Asymmetric Passive Components in Microwave Integrated Circuits

Asymmetric Passive Components in Microwave Integrated Circuits

HEE-RAN AHN

WILEY-INTERSCIENCE
A JOHN WILEY & SONS, INC., PUBLICATION

Published by John Wiley & Sons, Inc., Hoboken, New Jersey.
Published simultaneously in Canada.

Library of Congress Cataloging-in-Publication Data:

Ahn, Hee-Ran, 1956–
Asymmetric passive components in microwave integrated circuits / by Hee-Ran Ahn.
p. cm.
"A Wiley-Interscience publication."
Includes bibliographical references and index.
ISBN-13: 978-0-471-73748-3
ISBN-10: 0-471-73748-8
1. Microwave integrated circuits. I. Title.

TK7876.A38 2006
621.381'32–dc22

2005056772

10 9 8 7 6 5 4 3 2 1

Contents

Preface xi

1 Introduction 1

1.1 Asymmetric Passive Components 1

1.2 Circuit Parameters 2

1.3 Asymmetric Four-Port Hybrids 3

1.3.1 Asymmetric Ring Hybrids 3

1.3.2 Asymmetric Branch-Line Hybrids 4

1.4 Asymmetric Three-Port Power Dividers 5

1.5 Asymmetric Two-Port Components 6

References 6

2 Circuit Parameters 10

2.1 Scattering Matrix 10

2.1.1 Transmission-Line Theory 11

2.1.2 Basis-Dependent Scattering Parameters of a One-Port Network 12

2.1.3 Voltage- and Current-Basis Scattering Matrices of n-Port Networks 14

2.1.4 Complex Normalized Scattering Matrix 17

2.2 Scattering Parameters of Reduced Multiports 18

2.2.1 Examples of Reduced Multiports 21

2.3 Two-Port Network Analysis Using Scattering Parameters 23

2.4 Other Circuit Parameters 29
2.4.1 ABCD Parameters 29
2.4.2 Open-Circuit Impedance and Short-Circuit Admittance Parameters 36
2.4.3 Conversion Matrices of Two-Port Networks Terminated in Arbitrary Impedances 40
2.5 Analyses of Symmetric Networks 43
2.5.1 Analyses with Even- and Odd-Mode Excitations 43
2.5.2 Useful Symmetric Two-Port Networks 45
2.5.3 Properties of Symmetric Two-Port Networks 47
2.6 Analyses with Image Parameters 47
2.6.1 Image Impedances 47
2.6.2 Image Propagation Constants 49
2.6.3 Symmetrical and Common Structures 50
Exercises 52
References 54

3 Conventional Ring Hybrids 56

3.1 Introduction 56
3.2 Original Concept of the 3-dB Ring Hybrid 57
3.3 Conventional Ring Hybrids 62
3.3.1 Coupled Transmission Lines 62
3.3.2 Ring Hybrids with Coupled Transmission Lines 68
3.3.3 Wideband Ring Hybrids 71
3.3.4 Symmetric Ring Hybrids with Arbitrary Power Divisions 74
3.3.5 Conventional Lumped-Element Ring Hybrids 77
3.3.6 Mixed Small Ring Hybrids 80
3.4 Conventional 3-dB Uniplanar Ring Hybrids 84
3.4.1 Uniplanar T-Junctions 85
3.4.2 Transitions 86
3.4.3 Wideband Uniplanar Baluns 86
3.4.4 Uniplanar Ring Hybrids 88
Exercises 90
References 91

4 Asymmetric Ring Hybrids **93**

4.1 Introduction 93

4.2 Derivation of Design Equations of Asymmetric Ring Hybrids 93

4.3 Small Asymmetric Ring Hybrids 99

4.4 Wideband or Small Asymmetric Ring Hybrids 100

4.4.1 Microstrip Asymmetric Ring Hybrids 100

4.4.2 Uniplanar Asymmetric Ring Hybrids 102

4.5 Miniaturized Ring Hybrids Terminated in Arbitrary Impedances 106

4.5.1 Asymmetric Lumped-Element Ring Hybrids 106

Exercises 122

References 122

5 Asymmetric Branch-Line Hybrids **125**

5.1 Introduction 125

5.2 Origin of Branch-Line Hybrids 125

5.3 Multisection Branch-Line Couplers 127

5.4 Branch-Line Hybrids for Impedance Transforming 132

5.5 Asymmetric Four-Port Hybrids 139

5.5.1 Analyses of Asymmetric Four-Port Hybrids 139

5.5.2 Conventional–Direction Asymmetric Branch-Line Hybrids 140

5.5.3 Anti-Conventional-Direction Asymmetric Branch-Line Hybrids 147

Exercises 150

References 151

6 Conventional Three-Port Power Dividers **154**

6.1 Introduction 154

6.2 Three-Port 3-dB Power Dividers 155

6.3 Three-Port Power Dividers with Arbitrary Power Divisions 156

6.4 Symmetric Analyses of Asymmetric Three-Port Power Dividers 160

6.5 Three-Port 3-dB Power Dividers Terminated in Complex Frequency-Dependent Impedances 163
6.6 Three-Port 45° Power Divider/Combiner 167
Exercises 168
References 168

7 Three-Port 3-dB Power Dividers Terminated in Different Impedances 170
7.1 Introduction 170
7.2 Perfect Isolation Condition 171
7.3 Analyses 173
7.4 Scattering Parameters of Three-Port Power Dividers 177
7.5 Lumped-Element Three-Port 3-dB Power Dividers 186
7.6 Coplanar Three-Port 3-dB Power Dividers 188
Exercises 189
References 190

8 General Design Equations for *N*-Way Arbitrary Power Dividers 192
8.1 Introduction 192
8.2 General Design Equations for Three-Port Power Dividers 193
8.2.1 Coplanar Three-Port Power Divider Terminated in 50 Ω, 60 Ω, and 70 Ω 196
8.2.2 Determining Z_{Ad} 197
8.3 General Design Equations for N-Way Power Dividers 199
8.3.1 Analyses of N-Way Power Dividers 200
Exercises 204
References 204

9 Asymmetric Ring-Hybrid Phase Shifters and Attenuators 206
9.1 Introduction 206
9.2 Scattering Parameters of Asymmetric Ring Hybrids 207
9.3 Asymmetric Ring-Hybrid Phase Shifters 209
9.3.1 Uniplanar Asymmetric Ring-Hybrid −135° Phase Shifter 216

9.4 Asymmetric Ring-Hybrid Attenuator with Phase Shifts 216
9.4.1 Microstrip Asymmetric Ring-Hybrid 4-dB Attenuator with 45° Phase Shift 220
Exercises 222
References 223

10 Ring Filters and Their Use in a New Measurement Technique for Inherent Ring-Resonance Frequency 225

10.1 Introduction 225
10.2 Ring Filters 226
10.2.1 Analyses of Ring Filters 226
10.2.2 Measurements 230
10.3 New Measurement Technique for Inherent Ring-Resonance Frequency 230
10.3.1 Lossless Case 230
10.3.2 Loss Case 234
10.4 Conclusions 237
Exercises 238
References 238

11 Small Impedance Transformers, CVTs and CCTs, and Their Applications to Small Power Dividers and Ring Filters 240

11.1 Small Transmission-Line Impedance Transformers 240
11.2 Mathematical Approach for CVTs and CCTs 241
11.2.1 CVTs and CCTs 242
11.2.2 Microstrip CVTs and CCTs 247
11.2.3 Bounded Length of CVTs and CCTs 248
11.2.4 Phase Responses of CVTs and CCTs 251
11.3 CVT3PDs and CCT3PDs 253
11.3.1 Isolation Circuits of CVT3PDs and CVT3PDs 254
11.3.2 Design of CVT3PDs and CCT3PDs 256
11.4 Asymmetric Three-Port 45° Power Divider Terminated in Arbitrary Impedances 258
11.4.1 Asymmetric 45° Power Divider Terminated in 30 Ω, 60 Ω, and 50 Ω 259

11.5 CVT and CCT Ring Filters 261
11.5.1 Analyses of Ring Filters 262

Exercises 266

References 267

Appendix A: Symbols and Abbreviations 269

Appendix B: Conversion Matrices 272

Appendix C: Derivation of the Elements of a Small Asymmetric Ring Hybrid 276

Appendix D: Trigonometric Relations 279

Appendix E: Hyperbolic Relations 281

Index 283

Preface

This book was written primarily as an advanced text in microwave engineering for graduate students. Since it concentrates on the principle of circuit element designs, it is of value to engineers in industry who want to design advanced microwave circuits. To understand the text, transmission line theory, circuit matrices, and microwave circuit theory are the necessary background.

There are three classes of components in microwave integrated circuits: one is passive and the others are active and nonreciprocal ferrite components. This book treats the passive components intensively, especially asymmetric components. An epoch-making development was the even- and odd-mode excitation analyses suggested in 1956 by J. Reed and G. J. Wheeler, and their analyses were based on the symmetrical structures. Therefore, only symmetric passive designs have been developed until now. However, the passive components with arbitrary termination impedances are needed to reduce circuit size since they allow the elimination of the matching networks needed to obtain the desired output performances when the symmetric components are integrated with other elements. This results in components that are no longer symmetrical, so conventional design methods may not be used. Therefore, new design methods are needed. In this book, asymmetric design methods are illustrated for components such as asymmetric ring hybrids, asymmetric branch-line hybrids, asymmetric three-port power dividers, asymmetric ring-hybrid phase shifters and attenuators, asymmetric ring filters, and asymmetric impedance transformers, which are the basic and indispensable elements for integration with other active and/or passive devices.

The book is made up of eleven chapters: Chapter 1 provides a brief introduction to asymmetric passive components in microwave integrated circuits and their short history, starting with the asymmetric ring hybrids first described in 1994 by the author. Microwave circuit parameters such as scattering, *ABCD*, impedance, admittance, and image parameters are explained in Chapter 2. Basis-independent or basis-dependent scattering parameters are discussed, and an easy method of analyzing any network with multiports is described in more detail. A particular feature of this chapter is a conversion table between various circuits matrices characterizing two-port networks terminated in arbitrary impedances.

Chapters 3 and 4 cover ring hybrids. The design method of conventional ring hybrids, is given in Chapter 3, and asymmetric ring hybrids are treated in Chapter 4, where their design equations and a method to reduce their size are explained. Asymmetric branch-line hybrids are discussed in Chapter 5. Branch-line and ring hybrids are both four-port components, but they have been treated differently in conventional analyses. However, it is suggested in this book that they are not different and that the name *ring hybrid* or *branch-line hybrid* may be determined by the power division directions. Therefore, asymmetric branch-line hybrids may be designed using the same method as that used for asymmetric ring hybrids.

Chapters 6 to 8 treat three-port power dividers. Since they are very useful components in various applications, they are studied in more detail. Conventional three-port power dividers are described in Chapter 6. Asymmetric three-port power dividers are covered in Chapters 7 and 8: the case of equal power division in Chapter 7 and that of arbitrary power division in Chapter 8.

Chapters 9 to 11 cover asymmetric two-port components. They are asymmetric ring-hybrid phase shifters and attenuators, ring filters, and asymmetric impedance transformers. The asymmetric impedance transformers are not only small but can also produce arbitrary phase shifters of less than 90°. Due to these distinct properties, they can be used for various applications, such as small three-port power dividers, small ring filters, and asymmetric three-port 45° power dividers. Several applications are described in the latter part of Chapter 11; other applications are left for the readers.

Several people deserve acknowledgment for their help in completing the book. Professor Ingo Wolff at Duisburg–Essen University in Germany encouraged me to write the habilitation thesis that has served as a source for the book; Professor Tatsuo Itoh at UCLA in the United States, Professor Wolfgang Menzel at Ulm University in Germany, and Professor Kwyro Lee at KAIST (Korea Advanced Institute of Science and Technology) in Korea reviewed the book despite their busy schedules; and Professor Bumman Kim at POSTECH (Pohang University of Science and Technology) in Korea helped me concentrate on completing the book. I am especially appreciative of the interest and help of Professor Kai Chang at Texas A&M University. Without their interest and assistance, this task could not have been accomplished.

HEE-RAN AHN

Pohang University Science and Technology (POSTECH)

CHAPTER ONE

Introduction

1.1 ASYMMETRIC PASSIVE COMPONENTS

In microwave integrated circuits, there are three classes of components: one passive, the others active and nonreciprocal ferrite components. Power dividers, phase shifters, impedance transformers, and filters are typical passive components, and their conventional structures have been symmetric. However, asymmetric structures, with arbitrary termination impedances, are strongly preferable since they allow elimination of the matching networks needed to obtain the output performances desired when symmetric components are integrated with other elements.

The study of asymmetric passive components started with asymmetric ring hybrids in 1991 [1–5], and the design equations of asymmetric 3-dB ring hybrids were derived in 1994 by Ahn et al. [6]. Shortly after, those of asymmetric ring hybrids with arbitrary power divisions were synthesized assuming conditions of perfect isolation [7,8]. In 2000, a new concept [9], which stated that ring hybrids are not different from branch-line hybrids [10–13], allowed two types of asymmetric branch-line hybrids to be investigated and their design equations derived. In this book, asymmetric ring hybrids and branch-line hybrids are discussed in Chapters 3 to 5.

For asymmetric three-port power dividers, Ahn and Wolff [14,15] derived a perfect isolation condition that had never been analyzed for these types of conventional three-port power dividers. Using perfect isolation conditions 3-dB three-port power dividers terminating in different impedances could be constructed and their design equations derived. However, the design equations were available only for three-port power dividers with equal power division. In 2000, general design equations were derived for three-port power dividers with both arbitrary termination impedances and arbitrary power divisions [16,17]. In this book, the design of asymmetric three-port power dividers with arbitrary power divisions is treated in Chapters 7 and 8.

Development of the monolithic microwave integrated circuit (MMIC) technique has created a strong incentive to reduce circuit size. Three-port power dividers are key components in microwave integrated circuits, and there have been many trials to reduce their size. One of them was to adapt lumped-element circuits to be equivalent to transmission-line sections [1–6,14]. However, the bandwidths of three-port power dividers with lumped elements were not sufficient, so wideband impedance transformers were required. For this, two types of small wideband impedance transformers were introduced: the constant-voltage standing-wave-ratio transmission-line impedance transformer (CVT) and the constant-conductance transmission-line impedance transformer (CCT) [17–19]. CVTs and CCTs have arbitrary phase shifts of less than 90°, which differs considerably from conventional impedance transformers, which have only odd multiples of 90° phase shifts. Thus, the distinct characteristics of CVTs and CCTs allow construction of small CVT and CCT 3-dB power dividers (CVT3PD and CCT3PDs), three-port 45° power dividers, and CVT and CCT ring filters. One CCT3PD has the smallest recorded, and three-port 45° power dividers are very important in getting rid of third harmonics of active devices. The key components of CVTs and CCTs are treated in Chapter 11 as asymmetric two-port components.

In addition to CVTs and CCTs, asymmetric ring-hybrid phase shifters and attenuators [20–22] and ring filters are asymmetric two-port components and are considered in Chapters 9 and 10. Since asymmetric phase shifters and attenuators consist of asymmetric ring hybrids, they may be used as impedance transformers together with their original functions. Conventional 180° phase shifters were used for building linearizers in high-power systems, but they had narrowband properties. For this, ring filters [23] were introduced as wideband 180° transmission lines. Each ring filter consists of a ring and two short stubs, and it is possible to measure an inherent ring resonance frequency by having the two short stubs differ in length. This measurement technique is treated in Chapter 10.

1.2 CIRCUIT PARAMETERS

The measured quantities of passive components at microwave frequencies are almost always the scattering parameters. The scattering matrix discussed in Chapter 2 is admirably suitable for the description of a large class of passive microwave components and is used as much as possible throughout the book. In many cases it leads to a complete understanding of a microwave device while avoiding the need to construct a formal electromagnetic boundary-value problem for the structure. The entries of the scattering matrix of an n-port junction are a set of quantities related to incident and reflected waves at the ports of the junction that describe the performance of a network under any termination conditions specified. The coefficients along the main diagonal of the scattering matrix are reflection coefficients, whereas those along the off-diagonal are transmission coefficients.

A scattering matrix exists for every linear, passive, and time-invariant network, and it is possible to deduce important general properties of junctions containing a number of ports by invoking such junction properties as reciprocity and power conservation. Since the entries of the scattering matrix $\boldsymbol{S}$, impedance matrix $\boldsymbol{Y}$, and admittance matrix $\boldsymbol{Z}$ of a symmetric network are linear combinations of the circuit eigenvalues, their direct evaluation or measurement provides an alternative formulation of network parameters. Therefore, the relation between the scattering matrix and other circuit matrices is important and is described in Chapter 2.

1.3 ASYMMETRIC FOUR-PORT HYBRIDS

Many different types of power dividers, with and without isolation between output ports, are used for various applications. They perform a variety of functions, such as splitting and combining power in mixers (hybrids), sampling power from sources for level control, separating incident and reflected signals in network analyzers, and dividing power among a number of loads. Certain power dividers, which provide isolation between their output ports, are branch-line hybrids, ring hybrids, and parallel-coupled directional couplers, and their two outputs are in phase or out of phase by 90° or 180°. These power dividers are shown in Fig. 1.1(a), where the direction of power flow is indicated when power is fed into port ①. As shown, the direction of the ring hybrid is the same as that of the parallel-coupled directional coupler, but the two output signals of the ring hybrid are in phase or 180° out of phase, whereas those of the parallel-coupled directional coupler are 90° out of phase. The branch-line hybrid in Fig. 1.1(b) is same as that in Fig. 1.1(c), in that the two output signals are out of phase by 90°, but the power division directions are different from each other. Thus, the branch-line coupler (hybrid) is called a *forward coupler*, whereas the parallel-coupled directional coupler is called a *backward coupler*.

1.3.1 Asymmetric Ring Hybrids

The first conventional ring hybrid to be treated in Chapter 3 was investigated by Tyrrel in 1947 [24]. Tyrrel tried to explain ring hybrids using the concept of waveguide T-junctions, and described two types of hybrid circuits, one involving a ring or loop transmission line and the other relying on the symmetry properties of certain four-arm junctions. After he described the fundamental characteristics of distributed circuit hybrids, a number of workers discussed the performance of practical wideband realizations constructed in coaxial line [25–27] and stripline [28]. One of them was that two coupled-line filters were used for a wideband ring hybrid in the 1950s. In 1961, Pon [29] derived design equations for ring hybrids with arbitrary power divisions. In 1968, March [28] developed a wideband ring hybrid, adapting one coupled-line filter instead of a three-quarter-wavelength transmission line, which causes narrowband responses.

In the 1980s, as uniplanar techniques emerged for MMIC applications, there were several researchers developed small broadband ring hybrids which employed

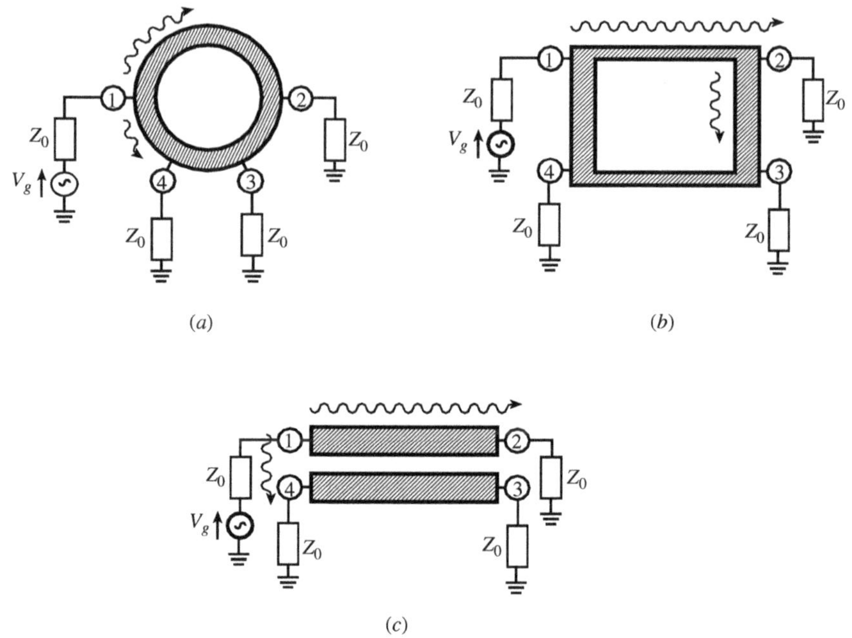

FIGURE 1.1 Four-port power dividers and their power flows: (*a*) ring hybrid; (*b*) branch-line hybrid; (*c*) parallel-coupled transmission-line directional coupler.

a combination of coplanar waveguides and slotlines using only one-sided substrates [30–32]. Since the first ring hybrid was introduced, ring hybrids have been studied and used for various applications in microwave equipment. Thus, they are indispensable components in various MICs (microwave integrated circuits) or MMICs (monolithic microwave integrated circuits), such as balanced mixers, balanced amplifiers, frequency discriminators, phase shifters, feeding networks in antenna arrays, and so on. The important conventional ring hybrids are introduced and discussed in Chapter 3.

In practice, ring hybrids are used together with other active and passive devices. Thus, to obtain a desired performance, additional matching networks are necessary for conventional ring hybrids. Therefore, ring hybrids terminating in arbitrary impedances can reduce the size of MICs significantly. Since no symmetry plane is available for asymmetric ring hybrids terminating in arbitrary impedance, the conventional method of even- and odd-mode excitation analyses cannot be used, so new design methods are required. In Chapter 4 we discuss how to derive design equations for asymmetric ring hybrids [8] and how to reduce the loss produced by inductors of asymmetric lumped-element ring hybrids [6].

1.3.2 Asymmetric Branch-Line Hybrids

Branch-line directional couplers, which originated with Mumford [33], consist of two adjacent transmission lines with one or more coupling elements between them. One of these two lines is the main or primary line and the other one is the secondary line. A small fraction of the energy in the main line is transferred, through the coupling elements, to the secondary line. The mechanisms of the power transmission and isolation from the primary line to the secondary line are discussed in Chapter 5. If the coupling elements are branch lines and hybrid T-junctions are used for the directional couplers, they are called *branch-line hybrids*, and one-stage branch-line hybrids are narrowband. So to increase the bandwidth, multiple branch-line hybrids [34] are needed and the design of such a directional coupler and the calculation of its frequency response are also covered in Chapter 5. The branch-line directional coupler is particularly desirable since the design constants are readily found and its frequency response can be calculated. Branch-line hybrids may be used for impedance transformers with one or two sections [35,36]; their design methods are treated in Chapter 5.

Branch-line hybrids have been studied for a long time. However, these studies have focused on symmetric branch-line hybrids [33–40]. If branch-line hybrids are terminated in arbitrary impedances, they are no longer symmetric and a new design method is needed. In the latter part of Chapter 5, the isolation mechanism of asymmetric four-port hybrids and design equations for asymmetric branch-line hybrids are treated.

1.4 ASYMMETRIC THREE-PORT POWER DIVIDERS

The history of three-port power dividers began in 1960 with Wilkinson [41], who described a device that separated one signal into n equiphase–equiamplitude signals. Theoretically perfect isolation between all output ports is achieved at one frequency. With $n = 2$, his circuit may be reduced to a three-port power divider. In 1965, Parad and Moynihan [42] presented a hybrid with the output signals in phase and an arbitrary amplitude difference. The perfect three-port hybrid property was again achieved at one frequency. In 1968, Cohn [43] presented a class of equal-power dividers with isolation and impedance matching at any number of frequencies. In 1971, Ekinge [44] described a three-port hybrid consisting of n sections in cascade, each section composed of two coupled lossless transmission lines with a certain electrical length and an intermediate resistor. His design seems to be similar to that of Cohn [43]. However, Cohn treated equal-power-split three-port hybrids, whereas Ekinge dealt with three-port hybrids with arbitrary power split.

For the design of three-port power dividers, it is very important to determine the isolation conditions or the values of isolation resistors. A number of papers have dealt with how to get isolation between output signals. Parad and Moynihan [42] suggested one of many isolation conditions, and Cohn, Ekinge, et al. [43–45] had to use intensive optimizing methods to derive the isolation resistances. In this book, a perfect isolation condition without optimization is discussed in Chapter 7.

In addition to the above, an optimization method for 3-dB three-port power dividers terminated in complex frequency-dependent impedances was suggested [45] and is discussed in Chapter 6. The output signals of conventional three-port power dividers are mostly in phase. However, if they are out of phase by 45° power dividers are particularly important to reduce unwanted intermodulation harmonic frequencies [46]. Such three-port power dividers can be built in two ways: The first is simply to use a delay line, and the second uses the small impedance transformers, CVTs and CCTs. These two ways are introduced in Chapters 6 and 11, respectively.

Since Wilkinson, studies on three-port hybrids have continued [42–48] to focus on symmetrical structures for which conventional even- and odd-mode excitation methods [49] can be used. If three-port power dividers are terminated in arbitrary impedances, they are no longer symmetric, and new ideas to derive design equations are needed. In earlier work [44], Ekinge mentioned that all the termination impedances had to be equal to each other, or at least the two output termination impedances be the same for the equal power division. However, Ahn and Wolff [14] showed that all the termination impedances could differ from each other despite equal power division. This is described in Chapter 7. General design equations that are available to any type of N-way power dividers with arbitrary power divisions and arbitrary termination impedances are discussed in Chapter 8.

1.5 ASYMMETRIC TWO-PORT COMPONENTS

Phase shifters, filters, and impedance transformers are two-port components. The phase shifter, as a general-purpose device in microwave components, finds use in a variety of communication and radar systems, microwave instrumentation, measurement systems, and industrial applications. Several different forms of phase shifters have been suggested, but most applications use a 3-dB hybrid power divider with symmetric reflection terminations. However, conventional ring-hybrid phase shifters require an additional transmission-line section to utilize symmetrical reflection terminations [50]. To reduce the size, a new design method for ring-hybrid phase shifters was presented without this transmission-line section [20]. In Chapter 9, the method is introduced and asymmetric ring-hybrid phase shifters and attenuators are presented and discussed.

Conventional impedance transformers have only odd multiples of 90° phase shift [51]. However, impedance transformers with arbitrary phase shifts are needed to reduce the size of microwave integrated circuits. The ring filters discussed in Chapter 10 have 180° phase shifts, and the CVTs and CCTs in Chapter 11 have arbitrary phase shifts of less than 90°. Asymmetric phase shifters, attenuators, impedance transformers, and ring filters are treated as asymmetric two-port components in Chapters 10 and 11.

REFERENCES

1. H.-R. Ahn, I.-S. Chang, and S.-W. Yun, A Study on the Miniaturization of 3-dB Ring Hybrid and Power Divider Using Lumped-Element Circuit, *J. KITE*, Vol. 28-A, No. 1, January 1991, pp. 15–22.

2. H.-R. Ahn, I.-S. Chang, and S.-W. Yun, A Study on the Miniaturization of 3-dB Ring Hybrid Having Arbitrary Termination Impedances Using Lumped-Element Circuit, *MTT Korean Chapter KITE-S Dig.*, Vol. 14, No. 2, October 1991, pp. 104–109.
3. H.-R. Ahn, I.-S. Chang, and S.-W. Yun, A Study on the Miniaturization of 3-dB Ring Hybrid Having Arbitrary Termination Impedances Using Lumped Equivalent Circuit, *J. KITE*, Vol. 29-A, No. 3, March 1992, pp. 25–32.
4. H.-R. Ahn, I.-S. Chang, and S.-W. Yun, A Study on the Miniaturization of 3-dB Ring Hybrid Having Arbitrary Termination Impedances Using Lumped Equivalent Circuit, *J. Telecommun. Rev.*, No. 5, May 1992, pp. 112–125.
5. H.-R. Ahn, I.-S. Chang, and S.-W. Yun, Lumped Element 3-dB 180° Hybrid with Asymmetrically Terminated Impedances, *J. KITE*, Vol. 31-A, No. 6, June 1994, pp. 18–25.
6. H.-R. Ahn, I.-S. Chang, and S.-W. Yun, Miniaturized 3-dB Ring Hybrid Terminated by Arbitrary Impedances, *IEEE Trans. Microwave Theory Tech.*, Vol. 42, December 1994, pp. 2216–2221.
7. H.-R. Ahn, I. Wolff, and I.-S. Chang, Arbitrary Termination Impedances, Arbitrary Power Division and Small-Sized Ring Hybrids, *IEEE MTT-S Dig.*, June 1997, pp. 285–288.
8. H.-R. Ahn, I. Wolff, and I.-S. Chang, Arbitrary Termination Impedances, Arbitrary Power Division and Small-Sized Ring Hybrids, *IEEE Trans. Microwave Theory Tech.*, Vol. 44, December 1997, pp. 2241–2247.
9. H.-R. Ahn and I. Wolff, Asymmetric Four-Port and Branch-Line Hybrids, *IEEE Trans. Microwave Theory Tech.*, Vol. 48, September 2000, pp. 1585–1588.
10. H.-R. Ahn and I. Wolff, 3-dB Branch-Line Hybrids with Arbitrary Termination Impedance Values, *IEICE Trans. Electrons*, Vol. E82-C, No. 7, July 1999, pp. 1324–1326.
11. H.-R. Ahn and I. Wolff, Arbitrary Power Division Branch-Line Hybrid Terminated by Arbitrary Impedances, *IEE Electron. Lett.*, Vol. 35, No. 7, April 1999, pp. 572–273.
12. H.-R. Ahn and I. Wolff, 3-dB Branch-Line Hybrid Terminated by Arbitrary Impedances, *IEE Electron. Lett.*, Vol. 34, No. 11, May 1998, pp. 1109–1110.
13. H.-R. Ahn and I. Wolff, Asymmetric Four-Port Hybrids, Asymmetric 3-dB Branch-Line Hybrids, *Asia–Pacific Microwave Conf. Proc.*, Yokohama, Japan, December 1998, pp. 677–680.
14. H.-R. Ahn and I. Wolff, Three-Port 3-dB Power Divider Terminated by Different Impedances and Its Application to MMIC's, *IEEE Trans. Microwave Theory Tech.*, Vol. 47, June 1999, pp. 786–794.
15. H.-R. Ahn and I. Wolff, Three-Port 3-dB Power Divider Terminated by Arbitrary Impedances, *IEEE MTT-S Dig.*, Baltimore, June 1998, pp. 781–784.
16. H.-R. Ahn and I. Wolff, General Design Equations of Three-Port Power Dividers, *IEEE MTT-S Dig.*, Boston, June 2000, pp. 1137–1140.
17. H.-R. Ahn and I. Wolff, General Design Equations of Three-Port Power Dividers, Small-Sized Impedance Transformers, and Their Applications to Small-Sized Three-Port 3-dB Power Divider, *IEEE Trans. Microwave Theory Tech.*, Vol. 49, July 2001, pp. 1277–1288.
18. H.-R. Ahn and I. Wolff, Small-Sized Impedance Transformers, *IEEE MTT/AP/EMC Korea Chapter KEES Dig.*, Vol. 23, No. 2, September 2000, pp. 157–160.

19. H.-R. Ahn and I. Wolff, Miniaturized Impedance Transformers, *MIOP Dig.*, May 2001, pp. 274–278.
20. H.-R. Ahn and I. Wolff, Asymmetric Ring Hybrid Phase-Shifters and Attenuators, *IEEE Trans. Microwave Theory Tech.*, Vol. 50, April 2002, pp. 1146–1155.
21. H.-R. Ahn and I. Wolff, Asymmetric Ring-Hybrid Phase Shifters, *IEEE MTT/AP/EMC Korea Chapter KEES Dig.*, Vol. 23, No. 2, September 2000, pp. 165–168.
22. H.-R. Ahn and I. Wolff, Small-Sized Ring-Hybrid Phase Shifters, *MIOP Dig.*, May 2001, pp. 274–278.
23. H.-R. Ahn and I. Wolff, Novel Ring Filter as a Wide-Band 180° Transmission-Line, *EUMC Proc.*, Vol. III, October 1999, pp. 95–98.
24. W. A. Tyrrel, Hybrid Circuits for Microwaves, *Proc. IRE*, Vol. 35, November 1947, pp. 1294–1306.
25. T. Morita and L. S. Sheingold, A Coaxial Magic-T, *IRE Trans. Microwave Theory Tech.*, Vol. 1, November 1953, pp. 17–23.
26. V. I. Albanese and W. P. Peyser, An Analysis of a Broad-Band Coaxial Hybrid Ring, *IRE Trans. Microwave Theory Tech.*, Vol. 6, October 1958, pp. 369–373.
27. W. V. Tyminski and A. E. Hylas, A Wide-Band Hybrid Ring for UHF, *Proc. IRE*, Vol. 41, January 1953, pp. 81–87.
28. S. March, Wideband Stripline Hybrid Ring, *IEEE Trans. Microwave Theory Tech.*, Vol. 16, June 1968, pp. 361–362.
29. C. Y. Pon, Hybrid-Ring Directional Coupler for Arbitrary Power Divisions, *IRE Trans. Microwave Theory Tech.*, Vol. 9, November 1961, pp. 529–535.
30. T. Hirota, Y. Tarusawa, and H. Ogawa, Uniplanar MMIC Hybrids: A Proposed New MMIC Structure, *IEEE Trans. Microwave Theory Tech.*, Vol. 35, June 1987, pp. 576–581.
31. C.-H. Ho, L. Fan, and K. Chang, Broad-Band Uniplanar Hybrid-Ring and Branch-Line Couplers, *IEEE Trans. Microwave Theory Tech.*, Vol. 41, December 1993, pp. 2116–2124.
32. C.-H. Ho, L. Fan, and K. Chang, New Uniplanar Coplanar Waveguide Hybrid-Ring Couplers and Magic-T's, *IEEE Trans. Microwave Theory Tech.*, Vol. 42, December 1994, pp. 2440–2448.
33. W. W. Mumford, Directional Couplers, *Proc. IRE,* Vol. 35, February 1947, pp. 159–165.
34. J. Reed, The Multiple Branch Waveguide Coupler, *IRE Trans. Microwave Theory Tech.*, Vol. 6, October 1958, pp. 398–403.
35. R. K. Gupta, S. E. Anderson, and W. Getsinger, Impedance-Transforming 3-dB Hybrids, *IEEE Trans. Microwave Theory Tech.*, Vol. 35, December 1987, pp. 1303–1307.
36. S. Kumar, C. Tannous, and T. Danshin, A Multisection Broadband Impedance Transforming Branch-Line Hybrid, *IEEE Trans. Microwave Theory Tech.*, Vol. 43, November 1995, pp. 2517–2523.
37. R. Levy, Analysis of Practical Branch-Guide Directional Couplers, *IEEE Trans. Microwave Theory Tech.*, Vol. 17, May 1969, pp. 289–290.
38. H. J. Riblet, Mathematical Theory of Directional Couplers, *Proc. IRE,* Vol. 35, December 1947, pp. 1307–1313.

39. R. Levy and L. F. Lind, Synthesis of Symmetrical Branch-Guide Directional Couplers, *IEEE Trans. Microwave Theory Tech.*, Vol. 4, February 1968, pp. 80–89.

40. H. J. Riblet, The Application of a New Class of Equal-Ripple Functions to Some Familiar Transmission-Line Problems, *IEEE Trans. Microwave Theory Tech.*, Vol. 12, July 1964, pp. 415–421.

41. E. J. Wilkinson, An *n*-Way Hybrid Power Divider, *IRE Trans. Microwave Theory Tech.*, Vol. 8, January 1960, pp. 116–118.

42. L. I. Parad and R. L. Moynihan, Split-Tee Power Divider, *IRE Trans. Microwave Theory Tech.*, Vol. 8, January 1965, pp. 91–95.

43. S. B. Cohn, A Class of Broadband Three-Port TEM-Mode Hybrids, *IRE Trans. Microwave Theory Tech.*, Vol. 16, February 1968, pp. 110–116.

44. R. B. Ekinge, A New Method of Synthesizing Matched Broad-Band TEM-Mode Three-Ports, *IEEE Trans. Microwave Theory Tech.*, Vol. 19, January 1971, pp. 81–88.

45. S. Rosloniec, Three-Port Hybrid Power Dividers Terminated in Complex Frequency-Dependent Impedances, *IEEE Trans. Microwave Theory Tech.*, Vol. 44, August 1996, pp. 1490–1493.

46. H. Hayashi, H. Okazaki, A. Kanda, T. Hirota, and M. Muraguch, Millimeter-Wave-Band Amplifier and Mixer MMIC's Using a Broad-Band 45° Power Divider/Combiner, *IEEE Trans. Microwave Theory Tech.*, Vol. 46, June 1998, pp. 811–818.

47. B. Kopp, Asymmetric Lumped Element Power Splitters, in *IEEE MTT-S Dig.*, 1989, pp. 333–336.

48. D. Köther, B. Hopf, Th. Sporkmann, and I. Wolff, MMIC Wilkinson Couplers for Frequencies Up to 110 GHz, in *IEEE MTT-S Dig.*, 1995, pp. 663–665.

49. J. Reed and G. J. Wheeler, A Method of Analysis of Symmetrical Four-Port Networks, *IRE Trans. Microwave Theory Tech.*, Vol. 4, October 1956, pp. 346–352.

50. J. F. White, Diode Phase Shifters for Array Antennas, *IEEE Trans. Microwave Theory Tech.*, Vol. 22, June 1974, pp. 658–674.

51. L. Young, The Quarter-Wave Transformer Prototype Circuit, *IRE Trans. Microwave Theory Tech.*, Vol. 8, September 1960, pp. 483–489.

CHAPTER TWO

Circuit Parameters

2.1 SCATTERING MATRIX

A large class of passive microwave components may be characterized by any of various sets of parameters, such as short-circuit admittance parameters, open-circuit impedance parameters, hybrid parameters, transmission parameters, scattering parameters, and so on. However, not all of these parameters will always exist because most parameters are partially defined with respect to zero or infinite loading at the ports. The scattering parameters, on the other hand, are defined in terms of some finite stable loadings at the ports. Because of this, they always exist for all linear passive networks.

The scattering parameters originated from the theory of transmission lines and form a matrix of transformation between variables which are linear combinations of the voltages and currents in a network. Since they are particularly suitable for problems of power transfer in networks designed to be terminated by complex loads, the scattering formalism finds convenient application in problems that involve insertion loss (e.g., filters, attenuators, power dividers, hybrids) and matching networks. Because the scattering parameters are closely associated with the power transfer properties of a network, they are indispensable in the design of microwave networks and permit the formulation of concise and useful expressions for energy constraints in passive structures. They are, therefore, particularly suitable for realization in the frequency domain. In addition, the scattering parameters possess another important property useful in general passive synthesis studies.

For the study of the scattering parameters, in this chapter we first treat how to get reflection coefficients from transmission-line theory and then apply the concepts to those of an n-port network. Since the scattering parameters are normalized to n complex loads on the complex plane, it will be shown that this

Asymmetric Passive Components in Microwave Integrated Circuits, by Hee-Ran Ahn

normalization can be extended to entire complex loads and to general n-port loads on the imaginary axis.

2.1.1 Transmission-Line Theory

The most important property of a transmission line is that electromagnetic fields can be uniquely related to voltages and currents. For this reason, circuit theory concepts are used for analyses of the structures, and the lumped-element equivalent circuit of a small section of transmission line with length dz is derived as shown in Fig. 2.1. It consists of series inductance per unit length L, series resistance per unit length R, shunt conductance per unit length G, and shunt capacitance per unit length C. Wave equations for $V(z)$ and $I(z)$ in Fig. 2.1 are obtained as

$$\frac{d^2V(z)}{dz^2} - \gamma^2 V(z) = 0, \tag{2.1a}$$

$$\frac{d^2I(z)}{dz^2} - \gamma^2 I(z) = 0, \tag{2.1b}$$

where $\gamma = \alpha + j\beta = \sqrt{(R + j\omega L)(G + j\omega C)}$ is the complex propagation constant, and α and β are phase and attenuation constants, respectively.

Assuming that $\alpha = 0$, the solutions for $V(z)$ and $I(z)$ in (2.1) are

$$V(z) = V_0^+ e^{-j\beta z} + V_0^- e^{j\beta z} = V_i(z) + V_r(z), \tag{2.2a}$$

$$I(z) = I_0^+ e^{-j\beta z} - I_0^- e^{j\beta z} = I_i(z) - I_r(z), \tag{2.2b}$$

where V_0^+, V_0^-, I_0^+, and I_0^- are voltage and current amplitudes traveling in positive and negative z-directions at $z = 0$. The total voltage $V(z)$ or current $I(z)$ along the line may be considered as the sum of an incident wave traveling in the positive z-direction and a reflected wave traveling in the negative z-direction, and the negative sign associated with the reflected current $I_r(z)$ indicates that the positive direction for $I_r(z)$ is opposite that for $I_i(z)$.

Figure 2.2 shows a uniform lossless transmission line that is connected between a load impedance Z_L and a voltage source V_g with an internal impedance

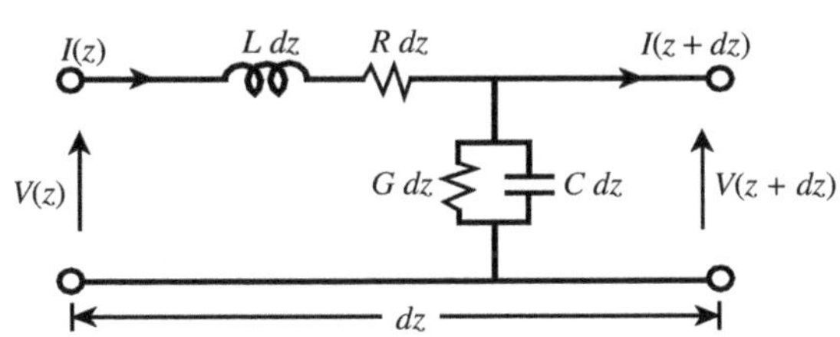

FIGURE 2.1 Lumped-element equivalent circuit of a transmission line of length dz.

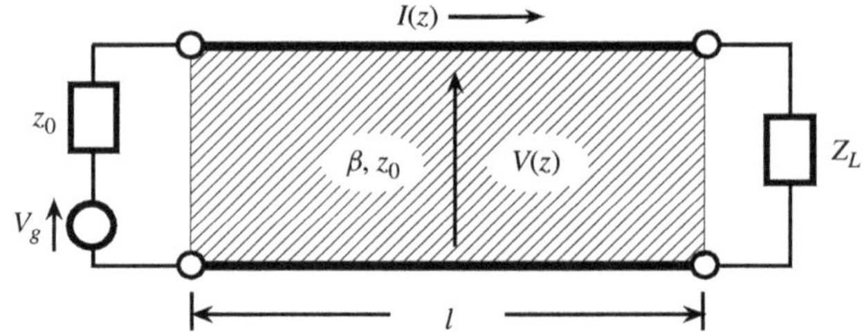

FIGURE 2.2 Uniform lossless transmission line.

z_0. When $Z_L \neq z_0$, reflections occur at $z = 0$, and the relations between voltages and currents and for z_0 and Z_L are given as

$$V(0) = V_i(0) + V_r(0), \tag{2.3a}$$

$$I(0) = I_i(0) - I_r(0), \tag{2.3b}$$

$$z_0 = \frac{V_r(0)}{I_r(0)} = \frac{V_i(0)}{I_i(0)}, \tag{2.3c}$$

$$Z_L = \frac{V(0)}{I(0)}. \tag{2.3d}$$

The ratio of reflected voltage to incident voltage at $z = 0$ is defined as the *voltage-basis reflection coefficient* $\Gamma^V = V_r(0)/V_i(0)$, and the ratio of reflected current to incident current at $z = 0$ is defined as the *current-basis reflection coefficient* $\Gamma^I = I_r(0)/I_i(0)$. By (2.2) and (2.3), these coefficients can easily be found and they are given by

$$\Gamma^I = (Z_L + z_0)^{-1}(Z_L - z_0), \tag{2.4a}$$

$$\Gamma^V = -(Y_L + y_0)^{-1}(Y_L - y_0), \tag{2.4b}$$

where $y_0 = 1/z_0$ and $Y_L = 1/Z_L$. For real z_0, $\Gamma^I = \Gamma^V$.

The most interesting conclusion from the relations in (2.4) is that reflections occur when $Z_L \neq z_0$ and that there is no reflected voltage or current wave when $Z_L = z_0$. Thus, when $Z_L \neq z_0$, Γ^I and Γ^V are finite, and standing waves of voltage and current exist along the transmission line. When $Z_L = z_0$, all the energy of the incident wave is transferred to the load, which cannot be distinguished from a transmission line with infinite length and characteristic impedance z_0. This concept is very important for the design of matching circuits.

2.1.2 Basis-Dependent Scattering Parameters of a One-Port Network

Since an equivalent impedance is found at $z = -l$ in Fig. 2.2, a one-port equivalent network is obtained from the uniform transmission line in Fig. 2.2. Thus, the concepts related to the reflection coefficients of the transmission line can be applied to those of the one-port network N. The one-port network N of Fig. 2.3(*a*) is characterized by its driving point impedance $Z(p)$, where p denotes $j\omega$. It is driven by a voltage source $V_g(p)$ in series with its *reference impedance* $z(p)$. The amount of power transferred from the source into the one-port network N depends on the impedance $Z(p)$, and the maximum power absorbed by the one-port network N is obtained with $Z(j\omega) = z_*(j\omega)$, where $z_*(j\omega)$ is the complex conjugate of $z(j\omega)$. When $Z(j\omega) = z_*(j\omega)$, the one-port network N is said to be *conjugately matched* to the load.

Like the case of transmission-line theory, the actual terminal voltage $V(p)$ in Fig. 2.3(*a*) is the sum of an *incident voltage* $V_i(p)$ and a *reflected voltage* $V_r(p)$, and the actual terminal current $I(p)$ is that of an *incident current* $I_i(p)$ and a *reflected current* $I_r(p)$. The incident current and voltage are those that appear under the optimal power-matching conditions. Thus, they are completely dependent on the loads, and Fig. 2.3(*b*) shows that the one-port network N is conjugately matched. From the Fig. 2.3(*b*), the relations between voltages and currents are obtained as

$$V_i(p) = [z(p) + z_*(p)]^{-1} V_g(p) z_*(p), \tag{2.5a}$$

$$I_i(p) = [z(p) + z_*(p)]^{-1} V_g(p). \tag{2.5b}$$

The matched load $z_*(p)$ is found from (2.5) as

$$V_i(p) = z_*(p) I_i(p). \tag{2.6}$$

As in (2.2), the reflected voltage $V_r(p)$ and current $I_r(p)$ are given as

$$V_r(p) = V(p) - V_i(p), \tag{2.7a}$$

$$-I_r(p) = I(p) - I_i(p). \tag{2.7b}$$

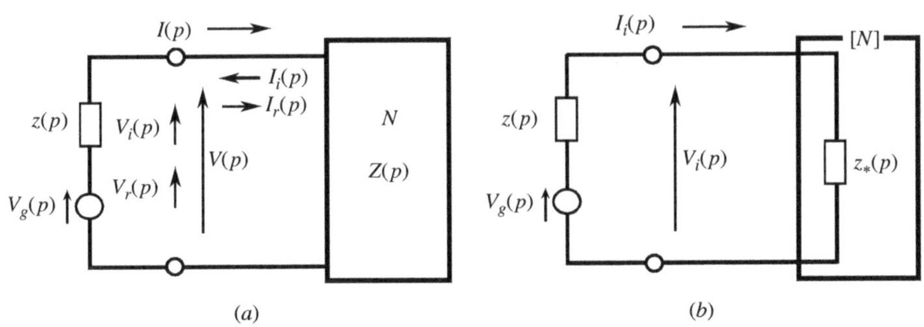

FIGURE 2.3 One-port networks: (*a*) characterized by its impedance $Z(p)$; (*b*) with the optimal power-matching condition.

The *voltage- and current-basis reflection coefficients* $\Gamma^V(p)$ and $\Gamma^I(p)$ are defined according to the relations

$$V_r(p) = \Gamma^V(p)V_i(p), \tag{2.8a}$$

$$I_r(p) = \Gamma^I(p)I_i(p). \tag{2.8b}$$

The reflection coefficients $\Gamma^V(p)$ and $\Gamma^I(p)$ are also referred to as *voltage-* and *current-basis scattering parameters,* respectively, of the one-port network N. In a similar way, $\Gamma^I(p)$ and $\Gamma^V(p)$ are

$$\Gamma^I(p) = [Z(p) + z(p)]^{-1}[Z(p) - z_*(p)], \tag{2.9}$$

$$\Gamma^V(p) = -[Y(p) + y(p)]^{-1}[Y(p) - y_*(p)]. \tag{2.10}$$

Comparing (2.9) with (2.10) yields

$$\Gamma^V(p)z_*(p) = z(p)\Gamma^I(p). \tag{2.11}$$

$\Gamma^I(p)$ is the same as $\Gamma^V(p)$ only when the reference impedance $z(p)$ is real, but they are in general different. On the real-frequency axis, they differ only by the phase, which is equal to twice the angle of the reference impedance $z(p)$. Both reflection coefficients are zero under optimal power-matching conditions, $Z(p) = z_*(p)$, and are related to the reference impedance $z(p)$ itself. It means that the incident waves see the impedance $z_*(p)$ and that the reflected waves see the reference impedance $z(p)$ itself.

2.1.3 Voltage- and Current-Basis Scattering Matrices of *n*-Port Networks

The reflection coefficients or scattering parameters of a one-port network have been discussed in detail. These concepts can be extended easily to an n-port network, and the scattering matrix of an n-port network is merely the matrix version of reflection coefficients of the one-port network. Figure 2.4 shows an n-port network N and its open-circuit impedance matrix $Z(p)$. Each of the n ports of N is loaded by a passive impedance $z_k(p)$ in series with a voltage source $V_{gk}(p)$. Since all elements of $z_k(p)$ are strictly passive, $z_k(p) + z_{k*}(p)$ cannot be identically zero.

The voltages, currents, and sources in Fig. 2.4 are represented by the vectors and related by the equation

$$\mathbf{V}_g(\mathbf{p}) = \mathbf{V}(\mathbf{p}) + \mathbf{z}(\mathbf{p})\mathbf{I}(\mathbf{p}) = [\mathbf{Z}(\mathbf{p}) + \mathbf{z}(\mathbf{p})]\mathbf{I}(\mathbf{p}), \tag{2.12}$$

where

$$\mathbf{V}(\mathbf{p}) = \begin{bmatrix} V_1(p) \\ V_2(p) \\ \vdots \\ V_n(p) \end{bmatrix}, \qquad \mathbf{I}(\mathbf{p}) = \begin{bmatrix} I_1(p) \\ I_2(p) \\ \vdots \\ I_n(p) \end{bmatrix}, \qquad \mathbf{V}_g(\mathbf{p}) = \begin{bmatrix} V_{g1}(p) \\ V_{g2}(p) \\ \vdots \\ V_{gn}(p) \end{bmatrix}, \tag{2.13}$$

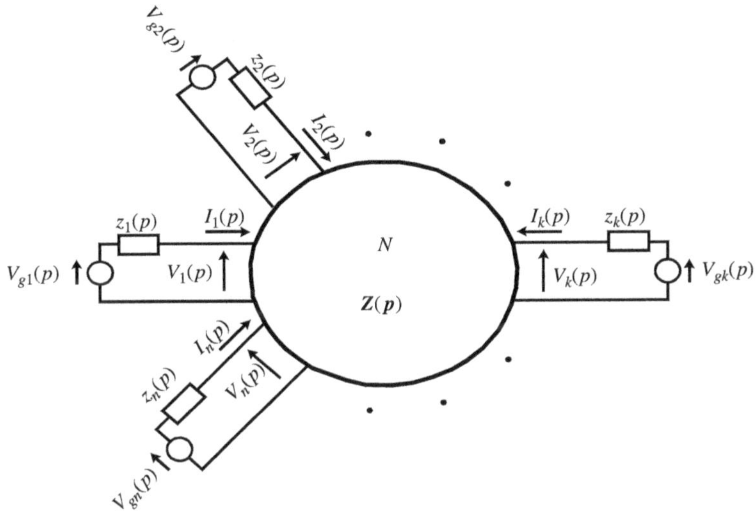

FIGURE 2.4 An n-port network N for current quantities.

and the *reference impedance matrix* $\boldsymbol{z}(\boldsymbol{p})$ of N is given as

$$\boldsymbol{z}(\boldsymbol{p}) = \begin{bmatrix} z_1(p) & 0 & \cdots & 0 \\ 0 & z_2(p) & 0 & 0 \\ \vdots & \vdots & \ddots & \vdots \\ 0 & 0 & \cdots & z_n(p) \end{bmatrix}, \tag{2.14}$$

whose kkth element is the reference impedance $z_k(p)$ of the kth port.

The concepts used for a one-port network can be applied to the scattering matrix of n-port network N. The *incident voltage vector* and *incident current vector* represent voltages $V_{i1}(p)$, $V_{i2}(p), \ldots, V_{in}(p)$ and currents $I_{i1}(p)$, $I_{i2}(p), \ldots, I_{in}(p)$, respectively, and they would appear at the terminals of the n-ports under the optimal power-matching conditions. Figure 2.5 shows the n-port network N terminated in optimal loads. Like the one-port case, the incident-voltage and incident-current vectors are represented by the equations

$$\boldsymbol{V}_i(\boldsymbol{p}) = \boldsymbol{z}_*(\boldsymbol{p})\boldsymbol{I}_i(\boldsymbol{p}), \tag{2.15a}$$

$$\boldsymbol{V}_g(\boldsymbol{p}) = [\boldsymbol{z}(\boldsymbol{p}) + \boldsymbol{z}_*(\boldsymbol{p})]\boldsymbol{I}_i(\boldsymbol{p}), \tag{2.15b}$$

where $\boldsymbol{z}(\boldsymbol{p}) + \boldsymbol{z}_*(\boldsymbol{p})$ is not identically singular with the passive loads. The *reflected voltage vector* $\boldsymbol{V}_r(\boldsymbol{p})$ and the *reflected current vector* $\boldsymbol{I}_r(\boldsymbol{p})$, defined by the difference between the actual quantities and incident quantities, are given as

$$\boldsymbol{V}_r(\boldsymbol{p}) = \boldsymbol{V}(\boldsymbol{p}) - \boldsymbol{V}_i(\boldsymbol{p}), \tag{2.16a}$$

$$-\boldsymbol{I}_r(\boldsymbol{p}) = \boldsymbol{I}(\boldsymbol{p}) - \boldsymbol{I}_i(\boldsymbol{p}). \tag{2.16b}$$

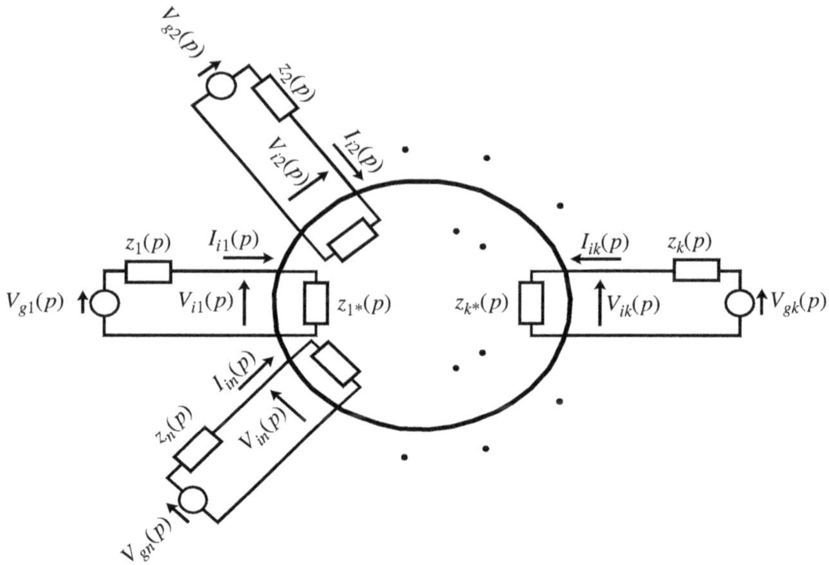

FIGURE 2.5 Matched n-port network for current quantities.

The matrix $\boldsymbol{S}^V$ relating the ratio of the reflected-voltage vector $\boldsymbol{V}_r(\boldsymbol{p})$ to the incident-voltage vector $\boldsymbol{V}_i(\boldsymbol{p})$,

$$\boldsymbol{V}_r(\boldsymbol{p}) = \boldsymbol{S}^V(\boldsymbol{p})\boldsymbol{V}_i(\boldsymbol{p}), \tag{2.17}$$

is called the *voltage-basis scattering matrix.* Similarly, the matrix $\boldsymbol{S}^I$ relating the reflected-current vector $\boldsymbol{I}_r(\boldsymbol{p})$ to the incident-current vector $\boldsymbol{I}_i(\boldsymbol{p})$,

$$\boldsymbol{I}_r(\boldsymbol{p}) = \boldsymbol{S}^I(\boldsymbol{p})\boldsymbol{I}_i(\boldsymbol{p}), \tag{2.18}$$

is called the *current-basis scattering matrix.* The elements of $\boldsymbol{S}^V(\boldsymbol{p})$ and $\boldsymbol{S}^I(\boldsymbol{p})$ are referred to as the current- and voltage-basis scattering parameters of the n-port network. Using (2.15), (2.16), and (2.18), the $\boldsymbol{S}^I(\boldsymbol{p})$ is, in terms of $\boldsymbol{Z}(\boldsymbol{p})$ and $\boldsymbol{z}(\boldsymbol{p})$, derived as

$$\begin{aligned}\boldsymbol{S}^I(\boldsymbol{p}) &= \boldsymbol{U}_n - [\boldsymbol{Z}(\boldsymbol{p}) + \boldsymbol{z}(\boldsymbol{p})]^{-1}[\boldsymbol{z}(\boldsymbol{p}) + \boldsymbol{z}_*(\boldsymbol{p})] \\ &= [\boldsymbol{Z}(\boldsymbol{p}) + \boldsymbol{z}(\boldsymbol{p})]^{-1}[\boldsymbol{Z}(\boldsymbol{p}) - \boldsymbol{z}_*(\boldsymbol{p})],\end{aligned} \tag{2.19}$$

based on

$$\boldsymbol{I}_r(\boldsymbol{p}) = -\boldsymbol{I}(\boldsymbol{p}) + \boldsymbol{I}_i(\boldsymbol{p}) = -[\boldsymbol{Z}(\boldsymbol{p}) + \boldsymbol{z}(\boldsymbol{p})]^{-1}\boldsymbol{V}_g(\boldsymbol{p}) + \boldsymbol{I}_i(\boldsymbol{p}), \tag{2.20}$$

where $\boldsymbol{U}_n$ denotes an identity matrix of order n.

Extension of the above to the dual situation is the voltage-basis scattering matrix, and the voltage-basis scattering matrix $S^V(p)$ is, in a similar way, derived as

$$S^V(p) = [Y(p) + y(p)]^{-1}[y(p) + y_*(p)] - U_n$$
$$= -[Y(p) + y(p)]^{-1}[Y(p) - y_*(p)], \tag{2.21}$$

based on

$$V_r(p) = V(p) - V_i(p) = [Y(p) + y(p)]^{-1}I_g(p) - V_i(p), \tag{2.22}$$

$$I_g(p) = [y(p) + y_*(p)]V_i(p), \tag{2.23}$$

where $Y(p) = Z(p)^{-1}$, $y(p) = z(p)^{-1}$, and $I_g(p)$ is a current source vector connected in parallel with the reference admittance matrix $y(p)$. From (2.19) and (2.21), the current- and voltage-basis scattering matrices are related by the equation

$$S^V(p)z_*(p) = z(p)S^I(p). \tag{2.24}$$

The voltage- and current-basis scattering matrices are defined with respect to a reference impedance matrix $z(p)$, which is quite arbitrary, and they are not necessarily symmetric even for a reciprocal network. Even though the voltage- and current-basis scattering matrices are represented as very simple formulas, their computation from the definitions is usually very cumbersome when n is more than 2. Therefore, many efforts have been made to simplify n-port networks. Moreover, when $z(p)$ is assumed to exist, $[Z(s) + z(s)]^{-1}$ and $[Y(s) + y(s)]^{-1}$ always exist even though the matrices $Z(s)$ and $Y(s)$ do not exist. Thus, the current- and voltage-basis scattering matrices always exist, and they represent the admittance and impedance matrices of some n-port networks derived from the original n-port network.

2.1.4 Complex Normalized Scattering Matrix

In Section 2.1.3, two scattering matrices were defined, one based on the voltage and the other on the current. However, the two are correct only with equal termination impedances, so they need to be normalized regardless of the termination impedances. That produces a complex normalized scattering matrix, which is one matrix for one network. The normalized scattering matrix is independent of both currents and voltages and can be obtained from either current- or voltage-basis quantities. To introduce the normalization for which the normalized scattering matrix becomes basis independent, a definition of *paraconjugate Hermitian impedance and admittance matrices* $r(p)$ and $g(p)$ are first given [1–3] as

$$r(p) = \tfrac{1}{2}[z(p) + z_*(p)] = h(p)h_*(p), \tag{2.25a}$$

$$g(p) = \tfrac{1}{2}[y(p) + y_*(p)] = k(p)k_*(p), \tag{2.25b}$$

from which the *normalized incident-wave vector* $\boldsymbol{a}(\boldsymbol{p})$ and the *normalized reflected-wave vector* $\boldsymbol{a}(\boldsymbol{p})$ are defined as

$$\boldsymbol{a}(\boldsymbol{p}) = \boldsymbol{h}_*(\boldsymbol{p})\boldsymbol{I}_i(\boldsymbol{p}) = \boldsymbol{k}_*(\boldsymbol{p})\boldsymbol{V}_i(\boldsymbol{p}), \tag{2.26a}$$

$$\boldsymbol{b}(\boldsymbol{p}) = \boldsymbol{h}(\boldsymbol{p})\boldsymbol{I}_r(\boldsymbol{p}) = \boldsymbol{k}(\boldsymbol{p})\boldsymbol{V}_r(\boldsymbol{p}). \tag{2.26b}$$

The complex normalized scattering matrix $\boldsymbol{S}(\boldsymbol{p})$ is defined by the relation

$$\boldsymbol{b}(\boldsymbol{p}) = \boldsymbol{S}(\boldsymbol{p})\boldsymbol{a}(\boldsymbol{p}). \tag{2.27}$$

The elements of $\boldsymbol{S}(\boldsymbol{p})$ are called the (*complex*) *normalized scattering parameters* of the n-port network. Like the basis-dependent scattering matrices, $\boldsymbol{S}(\boldsymbol{p})$ can also be expressed in terms of the impedance matrix $\boldsymbol{Z}(\boldsymbol{p})$ and the reference impedance matrix $\boldsymbol{z}(\boldsymbol{p})$ from the current quantities in (2.18) and (2.26):

$$\boldsymbol{b}(\boldsymbol{p}) = \boldsymbol{h}(\boldsymbol{p})\boldsymbol{I}_r(\boldsymbol{p}) = \boldsymbol{h}(\boldsymbol{p})\boldsymbol{S}^I(\boldsymbol{p})\boldsymbol{I}_i(\boldsymbol{p}) = \boldsymbol{h}(\boldsymbol{p})\boldsymbol{S}^I(\boldsymbol{p})\boldsymbol{h}_*^{-1}(\boldsymbol{p})\boldsymbol{a}(\boldsymbol{p}), \tag{2.28}$$

giving

$$\boldsymbol{S}(\boldsymbol{p}) = \boldsymbol{h}(\boldsymbol{p})\boldsymbol{S}^I(\boldsymbol{p})\boldsymbol{h}_*^{-1}(\boldsymbol{p}), \tag{2.29a}$$

$$\boldsymbol{S}(\boldsymbol{p}) = \boldsymbol{h}(\boldsymbol{p})[\boldsymbol{Z}(\boldsymbol{p}) + \boldsymbol{z}(\boldsymbol{p})]^{-1}[\boldsymbol{Z}(\boldsymbol{p}) - \boldsymbol{z}_*(\boldsymbol{p})]\boldsymbol{h}_*^{-1}(\boldsymbol{p}). \tag{2.29b}$$

Alternatively, the normalized scattering matrix can be found in a similar way from voltage quantities as

$$\boldsymbol{S}(\boldsymbol{p}) = \boldsymbol{k}(\boldsymbol{p})\boldsymbol{S}^V(\boldsymbol{p})\boldsymbol{k}_*(\boldsymbol{p}), \tag{2.30a}$$

$$\boldsymbol{S}(\boldsymbol{p}) = -\boldsymbol{k}(\boldsymbol{p})[\boldsymbol{Y}(\boldsymbol{p}) + \boldsymbol{y}(\boldsymbol{p})]^{-1}[\boldsymbol{Y}(\boldsymbol{p}) - \boldsymbol{y}_*(\boldsymbol{p})]\boldsymbol{k}_*^{-1}(\boldsymbol{p}), \tag{2.30b}$$

where $\boldsymbol{Y}(\boldsymbol{p}) = \boldsymbol{Z}^{-1}(\boldsymbol{p})$ and $\boldsymbol{y}(\boldsymbol{p}) = \boldsymbol{z}^{-1}(\boldsymbol{p})$.

Useful formulas for the normalized waves are obtained easily in terms of the port voltage vector $\boldsymbol{V}(\boldsymbol{p})$, and current vector $\boldsymbol{I}(\boldsymbol{p})$ in conjunction with impedance matrices $\boldsymbol{Z}(\boldsymbol{p})$ and $\boldsymbol{z}(\boldsymbol{p})$. Substituting $\boldsymbol{I}_i(\boldsymbol{p}) = [\boldsymbol{z}(\boldsymbol{p}) + \boldsymbol{z}_*(\boldsymbol{p})]^{-1}\boldsymbol{V}_g(\boldsymbol{p})$ and $\boldsymbol{V}_g = [\boldsymbol{z}(\boldsymbol{p}) + \boldsymbol{Z}(\boldsymbol{p})]\boldsymbol{I}(\boldsymbol{p})$ into current quantities in (2.26) results in

$$\boldsymbol{a}(\boldsymbol{p}) = \tfrac{1}{2}\boldsymbol{h}^{-1}(\boldsymbol{p})[\boldsymbol{V}(\boldsymbol{p}) + \boldsymbol{z}(\boldsymbol{p})\boldsymbol{I}(\boldsymbol{p})], \tag{2.31a}$$

$$\boldsymbol{b}(\boldsymbol{p}) = \tfrac{1}{2}\boldsymbol{h}_*^{-1}(\boldsymbol{p})[\boldsymbol{V}(\boldsymbol{p}) - \boldsymbol{z}_*(\boldsymbol{p})\boldsymbol{I}(\boldsymbol{p})], \tag{2.31b}$$

where $[\boldsymbol{z}(\boldsymbol{p}) + \boldsymbol{z}_*(\boldsymbol{p})]^{-1} = [2\boldsymbol{h}(\boldsymbol{p})]^{-1}$ and $\boldsymbol{Z}(\boldsymbol{p})\boldsymbol{I}(\boldsymbol{p}) = \boldsymbol{V}(\boldsymbol{p})$ are used.

2.2 SCATTERING PARAMETERS OF REDUCED MULTIPORTS

A microwave receiving system often contains a multiport, such as a directional coupler, a polarizer, or a hybrid junction that has been calibrated. The multiport properties calibrated are usually the scattering parameters which are defined for matched load conditions. However, when the multiport is installed in the system, the terminations seen out of various ports are not generally matched. In the analysis of microwave systems, it is often of interest to determine how variations of reflection coefficient at the loads affect the attenuation and phase shift of a signal that is transmitted through the multiport. If analysis is required only of signal transmission between two main ports, it is sometimes convenient to treat the terminated multiport as an equivalent two-port that is connected between a source and a load.

When some of the ports on a multiport are terminated with fixed loads so that these ports are not available, the multiport is sometimes referred to as a *reduced multiport* [4,5] rather than as an equivalent multiport of reduced size. The purpose of this section is to discuss how to derive the equivalent scattering parameters of the reduced multiport. Due to the inherent compactness of the matrix notation, all the scattering parameters can be derived in an efficient manner even when the size of the multiport matrix becomes large.

Figure 2.6 shows an $(n+k)$-port network that is to be reduced to an n-port network by placing terminations $Z_{L,n+1}, Z_{L,n+2}, \ldots, Z_{L,n+k}$ on the ports numbered $n+1$ through $n+k$. To derive the scattering matrix of the equivalent n-port network, begin by placing partitioning lines after the nth row and the nth column of the original $(n+k)$-port matrix, as expressed by

$$\left[\begin{array}{c} b_1 \\ b_2 \\ \vdots \\ b_n \\ \hline b_{n+1} \\ \vdots \\ b_{n+k} \end{array}\right] = \left[\begin{array}{cccc|ccc} S'_{11} & S'_{12} & \cdots & S'_{1n} & S'_{1,n+1} & \cdots & S'_{1,n+k} \\ S'_{21} & S'_{22} & \cdots & S'_{2n} & \vdots & \cdots & \vdots \\ \vdots & \vdots & \cdots & \vdots & \vdots & \cdots & \vdots \\ S'_{n1} & S'_{n2} & \cdots & S'_{n,n} & S'_{n,n+1} & \cdots & S'_{n,n+k} \\ \hline S'_{n+1,1} & \cdots & \cdots & S'_{n+1,n} & S'_{n+1,n+1} & \cdots & S'_{n+1,n+k} \\ \vdots & \cdots & \cdots & \vdots & \vdots & \cdots & \vdots \\ S'_{n+k,1} & \cdots & \cdots & S'_{n+k,n} & S'_{n+k,n+1} & \cdots & S'_{n+k,n+k} \end{array}\right] \left[\begin{array}{c} a_1 \\ a_2 \\ \vdots \\ a_n \\ \hline a_{n+1} \\ \vdots \\ a_{n+k} \end{array}\right], \tag{2.32}$$

which can then be rewritten as

$$\left[\begin{array}{c} \mathbf{B_1} \\ \hline \mathbf{B_2} \end{array}\right] = \left[\begin{array}{c|c} \mathbf{P_{11}} & \mathbf{P_{12}} \\ \hline \mathbf{P_{21}} & \mathbf{P_{22}} \end{array}\right] \left[\begin{array}{c} \mathbf{A_1} \\ \hline \mathbf{A_2} \end{array}\right], \tag{2.33}$$

where

$$\mathbf{B_1} = \begin{bmatrix} b_1 \\ b_2 \\ \vdots \\ b_n \end{bmatrix}, \quad \mathbf{A_1} = \begin{bmatrix} a_1 \\ a_2 \\ \vdots \\ a_n \end{bmatrix}, \quad \mathbf{B_2} = \begin{bmatrix} b_{n+1} \\ b_{n+2} \\ \vdots \\ b_{n+k} \end{bmatrix}, \quad \mathbf{A_2} = \begin{bmatrix} a_{n+1} \\ a_{n+2} \\ \vdots \\ a_{n+k} \end{bmatrix}.$$

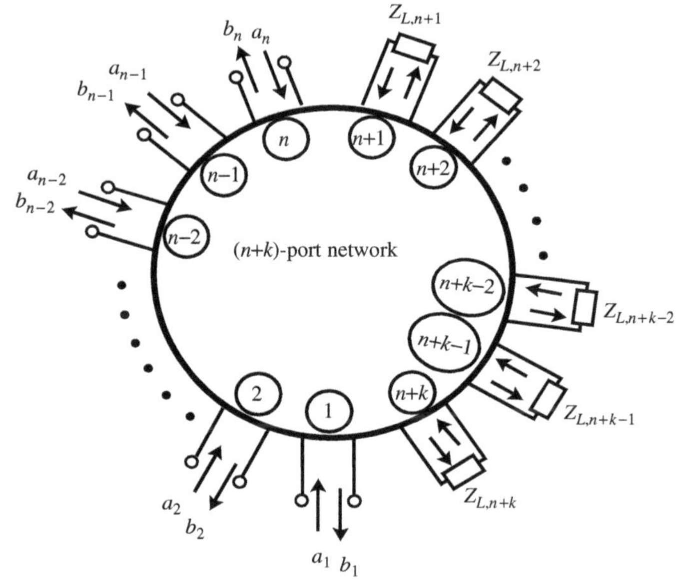

FIGURE 2.6 Incident and reflected waves associated with an $(n + k)$-port network.

$\boldsymbol{P}_{11}$, $\boldsymbol{P}_{12}$, $\boldsymbol{P}_{21}$, and $\boldsymbol{P}_{22}$ represent the partitioned submatrices of the original $(n + k)$-port scattering matrix. If the loads that terminate ports $n + 1$ through $n + k$ have reflection coefficients $\Gamma_{L,n+1}$ through $\Gamma_{L,n+k}$, respectively, then

$$\boldsymbol{A}_2 = \boldsymbol{\Gamma}_L \boldsymbol{B}_2, \tag{2.34}$$

where $\boldsymbol{\Gamma}_L$ is the load matrix expressed as

$$\boldsymbol{\Gamma}_L = \begin{bmatrix} \Gamma_{L,n+1} & 0 & 0 & 0 \\ 0 & \Gamma_{L,n+2} & \vdots & \vdots \\ \vdots & \vdots & \ddots & 0 \\ 0 & 0 & 0 & \Gamma_{L,n+k} \end{bmatrix}. \tag{2.35}$$

Substituting (2.34) for (2.33) and performing the matrix multiplication yields

$$\boldsymbol{S} = \begin{bmatrix} S_{11} & S_{12} & \cdot & \cdot & S_{1n} \\ S_{21} & S_{22} & \cdot & \cdot & \cdot \\ \cdot & \cdot & S_{ij} & \cdot & \cdot \\ \cdot & \cdot & \cdot & \cdot & \cdot \\ S_{n1} & \cdot & \cdot & \cdot & S_{nn} \end{bmatrix} = \boldsymbol{P}_{11} + (\boldsymbol{P}_{12}\Gamma_L)(\boldsymbol{U}_k - \boldsymbol{P}_{22}\Gamma_L)^{-1}\boldsymbol{P}_{21}, \tag{2.36}$$

where $\boldsymbol{U}_k$ is an identity matrix and $\boldsymbol{S}$ is the scattering matrix of the reduced multiport. The single-element relationship can be derived from (2.36) and expressed in the following convenient form [5]:

$$S_{ij} = \frac{\det\begin{vmatrix} S'_{ij} & (-\boldsymbol{P}_{12}\Gamma_L)i \\ (\boldsymbol{P}_{21})j & \boldsymbol{U}_k - \boldsymbol{P}_{22}\Gamma_L \end{vmatrix}}{\det|\boldsymbol{U}_k - \boldsymbol{P}_{22}\Gamma_L|}, \tag{2.37}$$

where $i = 1, 2, \ldots, n$, $j = 1, 2, \ldots, n$, $(-\boldsymbol{P}_{12}\Gamma_L)_i$ is a row matrix consisting of the ith row of the matrix $(-\boldsymbol{P}_{12}\Gamma_L)$, and $(\boldsymbol{P}_{21})_j$ is a column matrix consisting of the jth column of $\boldsymbol{P}_{21}$.

2.2.1 Examples of Reduced Multiports

Four-Port Network Terminated in Two Reflected Loads Formula (2.36) can be used in a variety of cases. One example is shown in Fig. 2.7, where a four-port network, two of whose ports are not perfectly matched, is depicted. Let the scattering matrix of the four-port network in Fig. 2.7 be given by

$$\boldsymbol{S}' = \begin{bmatrix} S'_{11} & S'_{12} & S'_{13} & S'_{14} \\ S'_{21} & S'_{22} & S'_{23} & S'_{24} \\ S'_{31} & S'_{32} & S'_{33} & S'_{34} \\ S'_{41} & S'_{42} & S'_{43} & S'_{44} \end{bmatrix}. \tag{2.38}$$

Referring to the notations in (2.33), the following are, obtained:

$$\boldsymbol{P}_{11} = \begin{bmatrix} S'_{11} & S'_{13} \\ S'_{31} & S'_{33} \end{bmatrix}, \tag{2.39a}$$

$$\boldsymbol{P}_{12} = \begin{bmatrix} S'_{12} & S'_{14} \\ S'_{32} & S'_{34} \end{bmatrix}, \tag{2.39b}$$

$$\boldsymbol{P}_{21} = \begin{bmatrix} S'_{21} & S'_{23} \\ S'_{41} & S'_{43} \end{bmatrix}, \tag{2.39c}$$

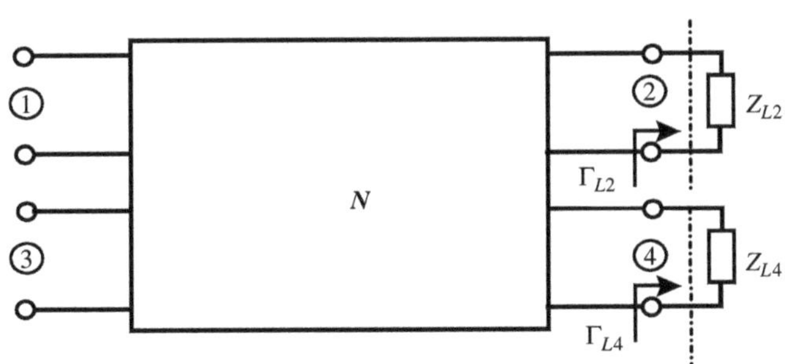

FIGURE 2.7 Four-port network with its two ports terminated in arbitrary loads.

$$\boldsymbol{P}_{22} = \begin{bmatrix} S'_{22} & S'_{24} \\ S'_{42} & S'_{44} \end{bmatrix}, \tag{2.39d}$$

$$\boldsymbol{\Gamma}_L = \begin{bmatrix} \Gamma_{L2} & 0 \\ 0 & \Gamma_{L4} \end{bmatrix}, \tag{2.39e}$$

where Γ_{L2} and Γ_{L4} are reflection coefficients at ports ② and ④.

By applying (2.39) to (2.36), the scattering matrix of the reduced two-port network is given by

$$S_{ij} = S'_{ij} + \frac{\Gamma_{L2}\begin{vmatrix} S'_{i2}S'_{2j} & -S'_{i2}S'_{24}\Gamma_{L4} \\ S'_{4j} & 1 - S'_{44}\Gamma_{L4} \end{vmatrix} + \Gamma_{L4}\begin{vmatrix} S'_{2j} & 1 - S'_{22}\Gamma_{L2} \\ -S'_{i4}S'_{4j} & -S'_{i4}S'_{42}\Gamma_{L2} \end{vmatrix}}{\begin{vmatrix} 1 - S'_{22}\Gamma_{L2} & -S'_{24}\Gamma_{L4} \\ -S'_{42}\Gamma_{L2} & 1 - S'_{44}\Gamma_{L4} \end{vmatrix}}, \tag{2.40}$$

where $i = 1, 2$ and $j = 1, 2$.

Three- and Two-Port Networks The use of formula (2.36) is demonstrated by finding the modified scattering matrix of three- and two-port networks, one of whose ports is not perfectly matched. The three-port network is terminated in Γ_{L3} at port ③ and the two-port network is terminated in Γ_{L2} at port ②, as depicted in Fig. 2.8(*a*) and (*b*). The scattering matrix of the reduced two-port network in Fig. 2.8(*a*) is given by

$$\boldsymbol{S} = \begin{bmatrix} S'_{11} + \dfrac{S'_{13}S'_{31}\Gamma_{L3}}{1 - S'_{33}\Gamma_{L3}} & S'_{12} + \dfrac{S'_{13}S'_{32}\Gamma_{L3}}{1 - S'_{33}\Gamma_{L3}} \\ S'_{21} + \dfrac{S'_{23}S'_{31}\Gamma_{L3}}{1 - S'_{33}\Gamma_{L3}} & S'_{22} + \dfrac{S'_{23}S'_{32}\Gamma_{L3}}{1 - S'_{33}\Gamma_{L3}} \end{bmatrix}, \tag{2.41}$$

where $\boldsymbol{S}'$ is the scattering matrix of the original three-port network, and that of the reduced one-port network in Fig. 2.8(*b*) is given by

$$S_{11} = S'_{11} + \frac{S'_{12}S'_{21}\Gamma_L}{1 - S'_{22}\Gamma_L}, \tag{2.42}$$

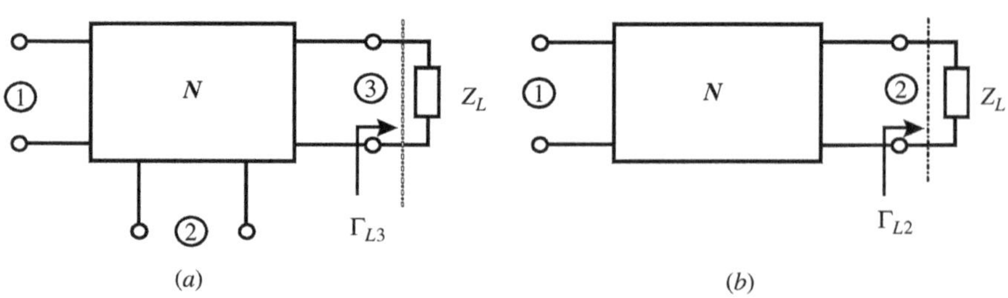

FIGURE 2.8 (*a*) Three-port network with one port terminated in an arbitrary load; (*b*) two-port network with one port terminated in an arbitrary load.

where S_{11} denotes the reflection coefficient at port ① of the reduced one-port network.

2.3 TWO-PORT NETWORK ANALYSIS USING SCATTERING PARAMETERS

Two-port subnetworks can be analyzed based on the scattering parameters given in various networks. The first objective of the analysis is to obtain expressions for the normalized incident and reflected waves, a_1, b_1, a_2, and b_2 in terms of network scattering parameters S_{11}, S_{12}, S_{21}, and S_{22}, the source quantities Γ_s and V_s, and the load reflection coefficient Γ_L. These parameters and quantities can be assumed to be known, as they will either be specified or will be obtainable by measurement. Once normalized incident and reflected waves have been found, they can be used to determine voltage and power gains for the network, and overall reflection coefficients for the network with load and source connected [6].

A two-port network and its one-port equivalent network are shown in Fig. 2.9(*a*) and (*b*), respectively. The standard $\boldsymbol{S}$ matrix relates a_1, a_2, b_1, and b_2 by

$$\begin{bmatrix} b_1 \\ b_2 \end{bmatrix} = \begin{bmatrix} S_{11} & S_{12} \\ S_{21} & S_{22} \end{bmatrix} \begin{bmatrix} a_1 \\ a_2 \end{bmatrix}. \tag{2.43}$$

From Fig. 2.9(*a*),

$$\Gamma_L = \frac{b_L}{a_L} = \frac{a_2}{b_2}, \tag{2.44}$$

which may be substituted in the relationship for b_2 in (2.43) to give

$$0 = S_{21}a_1 + \left(S_{22} - \frac{1}{\Gamma_L} \right) a_2. \tag{2.45}$$

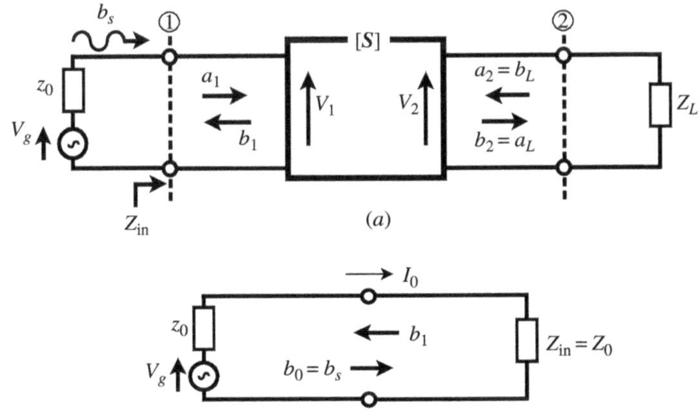

FIGURE 2.9 (*a*) Two-port network; (*b*) equivalent one-port network.

The incident and reflected waves at a port are defined in terms of the port voltage V_i, port current I_i, and a reference impedance as shown in (2.31). Assuming that all the reference impedances of the original scattering parameters are equally Z_0 and real, V_i and I_i at any port i can be written by normalized incident and reflected waves, a_i and b_i, defined such that

$$a_i + b_i = \frac{V_i}{\sqrt{Z_0}}, \qquad (2.46a)$$

$$a_i - b_i = I_i\sqrt{Z_0}. \qquad (2.46b)$$

The incident wave for the electromagnetic force (EMF) source V_g is b_1, as shown in Fig. 2.9(a), which will give rise to a reflected component $\Gamma_s b_1$, where Γ_s is a reflection coefficient of the source. In addition to this component, the V_g will contribute a component b_s so that the total reflected variable from the source is

$$a_1 = \Gamma_s b_1 + b_s. \qquad (2.47)$$

The component b_s in (2.47) may be found by arranging the circuit so that b_1 is zero, which can be done by making $Z_{\text{in}} = Z_0$ in Fig. 2.9(b). Making $Z_{\text{in}} = Z_0$ results in

$$I_0 = \frac{V_g}{z_0 + Z_0}. \qquad (2.48)$$

Assuming that the current flowing from the source is I_0 in Fig. 2.9(b), (2.46b) gives

$$a_1 - b_1 = I_0\sqrt{Z_0}. \qquad (2.49)$$

When $b_1 = 0$, $a_1 = I_0\sqrt{Z_0}$ in (2.49), with which b_s is obtained as

$$b_s = \frac{V_g\sqrt{Z_0}}{z_0 + Z_0}. \qquad (2.50)$$

Substituting (2.47) into the scattering equation for b_1 in (2.43) results in

$$\frac{-b_s}{\Gamma_s} = \left(S_{11} - \frac{1}{\Gamma_s}\right)a_1 + S_{12}a_2. \qquad (2.51)$$

Γ is often zero, so $1/\Gamma$ results in infinity and the mathematics are often difficult to handle. So (2.45) and (2.51) can be solved for a_1 and a_2 in convenient form as

$$-b_s = A_{11}a_1 + A_{12}a_2, \qquad (2.52a)$$

$$0 = A_{21}a_1 + A_{22}a_2, \qquad (2.52b)$$

where

$$A_{11} = \Gamma_s S_{11} - 1, \qquad A_{12} = \Gamma_s S_{12}, \qquad A_{21} = \Gamma_L S_{21}, \qquad A_{22} = \Gamma_L S_{22} - 1. \tag{2.53}$$

The complete set of equations for normalized incident and reflected waves are derived from (2.52), (2.53), and (2.43) as

$$a_1 = \frac{b_s A_{22}}{D}, \tag{2.54a}$$

$$a_2 = \frac{-b_s A_{21}}{D}, \tag{2.54b}$$

$$b_1 = \frac{b_s (A_{22} S_{11} - S_{12} A_{21})}{D}, \tag{2.54c}$$

$$b_2 = \frac{b_s (A_{22} S_{21} - S_{22} A_{21})}{D}, \tag{2.54d}$$

where $D = A_{21}A_{12} - A_{11}A_{22}$.

Voltage Gain The terminal voltage gain of the two-port network of Fig. 2.9(a) can be defined as

$$A_v = \frac{V_2}{V_1}, \tag{2.55}$$

where V_1 and V_2 are the terminal voltages at ports ① and ②, respectively. Applying (2.46a) to each port results in

$$A_v = \frac{a_2 + b_2}{a_1 + b_1}. \tag{2.56}$$

Now using (2.54) for the incident and reflected waves results in

$$A_v = \frac{-A_{21} + (A_{22} S_{21} - S_{22} A_{21})}{A_{22} + (A_{22} S_{11} - S_{12} A_{21})}. \tag{2.57}$$

The factor A_{11} is not included in (2.57). Thus, knowledge of the source parameters is not required.

Signal Flow Graph We have seen how the interconnection of sources, networks, and loads can be treated with various forms of scattering parameters. In this section, the signal flow graph is introduced as an additional technique that is very useful for the analysis of microwave networks in terms of scattering parameters or in terms of transmitted and reflected waves. The signal flow graph is a method of writing a set of equations in which the variables are represented by points and the interrelations by directed lines, giving a direct picture of signal flow [7,8]. When microwave network equations are written in scattering matrix

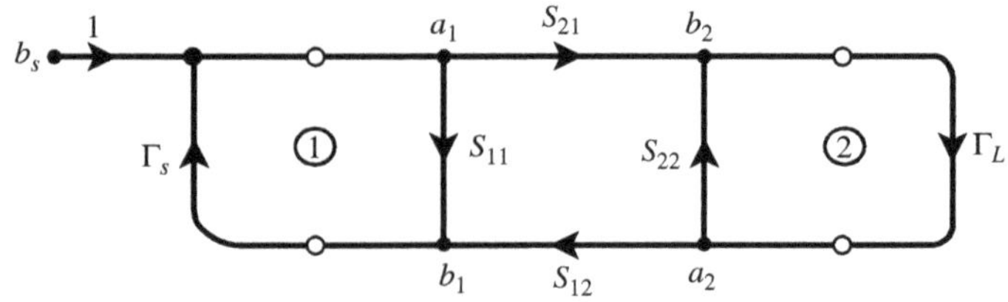

FIGURE 2.10 Signal flow graph of the two-port network in Fig. 2.9(*a*).

form, the corresponding flow graph is particularly useful because the flow graph of cascaded networks is constructed simply by joining together the flow graphs of the individual networks, and the solution is then available directly.

Figure 2.10 shows the signal flow graph of the two-port network in Fig. 2.9(*a*), where a_1 and a_2 are the normalized incident waves and b_1 and b_2 are normalized reflected waves at ports ① and ②. They are represented in the flow graph as points. The points are related to one another by directed lines marked with appropriate coefficients, which are the scattering coefficients S_{11}, S_{21}, S_{12}, and S_{22}, a source reflection coefficient Γ_S, and a load reflection coefficient Γ_L. The points and directed lines, called *nodes* and *branches*, are the primary components of a signal flow graph.

A normalized incident wave a_1 at port ① is split into two parts as shown in Fig. 2.10. One goes through S_{21} to node b_2, and the other goes through S_{11} to b_1 as a reflected wave. At node b_1, the wave goes out port ① and will be partially reflected and reenter port ① if a source with nonzero Γ_S is connected with the network. At node b_2, the wave goes out port ② and will also be partially reflected and reenter port ② if a load with nonzero Γ_L is connected. Part of the wave is reflected back through S_{22}, and part of the wave is transmitted through S_{12}. Once a microwave network has been represented in signal flow graph form, it is a relatively easy matter to solve for the ratio of any combination of wave amplitudes. There are two ways of doing this, but only one of them, Mason's rule, will be introduced.

Flow Graph Analysis Using Mason's Rule The signal flow graph can also be synthesized by Mason's rule [8–10]. It can be used to solve the relationship between any two nodes in the network as

$$T = \frac{\begin{array}{c} P_1\lfloor 1 - \sum L_1^{(1)} + \sum L_2^{(1)} - \cdots \rfloor \\ +P_2\lfloor 1 - \sum L_1^{(2)} + \sum L_2^{(2)} - \cdots \rfloor + \cdots \end{array}}{1 - \sum L_1 + \sum L_2 - \sum L_3 + \cdots}, \tag{2.58}$$

where P_1, P_2, ... are the various *paths*, L_1, L_2, ... are the first- and second-order *loops*, and $L_1^{(1)}$, $L_2^{(1)}$, ... are the first- and second-order loops that do not touch path P_1. A *path* is defined as a series of directed lines followed in sequence and in the same direction in such a way that no node is touched more than once.

The value of the path is the product of all coefficients encountered. A first-order loop is defined as a series of directed lines from a node back to the same node without crossing the same node twice. A second-order loop is the product of any two first-order loops that do not touch at any node.

If we are concerned with the ratio of b_1/b_s in Fig. 2.10, $P_1 = S_{11}$, $P_2 = S_{21}\Gamma_L S_{12}$, $\sum L_1 = S_{11}\Gamma_S + S_{22}\Gamma_L + S_{21}\Gamma_L S_{12}\Gamma_S$, $\sum L_2 = S_{11}\Gamma_S S_{22}\Gamma_L$, $\sum L_1^{(2)} = 0$, and $\sum L_2^{(1)} = 0$. So the ratio is calculated as

$$\frac{b_1}{b_s} = \frac{S_{11}(1 - S_{22}\Gamma_L) + S_{21}\Gamma_L S_{12}(1)}{1 - (S_{11}\Gamma_s + S_{22}\Gamma_L + S_{21}\Gamma_L S_{12}\Gamma_s) + S_{11}S_{22}\Gamma_s\Gamma_L}. \tag{2.59}$$

Other ratios are given similarly as

$$\frac{b_2}{b_s} = \frac{S_{21}}{1 - (S_{11}\Gamma_s + S_{22}\Gamma_L + S_{21}\Gamma_L S_{12}\Gamma_s) + S_{11}S_{22}\Gamma_s\Gamma_L}, \tag{2.60a}$$

$$\frac{a_1}{b_s} = \frac{1 - S_{22}\Gamma_L}{1 - (S_{11}\Gamma_s + S_{22}\Gamma_L + S_{21}\Gamma_L S_{12}\Gamma_s) + S_{11}S_{22}\Gamma_s\Gamma_L}, \tag{2.60b}$$

$$\frac{a_2}{b_s} = \frac{S_{21}\Gamma_L}{1 - (S_{11}\Gamma_s + S_{22}\Gamma_L + S_{21}\Gamma_L S_{12}\Gamma_s) + S_{11}S_{22}\Gamma_s\Gamma_L}. \tag{2.60c}$$

The results of (2.59) and (2.60) are the same as those from (2.54).

Scattering Transfer Parameters In dealing with circuits in cascade, as depicted in Fig. 2.11, the scattering formalism is not the best description of the network. To overcome this difficulty, a scattering transfer matrix is defined and this new matrix is known as a ***T*** matrix. It is obtained by rearranging the scattering relations so that input waves a_1 and b_1 are dependent variables and output waves a_2 and b_2 are independent ones. Rearranging (2.43) gives

$$\begin{bmatrix} b_1 \\ a_1 \end{bmatrix} = \begin{bmatrix} T_{11} & T_{12} \\ T_{21} & T_{22} \end{bmatrix} \begin{bmatrix} a_2 \\ b_2 \end{bmatrix}, \tag{2.61}$$

where

$$T_{11} = S_{12} - \frac{S_{11}S_{22}}{S_{21}},$$

$$T_{12} = \frac{S_{11}}{S_{21}},$$

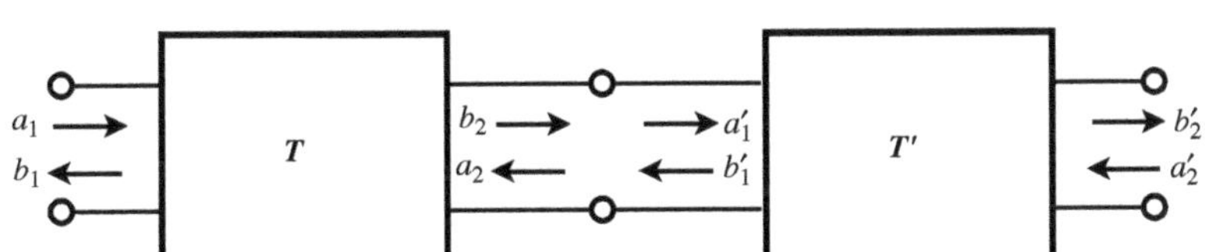

FIGURE 2.11 Two two-port networks connected in cascade.

$$T_{21} = -\frac{S_{22}}{S_{21}},$$

$$T_{22} = \frac{1}{S_{21}}.$$

A transfer matrix is also sometimes defined with the input waves as the independent variables and the output waves as the dependent variables. Referring to the notation in (2.61), the transfer matrix of the second network in Fig. 2.11 is

$$\begin{bmatrix} b'_1 \\ a'_1 \end{bmatrix} = \begin{bmatrix} T'_{11} & T'_{12} \\ T'_{21} & T'_{22} \end{bmatrix} \begin{bmatrix} a'_2 \\ b'_2 \end{bmatrix}. \tag{2.62}$$

Using

$$\begin{bmatrix} b'_1 \\ a'_1 \end{bmatrix} = \begin{bmatrix} a_2 \\ b_2 \end{bmatrix} \tag{2.63}$$

gives

$$\begin{bmatrix} b_1 \\ a_1 \end{bmatrix} = \begin{bmatrix} T_{11} & T_{12} \\ T_{21} & T_{22} \end{bmatrix} \begin{bmatrix} T'_{11} & T'_{12} \\ T'_{21} & T'_{22} \end{bmatrix} \begin{bmatrix} a'_2 \\ b'_2 \end{bmatrix}. \tag{2.64}$$

Taking the ratio b_1/a_1 gives S_{11} for the overall network. Since matrix multiplication is not commutative, these $\boldsymbol{T}$ matrices must be multiplied in the proper order. Using the alternative definition for the $\boldsymbol{T}$ parameters mentioned above, this matrix multiplication must be done in the reverse order.

Generalized Two-Port Scattering Parameters Although the scattering parameters of networks usually refer to 50 Ω, it is sometimes useful to define them in terms of specific load and generator impedances. For a two-port network, the new parameters for arbitrary load and generator impedances [11] may be given in terms of the measured 50-Ω parameters by

$$S_{11} = \frac{A_1^*}{A_1} \frac{(1 - r_2 S'_{22})(S'_{11} - r_1^*) + r_2 S'_{12} S'_{21}}{(1 - r_1 S'_{11})(1 - r_2 S'_{22}) - r_1 r_2 S'_{12} S'_{21}}, \tag{2.65a}$$

$$S_{12} = \frac{A_2^*}{A_1} \frac{S'_{12}\left[(1 - |r_1|^2)\right]}{(1 - r_1 S'_{11})(1 - r_2 S'_{22}) - r_1 r_2 S'_{12} S'_{21}}, \tag{2.65b}$$

$$S_{21} = \frac{A_1^*}{A_2} \frac{S'_{21}\left[(1 - |r_2|^2)\right]}{(1 - r_1 S'_{11})(1 - r_2 S'_{22}) - r_1 r_2 S'_{12} S'_{21}}, \tag{2.65c}$$

$$S_{22} = \frac{A_2^*}{A_2} \frac{(1 - r_1 S'_{11})(S'_{22} - r_2^*) + r_1 S'_{12} S'_{21}}{(1 - r_1 S'_{11})(1 - r_2 S'_{22}) - r_1 r_2 S'_{12} S'_{21}}, \tag{2.65d}$$

where

$$A_i = \frac{1 - r_i^*}{|1 - r_i|}\left(1 - |r_i|^2\right)^{1/2} \quad \text{and} \quad r_i = \frac{Z_i - Z_i'}{Z_i + Z_i'^*}, \qquad i = 1, 2$$

with normalized real impedance Z_i' and arbitrary real load impedance Z_i.

2.4 OTHER CIRCUIT PARAMETERS

If a network terminated in arbitrary impedances has more than two ports, derivation of the basis-independent scattering parameters of the network is cumbersome. They can be derived in several ways, including reduction of the port number and use of the multiport concept. For other approaches, knowledge of other commonly used circuit parameters is indispensable. In the rest of this chapter, the derivation of other circuit parameters and their relation to the scattering parameters are discussed.

2.4.1 ABCD Parameters

A typical microwave subsystem consists of a cascade of two-port networks such that the output of one network is connected to the input of the next, and so on. The two-port networks can be represented by their impedance, admittance, or scattering parameters. It is often more useful, however, to represent two-port networks by *ABCD* parameters because they allow computation of the matrix of the overall cascade network by multiplying the matrices of the individual networks.

Figure 2.12 shows two two-port networks. In Fig. 2.12(*a*) one network is expressed with incident and reflected waves, a_1, b_1, a_2, and b_2; two networks are connected in cascade in Fig. 2.12(*b*). The *ABCD* matrix of Fig. 2.12(*a*) is defined by the equations

$$\begin{bmatrix} V_1 \\ I_1 \end{bmatrix} = \begin{bmatrix} A & B \\ C & D \end{bmatrix} \begin{bmatrix} V_2 \\ I_2 \end{bmatrix}, \tag{2.66}$$

where I_1 is flowing into the two-port network and I_2 is flowing out of the network, as indicated in Fig. 2.12(*a*). These parameters are particularly useful, and the *ABCD* parameters for the two cascaded networks in Fig. 2.12(*b*) are given by

$$\begin{bmatrix} A & B \\ C & D \end{bmatrix} = \begin{bmatrix} A_a & B_a \\ C_a & D_a \end{bmatrix} \begin{bmatrix} A_b & B_b \\ C_b & D_b \end{bmatrix}. \tag{2.67}$$

By repeated application of this operation, the *ABCD* parameters can be computed for any number of two-port networks in cascade. Under certain conditions the *ABCD* parameters are interrelated in the special ways described below.

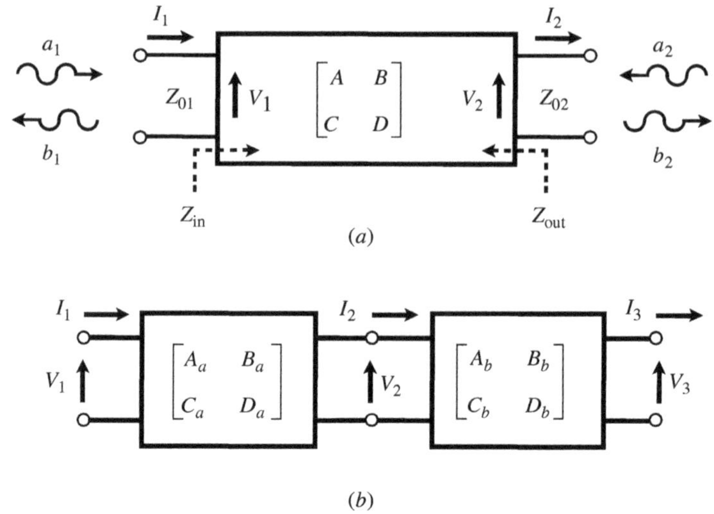

FIGURE 2.12 Networks for the *ABCD* parameters: (*a*) definition of currents and voltages; (*b*) two two-port networks in cascade.

Properties of ABCD Parameters If the network is reciprocal (passive),

$$AD - BC = 1. \tag{2.68}$$

If the network is symmetrical,

$$A = D. \tag{2.69}$$

If the network is lossless (i.e., without dissipative elements), the A and D terms will be purely real and the B and C terms will be purely imaginary for frequencies $j\omega$. If the network in Fig. 2.12(a) is turned around (i.e., ports ① and ② are interchanged), the square matrix in (2.66) is

$$\begin{bmatrix} A_t & B_t \\ C_t & D_t \end{bmatrix} = \begin{bmatrix} D & B \\ C & A \end{bmatrix}, \tag{2.70}$$

where the parameters with t subscripts are for the network when turned around, and the parameters without subscripts are for the network with its original orientation. In both cases, V_1 and I_1 are at the terminals at the left (port ①) and V_2 and I_2 are the terminals at the right (port ②).

ABCD Parameters for Useful Network Properties Consider the two-port network characterized by *ABCD* parameters in Fig. 2.12(a), where the characteristic impedances of the connecting lines at ports ① and ② are real Z_{01} and Z_{02}. The two-port network can be described with incident and reflected waves a_1, b_1, a_2, and b_2, which are, from (2.31), given as

$$a_1 = \frac{V_1 + Z_{01}I_1}{2\sqrt{Z_{01}}}, \tag{2.71a}$$

$$b_1 = \frac{V_1 - Z_{01}I_1}{2\sqrt{Z_{01}}}, \tag{2.71b}$$

$$a_2 = \frac{V_1 + Z_{02}I_1}{2\sqrt{Z_{02}}}, \tag{2.71c}$$

$$b_2 = \frac{V_2 + Z_{02}I_2}{2\sqrt{Z_{02}}}, \tag{2.71d}$$

where the defined current at port ② in Fig. 2.12(a) is the negative of that in (2.31).

The input reflection coefficient Γ_{in} is S_{11} in this case and is defined as the ratio of b_1 to a_1. Therefore, Γ_{in} is

$$\Gamma_{in} = \frac{b_1}{a_1} = \frac{V_1 - Z_{01}I_1}{V_1 + Z_{01}I_1} = \frac{AZ_{02} + B - CZ_{01}Z_{02} - DZ_{01}}{AZ_{02} + B + CZ_{01}Z_{02} + DZ_{01}}, \tag{2.72}$$

where $V_1 = AV_2 + BI_2$, $I_1 = CV_2 + DI_2$, and $Z_{02} = V_2/I_2$ are used.

The output reflection coefficient Γ_{out} is S_{22} in this case and is defined as the ratio of b_2 to a_2. Γ_{out} is found similarly as

$$\Gamma_{out} = \frac{-AZ_{02} + B - CZ_{01}Z_{02} + DZ_{01}}{AZ_{02} + B + CZ_{01}Z_{02} + DZ_{01}}. \tag{2.73}$$

The Γ_{in} and Γ_{out} are derived in other ways. They are given as

$$\Gamma_{in} = \frac{Z_{in} - Z_{01}}{Z_{in} + Z_{01}}, \tag{2.74a}$$

$$\Gamma_{out} = \frac{Z_{out} - Z_{02}}{Z_{out} + Z_{02}}, \tag{2.74b}$$

where

$$Z_{in} = \frac{V_1}{I_1} = \frac{AV_2 + BI_2}{CV_2 + DI_2} = \frac{AZ_{02} + B}{CZ_{02} + D},$$

$$Z_{out} = \frac{V_2}{I_2} = \frac{DV_1 + BI_1}{CV_1 + AI_1} = \frac{DZ_{01} + B}{CZ_{01} + A},$$

where the turned-around concept of (2.70) is used.

If the two-port network is reciprocal and passive, the A and D terms are purely real and the B and C terms are purely imaginary. Thus, the magnitude of Γ_{in} is equal to that of Γ_{out}, indicating that the magnitude of the reflection coefficient evaluated at port ① with port ② terminating in its reference impedance is equal to that at port ② with port ① terminating in its reference impedance.

A transmission coefficient T_{21} for the two-port network in Fig. 2.12(*a*) can be expressed in terms of the *ABCD* parameters as

$$T_{21} = \frac{b_2}{a_1} = \frac{(V_2 + Z_{02} I_2)/2\sqrt{Z_{02}}}{(V_1 + Z_{01} I_1)/2\sqrt{Z_{01}}} = \frac{2\sqrt{Z_{01}}\sqrt{Z_{02}}}{AZ_{02} + B + CZ_{01}Z_{02} + DZ_{01}}. \quad (2.75)$$

Note that $|T_{21}|^2$ is the ratio of power in the output wave to that in the input wave.

ABCD Parameters of Common Structures The common structures we meet frequently are two-port networks with a series impedance or shunt admittance, T-networks, π-networks, transmission lines, transformer circuits, and so on. They are connected in cascade and constitute more complicated microwave networks. Therefore, knowledge of their *ABCD* parameters is indispensable. Their derivation is treated in this section.

A two-port network with a series impedance is shown in Fig. 2.13(*a*), the situation with $I_2 = 0$ and $V_2 = 0$ is depicted in Fig. 2.13(*b*), and a two-port network with a shunt admittance and its *ABCD* parameters are represented in Fig. 2.13(*c*). The *ABCD* parameters can be found under the condition $I_2 = 0$ or $V_2 = 0$, as written in (2.66). $I_2 = 0$ or $V_2 = 0$ means that port ② is open- or short-circuited, respectively, as depicted in Fig. 2.13(*b*). When $I_2 = 0$ in Fig. 2.13(*b*), the following equations hold:

$$I_1 = 0, \quad (2.76a)$$

$$V_1 = V_2. \quad (2.76b)$$

When $V_2 = 0$ in Fig. 2.13(*b*), the voltages and currents at ports ① and ② are given as

$$I_1 = I_2, \quad (2.77a)$$

$$V_1 = I_1 Z. \quad (2.77b)$$

The *ABCD* parameters with the series impedance are given from (2.76) and (2.77), as

$$A = \left.\frac{V_1}{V_2}\right|_{I_2=0} = 1, \quad (2.78a)$$

$$B = \left.\frac{V_1}{I_2}\right|_{V_2=0} = \frac{I_1 Z}{I_1} = Z, \quad (2.78b)$$

$$C = \left.\frac{I_1}{V_2}\right|_{I_2=0} = 0, \quad (2.78c)$$

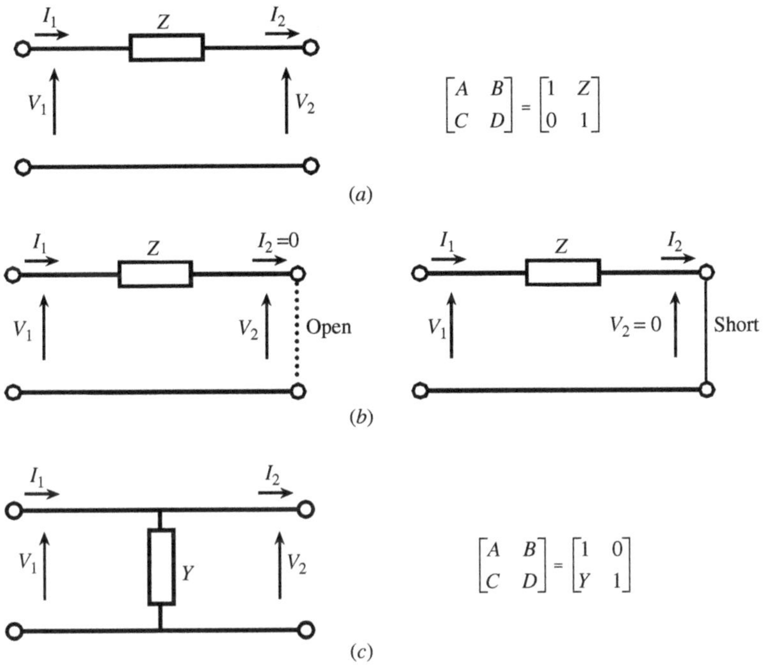

FIGURE 2.13 Two-port networks with a series impedance and a shunt admittance: (*a*) two-port network with a series impedance and its *ABCD* parameters; (*b*) situation with $I_2 = 0$ and $V_2 = 0$; (*c*) two-port network with a shunt admittance and its *ABCD* parameters.

$$D = \left.\frac{I_1}{I_2}\right|_{V_2=0} = 1. \tag{2.78d}$$

The two-port network with a shunt admittance is the dual network of that in Fig. 2.13(*a*). Therefore, the voltages and currents are interchangeable, so the following relations hold: When $V_2 = 0$,

$$V_1 = 0, \tag{2.79a}$$

$$I_1 = I_2, \tag{2.79b}$$

and when $I_2 = 0$,

$$V_1 = V_2, \tag{2.80a}$$

$$I_1 = V_1 Y. \tag{2.80b}$$

The *ABCD* parameters of the two-port network with a shunt admittance are easily found from (2.79) and (2.80) to be those in Fig. 2.13(*c*).

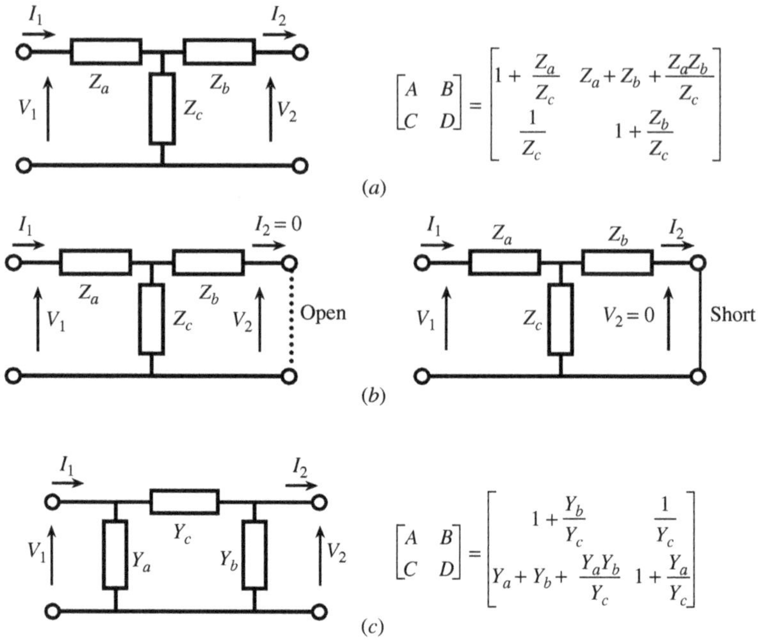

FIGURE 2.14 T- and Π-networks and their *ABCD* parameters: (*a*) T-network; (*b*) situation with $I_2 = 0$ and $V_2 = 0$; (*c*) Π-network.

A T-network with impedances Z_a, Z_b, and Z_c is depicted in Fig. 2.14(*a*), and the situation with $I_2 = 0$ and $V_2 = 0$ is shown in Fig. 2.14(*b*). The Π-network is the dual network of the T-network and is shown in Fig. 2.14(*c*). When $I_2 = 0$ in Fig. 2.14(*b*), the relations between voltages and currents are given as

$$V_1 = (Z_a + Z_c)I_1, \tag{2.81a}$$

$$V_2 = Z_c I_1. \tag{2.81b}$$

When $V_2 = 0$ in Fig. 2.14(*b*), they are

$$V_1 = Z_a I_1 + Z_b I_2, \tag{2.82a}$$

$$Z_c(I_1 - I_2) = Z_b I_2. \tag{2.82b}$$

The *ABCD* parameters of the T-network in Fig. 2.14(*a*) are calculated from (2.81) and (2.82), as written in Fig. 2.14(*a*). As the Π-network depicted in Fig. 2.14(*c*) is the dual network of the T-network, the voltages and currents are also interchangeable and the *ABCD* parameters are given similarly to those in Fig. 2.14(*c*).

One of the structures used most commonly at microwave frequencies is a uniform transmission line. Therefore, information about it is indispensable to the analysis of the entire microwave system. Here a method to obtain the *ABCD*

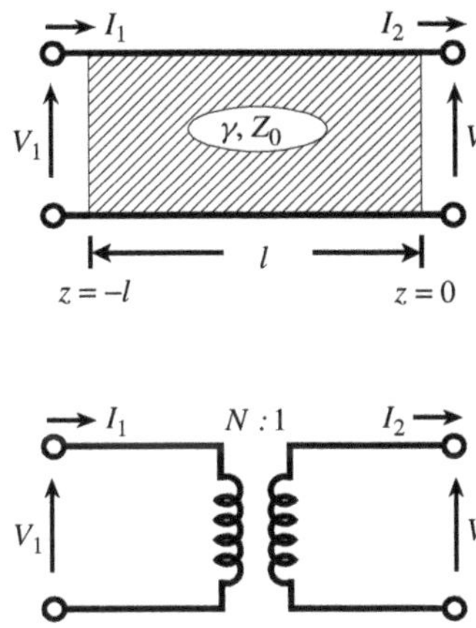

$$\begin{bmatrix} A & B \\ C & D \end{bmatrix} = \begin{bmatrix} \cosh\gamma l & Z_0 \sinh\gamma l \\ \dfrac{\sinh\gamma l}{Z_0} & \cosh\gamma l \end{bmatrix}$$

(*a*)

$$\begin{bmatrix} A & B \\ C & D \end{bmatrix} = \begin{bmatrix} N & 0 \\ 0 & \dfrac{1}{N} \end{bmatrix}$$

(*b*)

FIGURE 2.15 (*a*) Uniform transmission line and (*b*) transformer, and their *ABCD* parameters.

parameters of the uniform transmission line is introduced by which any of its circuit parameters can be derived. Figure 2.15(*a*) shows a uniform transmission line whose characteristic impedance, propagation constant, and length are Z_0, γ and l, respectively. From the wave equations in (2.1), the voltages and currents at z are

$$V(z) = V_0^+ e^{-\gamma z} + V_0^- e^{\gamma z}, \tag{2.83a}$$

$$I(z) = \frac{V_0^+}{Z_0} e^{-\gamma z} - \frac{V_0^-}{Z_0} e^{\gamma z}. \tag{2.83b}$$

The currents and voltages at ports ① and ② are obtained from (2.83) as

$$V_1 = V(-l) = V_0^+ e^{\gamma l} + V_0^- e^{-\gamma l}, \tag{2.84a}$$

$$I_1 = I(-l) = \frac{V_0^+ e^{\gamma l}}{Z_0} - \frac{V_0^- e^{-\gamma l}}{Z_0}, \tag{2.84b}$$

$$V_2 = V(0) = V_0^+ + V_0^-, \tag{2.84c}$$

$$I_2 = I(0) = \frac{V_0^+}{Z_0} - \frac{V_0^-}{Z_0}. \tag{2.84d}$$

When $I_2 = 0$ and $V_0^+ = V_0^-$, the A and C terms are

$$A = \frac{V_1}{V_2} = \frac{e^{\gamma l} + e^{-\gamma l}}{2} = \cosh\gamma l, \tag{2.85a}$$

$$C = \frac{I_1}{V_2} = \frac{1}{Z_0}\frac{e^{\gamma l} - e^{-\gamma l}}{2} = \frac{1}{Z_0}\sinh\gamma l. \tag{2.85b}$$

All *ABCD* parameters are derived similarly to those in Fig. 2.15(*a*). A transformer is also a useful two-port network, and is depicted in Fig. 2.15(*b*), where the turns ratio is $N : 1$. Since $V_1 = NV_2$ and $I_1 = 1/NI_2$ in this case, its *ABCD* parameters are derived similarly to those in Fig. 2.15(*b*).

2.4.2 Open-Circuit Impedance and Short-Circuit Admittance Parameters

Open-Circuit Impedance Parameters In Fig. 2.16, the *open-circuit impedances* of a two-port network may be defined by the equations

$$V_1 = Z_{11}I_1 + Z_{12}I_2, \tag{2.86a}$$

$$V_2 = Z_{21}I_1 + Z_{22}I_2. \tag{2.86b}$$

Physically, Z_{11} is the input impedance at port ① when port ② is open-circuited. The quantity Z_{12} could be measured as the ratio of the voltage V_1 to the current I_2, V_1/I_2, when port ① is open-circuited and current I_2 is flowing into port ②. The parameters Z_{22} and Z_{21} may be interpreted analogously.

Short-Circuit Admittance Parameters In a similar fashion, the *short-circuited admittances* may be defined as shown in Fig. 2.16, where the relations are given as

$$I_1 = Y_{11}V_1 + Y_{12}V_2, \tag{2.87a}$$

$$I_2 = Y_{21}V_1 + Y_{22}V_2. \tag{2.87b}$$

In this case, Y_{11} is the input admittance at port ① when port ② is short-circuited. The parameter Y_{12} may be computed as the ratio I_1/V_2 when port ① is short-circuited, where V_2 is the voltage at port ②. For reciprocal networks, $Z_{12} = Z_{21}$ and $Y_{12} = Y_{21}$, and they are all purely imaginary for a complex plane.

Conversions Between ABCD, Impedance, and Admittance Parameters Knowing the impedance parameters of a network, its *ABCD* parameters can

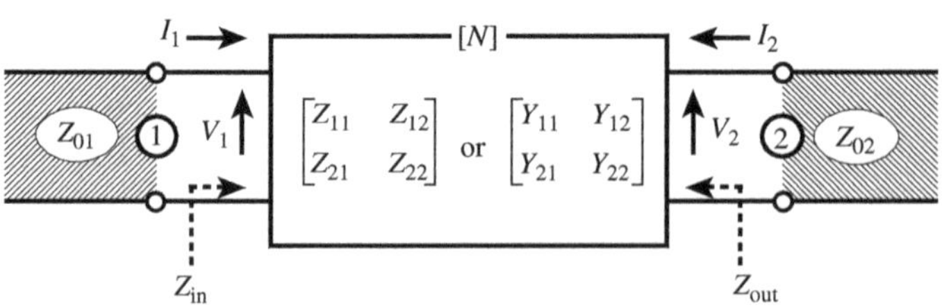

FIGURE 2.16 Two-port network.

easily be determined. For the definition of the *ABCD* parameters in Fig. 2.12(*a*), the relations between currents and voltages are

$$V_1 = Z_{11}I_1 - Z_{12}I_2, \tag{2.88a}$$

$$V_2 = Z_{21}I_1 - Z_{22}I_2. \tag{2.88b}$$

By rearranging (2.88) so that V_2 and I_2 are independent variables and V_1 and I_1 are dependent variables, the relation of the impedance to the *ABCD* parameters is obtained as

$$V_1 = \frac{Z_{11}}{Z_{21}}V_2 + \left(\frac{Z_{11}Z_{22}}{Z_{21}} - Z_{12}\right)I_2 = AV_2 + BI_2, \tag{2.89a}$$

$$I_1 = \frac{1}{Z_{21}}V_2 + \frac{Z_{22}}{Z_{21}}I_2 = CV_2 + DI_2. \tag{2.89a}$$

For the conversion of admittance parameters to *ABCD* parameters, a new relation can be written so that V_2 and I_2 are dependent variables and V_1 and I_1 are independent variables:

$$V_1 = -\frac{1}{Y_{21}}(Y_{22}V_2 + I_2) = AV_2 + BI_2, \tag{2.90a}$$

$$I_1 = \left(\frac{-Y_{11}Y_{22}}{Y_{21}} + Y_{12}\right)V_2 - \frac{Y_{11}}{Y_{21}}I_2 = CV_2 + DI_2. \tag{2.90b}$$

The conversion of *ABCD* parameters to impedance parameters or admittance parameters is easily derived and one example of the conversion of *ABCD* to admittance parameters is shown. Rearranging the relation for the *ABCD* parameters in (2.66) so that I_1 and I_2 are dependent variables and V_1 and V_2 are independent variables results in

$$V_1 = \frac{A}{C}I_1 + \frac{AD - BC}{C}I_2 = Z_{11}I_1 + Z_{12}I_2, \tag{2.91a}$$

$$V_2 = \frac{I_1}{C} + \frac{D}{C}I_2 = Z_{21}I_1 + Z_{22}I_2. \tag{2.91b}$$

Other conversions can be derived in a similar way. Based on (2.89)–(2.91), the conversions between *ABCD*, impedance, and admittance parameters are as listed in Table 2.1.

Common Structures The impedance and admittance parameters can be obtained from the *ABCD* parameters, but an additional computation is needed. Here, the useful common structures introduced above will be discussed and their derivation introduced from the viewpoint of impedance and admittance parameters. Consider the uniform transmission line shown in Fig. 2.17(*a*), where the currents and voltages at ports ① and ② are as indicated. From the wave equations of the

TABLE 2.1 Conversions Between *ABCD*, Impedance, and Admittance Parameters[a]

	$[A]$	$[Y]$	$[Z]$
$[A]$	$\begin{bmatrix} A & B \\ C & D \end{bmatrix}$	$\dfrac{1}{Y_{21}}\begin{bmatrix} -Y_{22} & -1 \\ -\lvert Y\rvert & -Y_{11} \end{bmatrix}$	$\dfrac{1}{Z_{21}}\begin{bmatrix} Z_{11} & \lvert Z\rvert \\ 1 & Z_{22} \end{bmatrix}$
$[Y]$	$\dfrac{1}{B}\begin{bmatrix} D & \Delta \\ -1 & A \end{bmatrix}$	$\begin{bmatrix} Y_{11} & Y_{12} \\ Y_{21} & Y_{22} \end{bmatrix}$	$\dfrac{1}{\lvert Z\rvert}\begin{bmatrix} Z_{22} & -Z_{21} \\ -Z_{12} & Z_{11} \end{bmatrix}$
$[Z]$	$\dfrac{1}{C}\begin{bmatrix} A & \Delta \\ 1 & D \end{bmatrix}$	$\dfrac{1}{\lvert Y\rvert}\begin{bmatrix} Y_{22} & -Y_{21} \\ -Y_{12} & Y_{11} \end{bmatrix}$	$\begin{bmatrix} Z_{11} & Z_{12} \\ Z_{21} & Z_{22} \end{bmatrix}$

[a] $\Delta = AD - BC$, $|Z| = Z_{11}Z_{22} - Z_{21}Z_{12}$, $|Y| = Y_{11}Y_{22} - Y_{12}Y_{21}$.

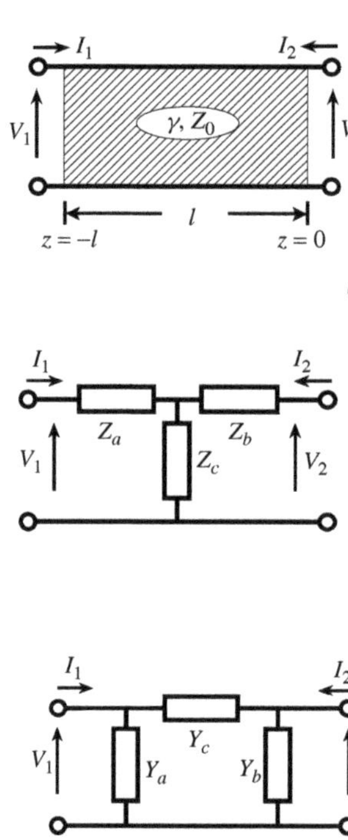

$$\begin{bmatrix} Z_{11} & Z_{12} \\ Z_{21} & Z_{22} \end{bmatrix} = \begin{bmatrix} Z_0 \coth \gamma l & \dfrac{Z_0}{\sinh \gamma l} \\ \dfrac{Z_0}{\sinh \gamma l} & Z_0 \coth \gamma l \end{bmatrix}$$

$$\begin{bmatrix} Y_{11} & Y_{12} \\ Y_{21} & Y_{22} \end{bmatrix} = \begin{bmatrix} \dfrac{\coth \gamma l}{Z_0} & -\dfrac{1}{Z_0 \sinh \gamma l} \\ \dfrac{1}{Z_0 \sinh \gamma l} & \dfrac{\coth \gamma l}{Z_0} \end{bmatrix}$$

(*a*)

$$\begin{bmatrix} Z_{11} & Z_{12} \\ Z_{21} & Z_{22} \end{bmatrix} = \begin{bmatrix} Z_a + Z_c & Z_c \\ Z_c & Z_b + Z_c \end{bmatrix}$$

$$\begin{bmatrix} Y_{11} & Y_{12} \\ Y_{21} & Y_{22} \end{bmatrix} = \frac{1}{\Delta_z}\begin{bmatrix} Z_b + Z_c & -Z_c \\ -Z_c & Z_a + Z_c \end{bmatrix}$$

where $\Delta_z = Z_aZ_b + Z_bZ_c + Z_cZ_a$

(*b*)

$$\begin{bmatrix} Y_{11} & Y_{12} \\ Y_{21} & Y_{22} \end{bmatrix} = \begin{bmatrix} Y_a + Y_c & -Y_c \\ -Y_c & Y_b + Y_c \end{bmatrix}$$

$$\begin{bmatrix} Z_{11} & Z_{12} \\ Z_{21} & Z_{22} \end{bmatrix} = \frac{1}{\Delta_Y}\begin{bmatrix} Y_b + Y_c & Y_c \\ Y_c & Y_a + Y_c \end{bmatrix}$$

where $\Delta_Y = Y_aY_b + Y_bY_c + Y_cY_a$

(*c*)

FIGURE 2.17 Impedance and admittance parameters of common structures: (*a*) transmission line; (*b*) T-network; (*c*) Π-network.

transmission line in (2.2), the currents and voltages at ports ① and ② are given as

$$V_1 = V(-l) = V_0^+ e^{\gamma l} + V_0^- e^{-\gamma l}, \tag{2.92a}$$

$$I_1 = I(-l) = \frac{V_0^+ e^{\gamma l}}{Z_0} - \frac{V_0^- e^{-\gamma l}}{Z_0}, \tag{2.92b}$$

$$V_2 = V(0) = V_0^+ + V_0^-, \tag{2.92c}$$

$$I_2 = -I(0) = -\left(\frac{V_0^+}{Z_0} - \frac{V_0^-}{Z_0}\right). \tag{2.92d}$$

In (2.86), Z_{11} is the ratio of V_1 to I_1 and Z_{21} that of V_2 to I_1 when $I_2 = 0$. For $I_2 = 0$ in (2.92*d*), $V_0^+ = V_0^-$ and Z_{11} and Z_{21} are easily derived as

$$Z_{11} = \frac{V_1}{I_1} = \frac{V(-l)}{I(-l)} = Z_0 \coth \gamma l, \tag{2.93a}$$

$$Z_{21} = \frac{V_2}{I_1} = \frac{V(0)}{I(-l)} = Z_0 \frac{2}{e^{\gamma l} - e^{-\gamma l}} = \frac{Z_0}{\sinh \gamma l}. \tag{2.93b}$$

The uniform transmission line is symmetric, so $Z_{11} = Z_{22}$ and $Z_{12} = Z_{21}$ are valid. Therefore, the impedance matrix is derived similarly to that in Fig. 2.17(*a*), as is the admittance matrix.

Since any microwave two-port network can be equivalent to a T- or Π-network, derivation of their admittance and impedance matrices is particularly useful. Thus, in the T-network depicted in Fig. 2.17(*b*), the relation between its voltages and currents is given by

$$V_1 = Z_a I_1 + Z_c(I_1 + I_2) = (Z_a + Z_c)I_1 + Z_c I_2, \tag{2.94a}$$

$$V_2 = Z_b I_2 + Z_c(I_1 + I_2) = Z_c I_1 + (Z_b + Z_c)I_2. \tag{2.94b}$$

Therefore, its impedance and admittance parameters are derived similarly to those in Fig. 2.17(*b*); and the impedance and admittance parameters of the Π-network are derived similarly to those in Fig. 2.17(*c*).

Input and Output Impedances and Admittances The calculation of input and output impedances is important for building matching networks. They can be computed from any general circuit matrix, and one example with *ABCD* parameters was shown in (2.72)–(2.74). Here, their derivation from impedance and admittance parameters is introduced.

Consider the two-port network shown in Fig. 2.16. When a transmission line with characteristic impedance Z_{02} is connected at port ② and assumed to be infinite, the current at port ②, I_2, is related to V_2 and Z_{02}, and their relation is $V_2 = -I_2 Z_{02}$. Since the input impedance Z_{in} is expressed as the ratio of V_1 to I_1, impedance and admittance parameters describing the two-port network can be used:

$$Z_{in} = \frac{V_1}{I_1} = Z_{11} - \frac{Z_{12}Z_{21}}{Z_{22} + Z_{02}}$$

$$= \frac{Y_{22} + Y_{02}}{Y_{11}(Y_{22} + Y_{02}) - Y_{12}Y_{21}}, \tag{2.95}$$

where $Y_{02} = 1/Z_{02}$.

Consider the two-port network in Fig. 2.16, which is connected with a transmission line at port ①. If its characteristic impedance is Z_{01} and the transmission line is assumed to be infinitively long, the output impedance Z_{out} is calculated as

$$Z_{out} = \frac{V_2}{I_2} = Z_{22} - \frac{Z_{12}Z_{21}}{Z_{11} + Z_{01}}$$

$$= \frac{Y_{11} + Y_{01}}{Y_{22}(Y_{11} + Y_{01}) - Y_{12}Y_{21}}, \tag{2.96}$$

where $Y_{01} = 1/Z_{01}$.

2.4.3 Conversion Matrices of Two-Port Networks Terminated in Arbitrary Impedances

Frequently in practice, a two-port network is not terminated in equal impedances, and the two termination impedances have a series effect on the scattering matrix. Conversions without regard to termination impedances were treated in Table 2.1, and those related to the two termination impedances are treated in this section. Since the scattering matrix is defined based on the reference and network immittance (impedance and admittance) matrices, conversions between scattering and immittance matrices are first considered and then those between scattering and *ABCD* matrices. All the conversions are listed in Table 2.2, and the detailed processes are described in Appendix *B*.

Scattering Matrix ↔ Impedance Matrix Consider the two-port network in Table 2.2 terminated in two different real characteristic impedances of the transmission lines, Z_{01} and Z_{02}. The conversion of an impedance matrix to a scattering matrix is obtained from the definition of scattering matrix in (2.29*b*) based on current quantities. It is again given as

$$\boldsymbol{S} = \boldsymbol{h}[\boldsymbol{Z} + \boldsymbol{z}]^{-1}[\boldsymbol{Z} - \boldsymbol{z}_*]\boldsymbol{h}_*^{-1}, \tag{2.97}$$

where

$$\boldsymbol{Z} = \begin{bmatrix} Z_{11} & Z_{12} \\ Z_{21} & Z_{22} \end{bmatrix}, \quad \boldsymbol{z} = \begin{bmatrix} Z_{01} & 0 \\ 0 & Z_{02} \end{bmatrix}, \quad \boldsymbol{h} = \begin{bmatrix} \sqrt{Z_{01}} & 0 \\ 0 & \sqrt{Z_{02}} \end{bmatrix},$$

$$\boldsymbol{h}_*^{-1} = \begin{bmatrix} \dfrac{1}{\sqrt{Z_{01}}} & 0 \\ 0 & \dfrac{1}{\sqrt{Z_{02}}} \end{bmatrix},$$

and $\boldsymbol{z} = \boldsymbol{z}_*$ for real Z_{01} and Z_{02}.

TABLE 2.2 Conversions Between Scattering and Other Circuit Parameters[a]

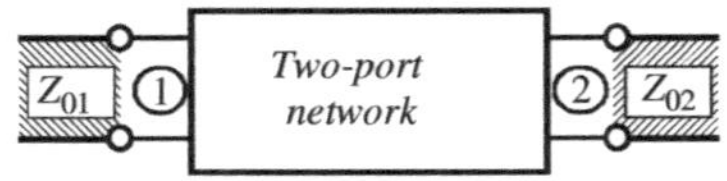

	[ABCD]	[Z]	[Y]
S_{11}	$\dfrac{AZ_{02}+B-CZ_{01}Z_{02}-DZ_{01}}{AZ_{02}+B+CZ_{01}Z_{02}+DZ_{01}}$	$\dfrac{(Z_{11}-Z_{01})(Z_{22}+Z_{02})-Z_{12}Z_{21}}{\Delta_Z}$	$-\dfrac{(Y_{11}-Y_{01})(Y_{22}+Y_{02})-Y_{12}Y_{21}}{\Delta_Y}$
S_{12}	$\dfrac{2\Delta\sqrt{Z_{01}}\sqrt{Z_{02}}}{AZ_{02}+B+CZ_{01}Z_{02}+DZ_{01}}$	$\dfrac{2\sqrt{Z_{01}}\sqrt{Z_{02}}Z_{12}}{\Delta_Z}$	$\dfrac{-2\sqrt{Y_{01}}\sqrt{Y_{02}}Y_{12}}{\Delta_Y}$
S_{21}	$\dfrac{2\sqrt{Z_{01}}\sqrt{Z_{02}}}{AZ_{02}+B+CZ_{01}Z_{02}+DZ_{01}}$	$\dfrac{2\sqrt{Z_{01}}\sqrt{Z_{02}}Z_{21}}{\Delta_Z}$	$\dfrac{-2\sqrt{Y_{01}}\sqrt{Y_{02}}Y_{21}}{\Delta_Y}$
S_{22}	$\dfrac{-AZ_{02}+B-CZ_{01}Z_{02}+DZ_{01}}{AZ_{02}+B+CZ_{01}Z_{02}+DZ_{01}}$	$\dfrac{(Z_{11}+Z_{01})(Z_{22}-Z_{02})-Z_{12}Z_{21}}{\Delta_Z}$	$-\dfrac{(Y_{11}+Y_{01})(Y_{22}-Y_{02})-Y_{12}Y_{21}}{\Delta_Y}$

$$\begin{bmatrix} Z_{11} & Z_{12} \\ Z_{21} & Z_{22} \end{bmatrix} = \frac{1}{(1-S_{11})(1-S_{22})-S_{12}S_{21}} \begin{bmatrix} Z_{01}[(1+S_{11})(1-S_{22})+S_{12}S_{21}] & 2\sqrt{Z_{01}}\sqrt{Z_{02}}S_{12} \\ 2\sqrt{Z_{01}}\sqrt{Z_{02}}S_{21} & Z_{02}[(1-S_{11})(1+S_{22})+S_{12}S_{21}] \end{bmatrix}$$

$$\begin{bmatrix} Y_{11} & Y_{12} \\ Y_{21} & Y_{22} \end{bmatrix} = \frac{1}{(1+S_{11})(1+S_{22})-S_{12}S_{21}} \begin{bmatrix} Y_{01}[(1-S_{11})(1+S_{22})+S_{12}S_{21}] & -2\sqrt{Y_{01}}\sqrt{Y_{02}}S_{12} \\ -2\sqrt{Y_{01}}\sqrt{Y_{02}}S_{21} & Y_{02}[(1+S_{11})(1-S_{22})+S_{12}S_{21}] \end{bmatrix}$$

$$\begin{bmatrix} A & B \\ C & D \end{bmatrix} = \frac{1}{2S_{12}} \begin{bmatrix} \dfrac{\sqrt{Z_{01}}[(1+S_{11})(1-S_{22})+S_{12}S_{21}]}{\sqrt{Z_{02}}} & \sqrt{Z_{01}}\sqrt{Z_{02}}[(1+S_{11})(1+S_{22})-S_{12}S_{21}] \\ \dfrac{(1-S_{11})(1-S_{22})-S_{12}S_{21}}{\sqrt{Z_{01}}\sqrt{Z_{02}}} & \dfrac{\sqrt{Z_{02}}[(1-S_{11})(1+S_{22})+S_{12}S_{21}]}{\sqrt{Z_{01}}} \end{bmatrix}$$

[a] $\Delta_Z = (Z_{11}+Z_{01})(Z_{22}+Z_{02}) - Z_{21}Z_{12}$, $\Delta_Y = (Y_{11}+Y_{01})(Y_{22}+Y_{02}) - Y_{21}Y_{12}$, $Z_{01} = 1/Y_{01}$, $Z_{02} = 1/Y_{02}$, $\Delta = AD - BC$.

Conversion of a scattering matrix to an impedance matrix is also possible. For that, $[\boldsymbol{Z} + z]\boldsymbol{h}^{-1}\boldsymbol{S}\boldsymbol{h}_* = [\boldsymbol{Z} - z_*]$ is derived from (2.97) and the impedance matrix $\boldsymbol{Z}$ is finally given as

$$\boldsymbol{Z} = (z\boldsymbol{h}^{-1}\boldsymbol{S}\boldsymbol{h}_* + z_*)(\boldsymbol{U}_n - \boldsymbol{h}^{-1}\boldsymbol{S}\boldsymbol{h}_*)^{-1}, \tag{2.98}$$

where $\boldsymbol{U}$ is an identity matrix of order 2,

$$\boldsymbol{h}^{-1} = \begin{bmatrix} \dfrac{1}{\sqrt{Z_{01}}} & 0 \\ 0 & \dfrac{1}{\sqrt{Z_{02}}} \end{bmatrix} \quad \text{and} \quad \boldsymbol{h}_* = \begin{bmatrix} \sqrt{Z_{01}} & 0 \\ 0 & \sqrt{Z_{02}} \end{bmatrix}$$

for real Z_{01} and Z_{02}.

Scattering Matrix ↔ Admittance Matrix The conversion of admittance to scattering matrices is also obtained from the definition of a scattering matrix in (2.30*b*). The scattering matrix based on voltage quantities is given as

$$\boldsymbol{S} = -\boldsymbol{k}[\boldsymbol{Y} + \boldsymbol{y}]^{-1}[\boldsymbol{Y} - \boldsymbol{y}_*]\boldsymbol{k}_*^{-1}, \tag{2.99}$$

where

$$\boldsymbol{Y} = \begin{bmatrix} Y_{11} & Y_{12} \\ Y_{21} & Y_{22} \end{bmatrix}, \quad \boldsymbol{y} = \begin{bmatrix} Y_{01} & 0 \\ 0 & Y_{02} \end{bmatrix}, \quad \boldsymbol{k} = \begin{bmatrix} \sqrt{Y_{01}} & 0 \\ 0 & \sqrt{Y_{02}} \end{bmatrix},$$

$$\boldsymbol{k}_*^{-1} = \begin{bmatrix} \dfrac{1}{\sqrt{Y_{01}}} & 0 \\ 0 & \dfrac{1}{\sqrt{Y_{02}}} \end{bmatrix},$$

and $\boldsymbol{y} = \boldsymbol{y}_*$ for real Y_{01} and Y_{02}, $Y_{01} = Z_{01}^{-1}$, and $Y_{02} = Z_{02}^{-1}$. Conversion of scattering to admittance matrices is also possible. For a calculation of the conversion, $[\boldsymbol{Y} + \boldsymbol{y}]\boldsymbol{k}^{-1}\boldsymbol{S}\boldsymbol{k}_* = -[\boldsymbol{Y} - \boldsymbol{y}_*]$ is first derived from (2.99) and the admittance matrix is finally given as

$$\boldsymbol{Y} = (\boldsymbol{y}_* - \boldsymbol{y}\boldsymbol{k}^{-1}\boldsymbol{S}\boldsymbol{k}_*)(\boldsymbol{k}^{-1}\boldsymbol{S}\boldsymbol{k}_* + \boldsymbol{U}_n)^{-1}, \tag{2.100}$$

where $\boldsymbol{U}$ is a identity matrix with order 2,

$$\boldsymbol{k}^{-1} = \begin{bmatrix} \dfrac{1}{\sqrt{Y_{01}}} & 0 \\ 0 & \dfrac{1}{\sqrt{Y_{02}}} \end{bmatrix} \quad \text{and} \quad \boldsymbol{k}_* = \begin{bmatrix} \sqrt{Y_{01}} & 0 \\ 0 & \sqrt{Y_{02}} \end{bmatrix}.$$

ABCD Matrix ↔ Scattering Matrix Since an *ABCD* matrix is not directly convertible to a scattering matrix, a conversion method can be used. From Table 2.1,

$$\begin{bmatrix} Z_{11} & Z_{12} \\ Z_{21} & Z_{22} \end{bmatrix} = \frac{1}{C}\begin{bmatrix} A & \Delta \\ 1 & D \end{bmatrix}, \tag{2.101}$$

where $\Delta = AD - BC$. Using (2.97) and (2.101) yields

$$\boldsymbol{S} = \boldsymbol{h}[\boldsymbol{A}_Z + \boldsymbol{z}]^{-1}[\boldsymbol{A}_Z - \boldsymbol{z}_*]\boldsymbol{h}_*^{-1}, \tag{2.102}$$

where

$$\boldsymbol{A}_Z = \frac{1}{C}\begin{bmatrix} A & \Delta \\ 1 & D \end{bmatrix}$$

and the conversion results are the same as those in (2.72) to (2.75).

For conversion of a scattering matrix to an *ABCD* matrix, the relation between the scattering matrix and the immittance matrices can be used. Table 2.1 gives suitable values: $A = Z_{11}/Z_{21}$, $B = -1/Y_{21}$, $C = 1/Z_{21}$, and $D = Z_{22}/Z_{21}$. Based on (2.72) to (2.75), (2.97) to (2.104), and Table 2.1, the conversions between the scattering matrix and other circuit matrices are as given in Table 2.2, where $S_{21} = S_{12}$ for passive two-port networks with unitary and reciprocal properties [12].

2.5 ANALYSES OF SYMMETRIC NETWORKS

In practical cases, asymmetric networks are common. However, if they have more than two ports, describing them with any circuit parameter is so difficult that easier methods are required. One of them is reduction of the port number, as already discussed. If, on the other hand, the networks are symmetric, analysis using the well-known even- and odd-mode excitation method is powerful.

The analyses of Reed and Wheeler [13] are needed for easy analysis of a symmetric three- or four-port network. The four-port network shown in Fig. 2.18(*a*) is assumed to be symmetrical about the plane S–S', so the impedances terminating various ports are the same. It is also assumed that the network is lossless and that the junction effects are negligible. A signal of amplitude of V_g is applied at port ① and divides in the network. The method of analysis to be described makes possible determination of the signals that appear at the four ports and how they vary (in phase and amplitude) with frequency.

2.5.1 Analyses with Even- and Odd-Mode Excitations

If two signals of amplitude $V_g/2$ and *in phase* (even-mode excitation) are applied at ports ① and ④, by symmetry a voltage maximum occurs at every point on the symmetry plane. That is, these points are all $Z = \infty$ and $Y = 0$. This is equivalent to an open circuit, as illustrated in Fig. 2.18(*b*). Similarly, if two

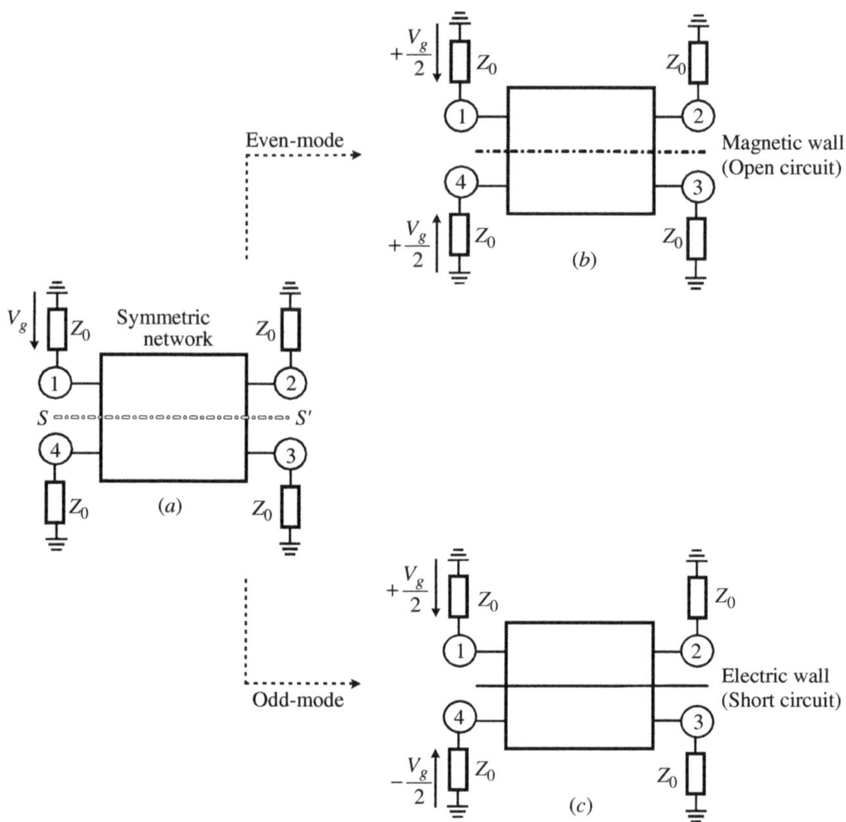

FIGURE 2.18 Even- and odd-mode excitation analyses for a symmetrical network.

signals of amplitude $V_g/2$ and *out of phase* (odd-mode excitation) are applied at ports ① and ④, a voltage minimum occurs at every point on the plane of symmetry. That is, these points are all $Z = \infty$ and $Y = 0$. This is equivalent to a short circuit, as explained in Fig. 2.18(*c*).

In each case, the problem is reduced to that of a two-port network. For even-mode excitation, a reflection coefficient S_{11e} and transmission coefficient S_{21e} are determined. Similarly, for odd-mode excitation, a reflection coefficient S_{11o} and transmission coefficient S_{21o} are determined. By superposition, the sum of the two cases is a signal V_g amplitude at port ①. The resulting signals out of the four ports are also superpositions of the results obtained from the even- and odd-mode excitation cases. Thus, the vector amplitudes of the signals emerging from the four ports are

$$A_1 = \tfrac{1}{2}S_{11e} + \tfrac{1}{2}S_{11o}, \tag{2.103a}$$

$$A_2 = \tfrac{1}{2}S_{21e} + \tfrac{1}{2}S_{21o}, \tag{2.103b}$$

$$A_3 = \tfrac{1}{2}S_{21e} - \tfrac{1}{2}S_{21o}, \qquad (2.103c)$$

$$A_4 = \tfrac{1}{2}S_{11e} - \tfrac{1}{2}S_{11o}. \qquad (2.103d)$$

2.5.2 Useful Symmetric Two-Port Networks

Inductive and capacitive microwave reactances are widely utilized in microwave engineering for the filter construction and impedance (admittance) matching of microwave components. Such reactances are generally obtained by introducing some physical discontinuity in the transmission line that perturbs either the electric or the magnetic fields of the structure. For the modeling of these structures, symmetric T- or Π-networks are useful in conjunction with transmission lines.

Figure 2.19(*a*) shows a symmetric Π-network. Applying node equations to the Π-network results in

$$I_1 = Y_a V_1 + Y_c(V_1 - V_2), \qquad (2.104a)$$

$$I_2 = Y_a V_2 + Y_c(V_2 - V_1). \qquad (2.104b)$$

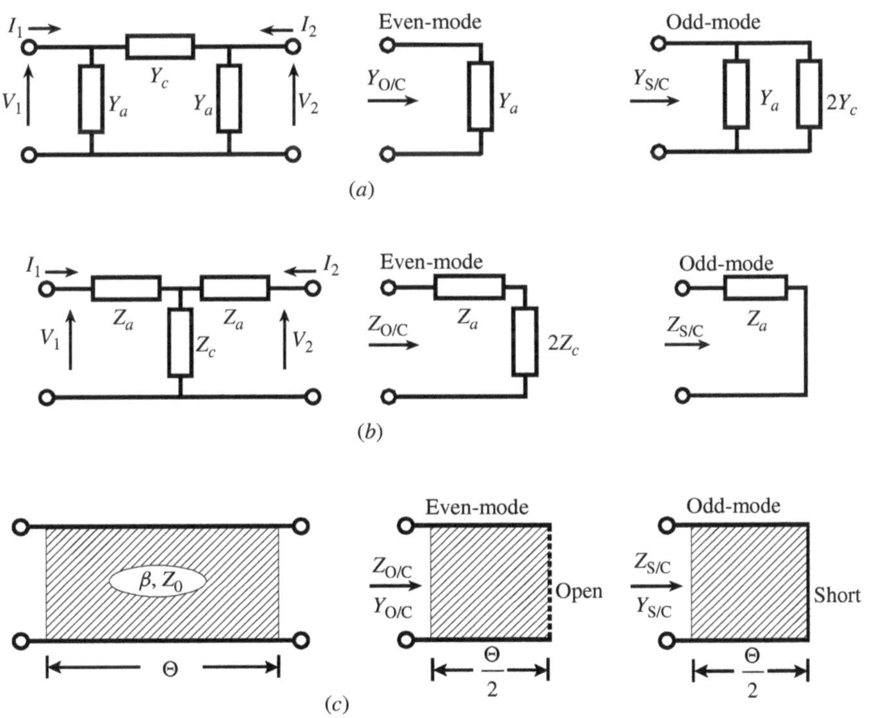

FIGURE 2.19 Useful symmetric structures and their equivalent circuits with even- and odd-mode excitations: (*a*) Π-network; (*b*) T-network; (*c*) transmission line.

Comparing two equations in (2.104) with those describing the admittance parameters in (2.87) gives

$$Y_{11} = Y_a + Y_c = Y_{22}, \tag{2.105a}$$

$$Y_{12} = Y_{21} = -Y_c. \tag{2.105b}$$

For the Π-network to be symmetric, $Y_{11} = Y_{22}$ as shown in (2.105), and the two half-sections associated with even- and odd-mode excitation are those shown in Fig. 2.19(*a*). The even- and odd-mode admittances are found and they are, with the admittance parameters, represented as

$$Y_{\mathrm{O/C}} = Y_a = Y_{11} + Y_{12}, \tag{2.106a}$$

$$Y_{\mathrm{S/C}} = Y_a + 2Y_c = Y_{11} - Y_{12}. \tag{2.106b}$$

The dual network of the Π-network is the T-network, as stated above. If all the relations between voltages and currents in a network are the same as those in another network when voltages and currents are interchanged, the two networks are said to be *dual networks*. Due to that, parallel and series connections, as well as short and open connections, are also interchangeable. The T-network shown in Fig. 2.19(*b*) is one suitable structure for the description of impedance parameters, and two half-sections associated with even- and odd-mode excitation are depicted in Fig. 2.19(*b*). If parallel and series connections are interchanged and impedances and admittances are interchanged, the even- and odd-mode Π-networks in Fig. 2.19(*a*) become the odd- and even-mode T-networks in Fig. 2.19(*b*), respectively. Using dual properties, the even- and odd-mode impedances of the T-network are easily found as

$$Z_{\mathrm{O/C}} = Z_a + 2Z_c = Z_{11} + Z_{12}, \tag{2.107a}$$

$$Z_{\mathrm{S/C}} = Z_a = Z_{11} - Z_{12}. \tag{2.107b}$$

As operating frequencies become higher than microwave frequencies, transmission lines should be considered as main elements because discontinuity effects are not negligible. A transmission-line section of length Θ and its two equivalent networks with the even- and odd-mode excitations are described in Fig. 2.19(*c*). The two even- and odd-mode networks are obtained by bisecting the network, and their even- and odd-mode impedances and admittances are found as

$$Z_{\mathrm{O/C}} = -jZ_0 \cot\frac{\Theta}{2}, \tag{2.108a}$$

$$Y_{\mathrm{O/C}} = \frac{j}{Z_0} \tan\frac{\Theta}{2}, \tag{2.108b}$$

$$Z_{\mathrm{S/C}} = jZ_0 \tan\frac{\Theta}{2}, \tag{2.108c}$$

$$Y_{\mathrm{S/C}} = \frac{-j}{Z_0} \cot\frac{\Theta}{2}. \tag{2.108d}$$

2.5.3 Properties of Symmetric Two-Port Networks

For symmetric structures, the impedance and admittance parameters by which the two-port networks can be characterized are represented with even- and odd-mode impedances and admittances as shown in (2.106) and (2.107). From the derivation above, a conclusion can be obtained as

$$Z_{11} = Z_{22} = \frac{Z_{\mathrm{O/C}} + Z_{\mathrm{S/C}}}{2}, \tag{2.109a}$$

$$Z_{12} = Z_{21} = \frac{Z_{\mathrm{O/C}} - Z_{\mathrm{S/C}}}{2}, \tag{2.109b}$$

$$Y_{11} = Y_{22} = \frac{Y_{\mathrm{O/C}} + Y_{\mathrm{S/C}}}{2}, \tag{2.109c}$$

$$Y_{12} = Y_{21} = \frac{Y_{\mathrm{O/C}} - Y_{\mathrm{S/C}}}{2}. \tag{2.109d}$$

Using (2.109), the impedance parameters of a uniform transmission line can be obtained in an alternative way. For the uniform transmission line,

$$Z_{11} = Z_{22} = \frac{Z_{\mathrm{O/C}} + Z_{\mathrm{S/C}}}{2} = -jZ_0 \cot\Theta, \tag{2.110a}$$

$$Z_{12} = Z_{21} = \frac{Z_{\mathrm{O/C}} - Z_{\mathrm{S/C}}}{2} = \frac{-jZ_0}{\sin\Theta}, \tag{2.110b}$$

which are same as those with $\alpha = 0$ in Fig. 2.17(a).

2.6 ANALYSES WITH IMAGE PARAMETERS

The alternative circuit analysis involves image parameters, *image impedances*, and *image propagation constants* [14]. The method is particularly useful for designing filters consisting of parallel-coupled transmission lines and for building matching networks. In the case of a uniform transmission line, the image impedance is the characteristic impedance of the transmission line, and the image propagation constant is its propagation constant per unit length. The method is not limited to symmetric networks and can be extended to general networks, including asymmetric networks. Therefore, the objective of this section is to provide the background for an understanding of the design techniques used in later chapters.

2.6.1 Image Impedances

In this section, a new notation system for impedances is used. Readers should take care not to confuse the new subscripts for $\boldsymbol{Z}$ and $\boldsymbol{Y}$, which consist of the letter I plus a numeral (i.e., $I1$ and $I2$), with numeral-only subscripts used to reference values of open- and short-circuited parameters (i.e., 11 and 12). Consider the two-port networks in Fig. 2.20, where two two-port networks are connected in cascade in Fig. 2.20(*a*) and a two-port network is perfectly matched in Fig. 2.20(*b*). Setting $Z_{\text{in}} = Z_L$ in Fig. 2.20(*a*) makes the network symmetric with respect to a plane $(S-S')$, and the input impedance Z_{in} is, under symmetric conditions, defined as an *image impedance* at port ①, Z_{I1}. The second network, N_B in Fig. 2.20(*a*), is turned around (i.e., ports ① and ② are interchanged) and the *ABCD* parameters of the total network $N_A - N_B$ in cascade are, referring to (2.70), easily found as

$$\begin{bmatrix} A_T & B_T \\ C_T & D_T \end{bmatrix} = \begin{bmatrix} AD+BC & 2AB \\ 2CD & AD+BC \end{bmatrix}. \tag{2.111}$$

The input impedance Z_{in} is given as

$$Z_{\text{in}} = \frac{V_1}{I_1} = \frac{A_T V_2 + B_T I_2}{C_T V_2 + D_T I_2} = \frac{A_T Z_L + B_T}{C_T Z_L + D_T}. \tag{2.112}$$

Now the image impedance Z_{I1} at port ① is, setting $Z_{I1} = Z_{\text{in}} = Z_L$, calculated as

$$Z_{I1} = \sqrt{\frac{AB}{CD}}. \tag{2.113}$$

Then, by switching the A and D terms because of the interchanged input and output ports, the image impedance at port ② is readily found as

$$Z_{I2} = \sqrt{\frac{DB}{CA}}. \tag{2.114}$$

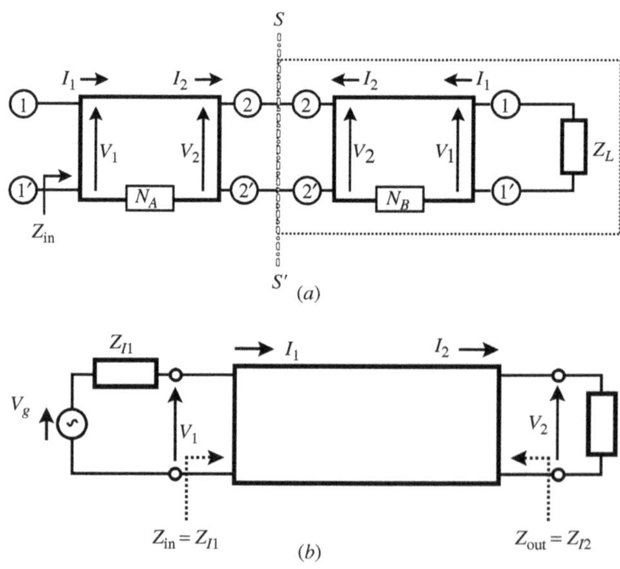

FIGURE 2.20 Two-port networks: (*a*) two networks in cascade; (*b*) two-port network perfectly matched.

Therefore, if a voltage source has a reference impedance (internal impedance) Z_{I1} and port ② is simultaneously terminated in Z_{I2}, it is said that the network is *matched.* This situation is shown in Fig. 2.20(*b*). For a two-port network terminated in Z_{I1} and Z_{I2} to be matched, scattering parameters S_{11} and S_{22} should be zero. The conditions for S_{11} and S_{22} to be zero are found with reference to (2.72) and (2.73), as

$$AZ_{I2} + B = CZ_{I1}Z_{I2} + DZ_{I1}, \tag{2.115a}$$

$$AZ_{I2} + CZ_{I1}Z_{I2} = B + DZ_{I1}. \tag{2.115b}$$

Since AZ_{I2} is already equal to DZ_{I1} from the ratio Z_{I2} to Z_{I2} in (2.113) and (2.114), the conditions for a two-port network to be matched are

$$AZ_{I2} = DZ_{I1}, \tag{2.116a}$$

$$B = CZ_{I1}Z_{I2}. \tag{2.116b}$$

2.6.2 Image Propagation Constants

Consider the voltage transfer function for a network terminated in its real image impedances in Fig. 2.20(*b*). From the definition of *ABCD* parameters in (2.66), the input voltage at port ① is expressed as

$$V_1 = AV_2 + BI_2 = AV_2 + B\frac{V_2}{Z_{I2}}, \tag{2.117}$$

where $V_2 = Z_{I2}I_2$ is used. Thus, we have the voltage ratio of V_1 to V_2 as

$$\frac{V_1}{V_2} = A + \frac{B}{Z_{I1}} = A + B\sqrt{\frac{AC}{BD}} = \sqrt{\frac{A}{D}}\left(\sqrt{AD} + \sqrt{BC}\right). \tag{2.118}$$

Similarly, the current ratio of I_1 to I_2 is given as

$$\frac{I_1}{I_2} = \sqrt{\frac{D}{A}}\left(\sqrt{AD} + \sqrt{BC}\right). \tag{2.119}$$

$\sqrt{A/D}$ and $\sqrt{D/A}$ are the magnitude of voltage and current ratios, and the propagation constant γ yields

$$e^{\gamma} = \sqrt{AD} + \sqrt{BC}, \tag{2.120a}$$

$$e^{-\gamma} = \sqrt{AD} - \sqrt{BC}, \tag{2.120b}$$

where a reciprocal property $AD - BC = 1$ is used.

TABLE 2.3 Image Parameters and Their Relation to General Circuit Parameters[a]

	[*ABCD*]	[*Z*]	[*Y*]	Mixed
		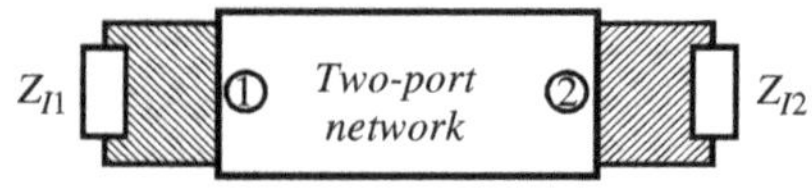		
Z_{I1}	$\sqrt{AB/CD}$	$\sqrt{(Z_{11}\|Z\|)/Z_{22}}$	$\sqrt{Y_{22}/(Y_{11}\|Y\|)}$	$\sqrt{Z_{11}/Y_{11}}$
Z_{I2}	$\sqrt{DB/CA}$	$\sqrt{(Z_{22}\|Z\|)/Z_{11}}$	$\sqrt{Y_{11}/(Y_{22}\|Y\|)}$	$\sqrt{Z_{22}/Y_{22}}$
γ	$\cosh^{-1}\sqrt{AD}$	$\cosh^{-1}[\sqrt{Z_{11}Z_{22}}/Z_{21}]$	$\cosh^{-1}[\sqrt{Y_{11}Y_{22}}/Y_{21}]$	
	$\sinh^{-1}\sqrt{BC}$	$\sinh^{-1}[\sqrt{\|Z\|}/Z_{21}]$	$\sinh^{-1}[\sqrt{\|Y\|}/Y_{21}]$	
	$\coth^{-1}\sqrt{AD/BC}$	$\coth^{-1}\sqrt{(Z_{11}Z_{22})/\|Z\|}$	$\coth^{-1}\sqrt{(Y_{11}Y_{22})/\|Y\|}$	$\coth^{-1}\sqrt{Z_{11}Y_{11}}$, $\coth^{-1}\sqrt{Z_{22}Y_{22}}$
	$A = \sqrt{Z_{I1}/Z_{I2}}\cosh\gamma$	$Z_{11} = Z_{I1}\coth\gamma$	$Y_{11} = Y_{I1}\coth\gamma$	
	$B = \sqrt{Z_{I1}Z_{I2}}\sinh\gamma$	$Z_{12} = \sqrt{Z_{I1}Z_{I2}}/\sinh\gamma$	$Y_{12} = -\sqrt{Y_{I1}Y_{I2}}/\sinh\gamma$	
	$C = \sinh\gamma/\sqrt{Z_{I1}Z_{I2}}$	$Z_{21} = Z_{12}$	$Y_{21} = Y_{12}$	
	$D = \sqrt{Z_{I2}/Z_{I1}}\cosh\gamma$	$Z_{22} = Z_{I2}\coth\gamma$	$Y_{22} = Y_{I2}\coth\gamma$	

[a] $|Z| = Z_{11}Z_{22} - Z_{21}Z_{12}$, $|Y| = Y_{11}Y_{22} - Y_{12}Y_{21}$, $Y_{I1} = Z_{I1}^{-1}$, $Y_{I2} = Z_{I2}^{-1}$.

Based on the background discussed above and Table 2.1, the image parameters and their relation to general circuit parameters are given in Table 2.3.

2.6.3 Symmetrical and Common Structures

Since the concept treated above is applied to general asymmetric networks, the image parameters can therefore be simplified for symmetric networks. In terms of impedance parameters, using the conversion matrix in Table 2.1, the image impedance Z_{I1} is, expressed as

$$Z_{I1} = \sqrt{\frac{Z_{11}(Z_{11}Z_{22} - Z_{12}Z_{21})}{Z_{22}}} = \sqrt{Z_{11}^2 - Z_{12}^2} = \sqrt{Z_{22}^2 - Z_{12}^2}, \qquad (2.121)$$

where $Z_{11} = Z_{22}$ is used for a symmetric network. Referring to (2.107), the impedances with the even- and odd-mode excitations $Z_{O/C}$ and $Z_{S/C}$ are $Z_{11} + Z_{12}$ and $Z_{11} - Z_{12}$, with which the image impedance Z_{I1} is, in different form, expressed as

$$Z_{I1} = \sqrt{Z_{O/C}Z_{S/C}} = Z_{I2}. \qquad (2.122)$$

Figure 2.21 shows common symmetrical structures. For the uniform transmission line in Fig. 2.21(*a*), $Z_{O/C} = -jZ_0\cot(\Theta/2)$ and $Z_{S/C} = jZ_0\tan(\Theta/2)$ with reference to (2.108). Thus, its image impedance is Z_0. For a symmetric network, $Z_{11} = Z_{22}$ and the A term is equal to the D term, as shown in (2.69). The propagation constant γ is expressed as $\cosh^{-1}\sqrt{AD}$ in Table 2.3, and $\cosh\gamma$ is, with $A = D$, given as

$$\cosh\gamma = A. \qquad (2.123)$$

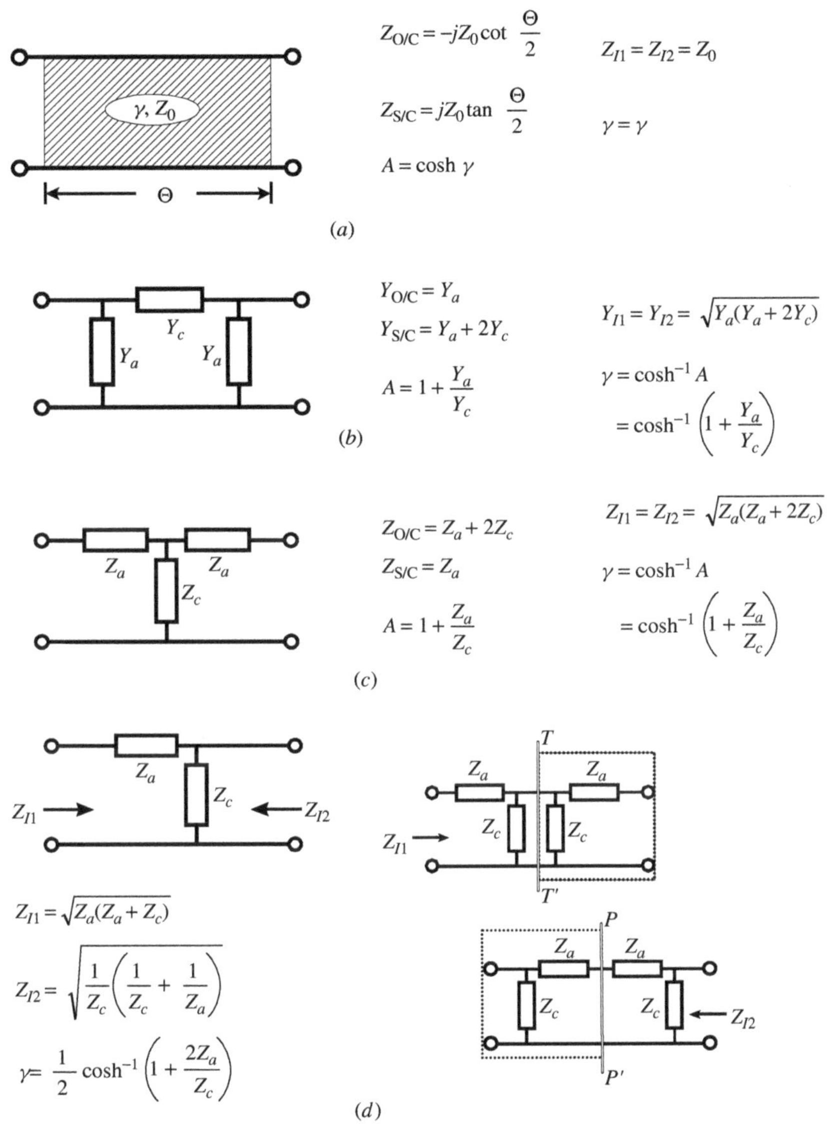

FIGURE 2.21 Image parameters of common structures: (*a*) transmission line; (*b*) Π-network; (*c*) T-network; (*d*) L-network.

Since the A term of the uniform transmission line is $A = \cosh \gamma l$ from Fig. 2.15, the propagation constant of the uniform transmission line per unit length becomes the image propagation constant γ. The image parameters of symmetric networks in Fig. 2.21(*b*) and (*c*) may be found similarly.

The L-network shown in Fig. 2.21(*d*) is not symmetric, but its image parameters may be obtained by use of the symmetrical properties. This situation is well depicted on the right side of Fig. 2.21(d) where symmetric T- and Π-networks are drawn with two planes, $T - T'$ and $P - P'$. When an image network is connected with the original L-network so that the two networks are symmetric with respect $T - T'$, the resulting network is a symmetric T-network. When the L-network is connected with its image network in such a way that the two networks are symmetrical with respect to $P - P'$, the resulting network is a symmetric π-network. So the image impedances are found similarly to those in Fig. 2.21(d).

As illustrated on the right side of Fig. 2.21(d), one T- or Π-network consists of two L-networks, so the image propagation constant of the L-network becomes half that of one T- or Π-network. Alternatively, using the *ABCD* parameters it may be obtained as

$$\gamma_L = \cosh^{-1}\sqrt{1 + \frac{Z_a}{Z_c}}, \tag{2.124a}$$

$$= \sinh^{-1}\sqrt{\frac{Z_a}{Z_c}}, \tag{2.124b}$$

$$= \coth^{-1}\sqrt{1 + \frac{Z_a}{Z_a}}, \tag{2.124c}$$

where $A = 1 + Z_a/Z_c$, $B = Z_a$, $C = 1/Z_c$, and $D = 1$ are used. Once the image propagation constant of the L-network is known, that of the T- or Π-network is simply twice that of the L-network.

EXERCISES

2.1 Consider the ideal transformer N with turns ratio $n : 1$ shown in Fig. E2.1. Compute the incident and reflected voltage and current vectors and the scattering matrices with respect to the reference impedance matrix

$$z(p) = \begin{bmatrix} R_1 & 0 \\ 0 & R_2 \end{bmatrix}. \tag{E2.1}$$

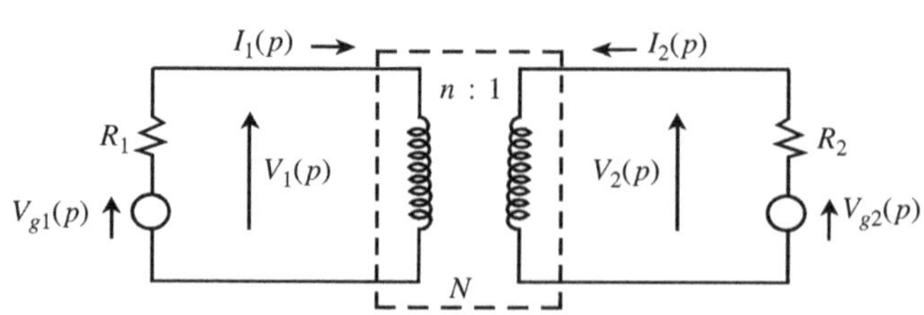

FIGURE E2.1 Ideal transformer N with turns ratio $n : 1$.

2.2 Consider the lossless two-port network N of Fig. E2.2. Compute its scattering matrix normalizing to the load impedances as shown in the figure. To this end, compute the following:

(a) The reference impedance matrix $z(p)$ and the impedance matrix $Z(p)$ of N

(b) The current-basis scattering matrix, $S^I(p)$

(c) The paraconjugate Hermitian part of $z(p)$

(d) The scattering matrix $S(p)$

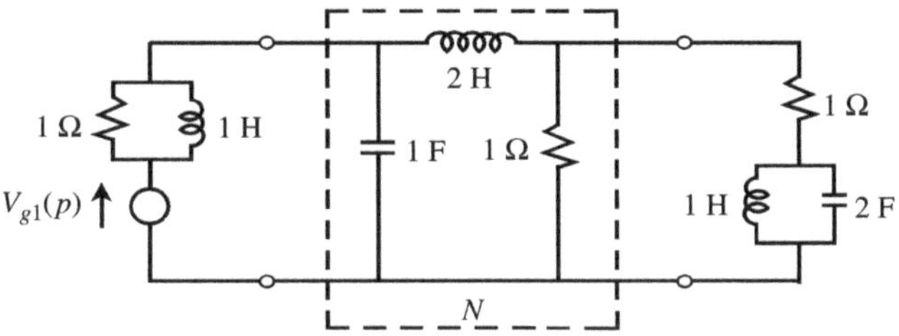

FIGURE E2.2 Lossless reciprocal two-port network together with its loading.

2.3 **(a)** For Fig. E2.3(a), find the relation between S_{12} and S_{22}.

(b) For Fig. E2.3(b), find the relation between S_{21} and S_{11}.

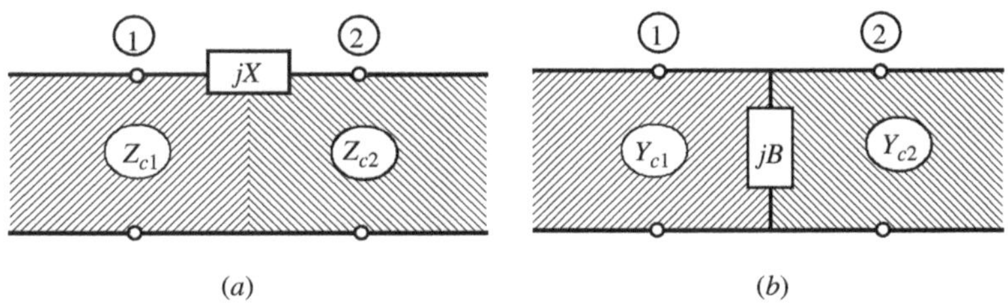

FIGURE E2.3 Two-port networks: (a) series stub; (b) parallel stub.

2.4 Scattering parameters are given as $S_{11} = 0.687 < 107^\circ$, $S_{21} = 1.72 < -59^\circ$, $S_{12} = 0.114 < -81^\circ$, and $S_{22} = 0.381 < 153^\circ$. Determine the incident and reflected waves, a_1, a_2, b_1, and b_2 for the following load conditions: **(a)** 50 Ω; **(b)** 100 Ω; **(c)** 200 Ω. The source is 3 mV EMF in series with a 50-Ω internal impedance.

2.5 Consider the Wilkinson three-port power divider N shown in Fig. E2.5. When it is terminated in equal impedances Z_0, its scattering parameters are

$$\begin{bmatrix} 0 & -\dfrac{j}{\sqrt{2}} & -\dfrac{j}{\sqrt{2}} \\ -\dfrac{j}{\sqrt{2}} & 0 & 0 \\ -\dfrac{j}{\sqrt{2}} & 0 & 0 \end{bmatrix}. \tag{E2.2}$$

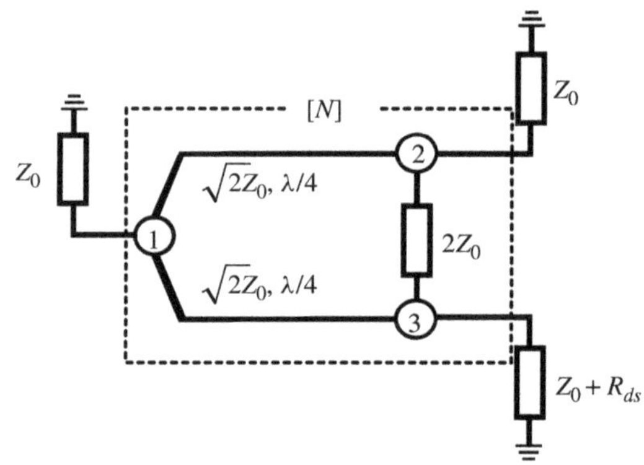

FIGURE E2.5 Wilkinson three-port power divider.

However, if it is terminated in the different impedances shown in Fig. E2.5, its scattering parameters should be changed. Find the changed scattering parameters of N.

2.6 Find the voltage gain between ports ① and ③ for the Wilkinson three-port power divider in Fig. E2.5.

2.7 Derive the scattering parameters shown in (E2.2) when the Wilkinson power divider is terminated in equal impedances.

REFERENCES

1. W.-K. Chen, *Theory and Design of Broadband Matching Networks*, Pergamon Press, New York, 1976, pp. 48–114.
2. R. A. Rohrer, The Scattering Matrix: Normalized to Complex n-Port Load Networks, *IEEE Trans. Circuit Theory*, Vol. 12, June 1965, pp. 223–230.
3. D. C. Youla, An Extension of the Concept of Scattering Matrix, *IEEE Trans. Circuit Theory*, Vol. 17, June 1970, pp. 46–54.
4. T. Nemoto and D. F. Wait, Microwave Circuit Analysis Using the Equivalent Generator Concept, *IEEE Trans. Microwave Theory Tech.*, Vol. 16, October 1968, pp. 866–873.
5. T. Y. Otoshi, On the Scattering Parameters of Reduced Multiport, *IEEE Trans. Microwave Theory and Tech.*, Vol. 17, September 1969, pp. 722–724.
6. D. Roddy, *Microwave Technology*, Prentice-Hall, Englewood Cliffs, NJ, 1986, pp. 66–69.
7. D. M. Pozar, *Microwave Engineering*, Addison-Wesley, Reading, MA, 1990, pp. 245–247.
8. J. K. Hunton, Analysis of Microwave Measurement Techniques by Means of Signal Flow Graphs, *IRE Trans. Microwave Theory Tech.*, Vol. 8, March 1960, pp. 206–212.
9. S. J. Mason, Feedback Theory: Some Properties of Signal Flow Graphs, *Proc. IRE*, Vol. 41, September 1953, pp. 1144–1156.

10. S. J. Mason, Feedback Theory: Further Properties of Signal Flow Graphs, *Proc. IRE,* Vol. 44, July 1956, pp. 920–926.
11. J. Helszajn, *Passive and Active Microwave Circuits*, Wiley, New York, 1978, pp. 19–20.
12. H.-R. Ahn, I. Wolff, and I.-S. Chang, Arbitrary Termination Impedances, Arbitrary Power Division and Small-Sized Ring Hybrids, *IEEE Trans. Microwave Theory Tech.*, Vol. 44, December 1997, pp. 2241–2247.
13. J. Reed and G. J. Wheeler, A Method of Analysis of Symmetrical Four-Port Networks, *IRE Trans. Microwave Theory Tech.*, Vol. 4, October 1956, pp. 346–352.
14. G. Matthaei, L. Young, and E. M. T. Jones, *Microwave Filters, Impedance-Matching Networks, and Coupled Structures*, Artech House, Dedham, MA, 1980, pp. 49–52.

CHAPTER THREE

Conventional Ring Hybrids

3.1 INTRODUCTION

The first ring hybrid was proposed by Tyrrel in 1947 [1]. Tyrrel tried to explain the ring hybrid with the concept of waveguide T-junctions and described two types of hybrid circuits, one involving a ring or loop transmission line and the other relying on the symmetry properties of certain four-arm junctions. After Tyrrel described the fundamental characteristics of distributed circuit hybrids, a number of workers discussed the performance of practical wideband realizations constructed in coaxial line and stripline [2–5]. One of them was that two coupled-line filters were used for a wideband ring hybrid in the 1950s. In 1961, Pon [6] derived design equations for ring hybrids with arbitrary power divisions. In 1968, March [5] developed a wideband ring hybrid, adapting one coupled-line filter instead of a three-quarter-wavelength transmission line, which causes narrowband responses. As new uniplanar techniques emerged for MMIC applications [7–13], there were several publications to realize small broadband ring hybrids which employed a combination of coplanar waveguides and slotlines using only one-sided substrates [9,11].

Since the first ring hybrid was introduced, ring hybrids have been studied and used for various applications in microwave equipment. Thus, they are indispensable components in various MICs (microwave integrated circuits) and MMICs (monolithic microwave integrated circuits), such as balanced mixers, balanced amplifiers, frequency discriminators, phase shifters, feeding networks in antenna arrays, and so on. In this chapter, conventional and symmetrical ring hybrids are treated.

3.2 ORIGINAL CONCEPT OF THE 3-dB RING HYBRID

As mentioned above, Tyrrel developed the first ring hybrid [1]. His idea began from the concept of a ring or loop of transmission lines. Figure 3.1 shows a ring waveguide structure. The electrical length around the ring is one and a half wavelengths and the wavelength corresponds to the guide wavelength of a hollow pipe. It is assumed that the electrical length around the ring is equal to the mean circumference. When dominant waves are sent into the ring from the sidearm, they will split at the junction into two sets of waves of equal amplitude traveling around the ring in opposite directions. A pure standing wave will therefore be set up within the ring. The two sets of waves spreading out from the junction have a 180° phase difference, due to the nature of an electric-plane T-junction [14]. When they reach the point diametrically opposite the junction, they have traversed paths of equal length and are still 180° out of phase.

This point is therefore a voltage minimum (zero). With this as a starting point, the standing wave within the ring can be mapped by marking off alternate voltage maxima and minima at quarter-wavelength intervals, as indicated on Fig. 3.1. If identical ports are connected symmetrically to the ring at points ② and ④ so as to form additional electric-plane T-junctions, and if these sidearms are terminated in their characteristic impedance, they will receive equal amounts of power, since they are series connections at voltage minima. The amplitude of waves proceeding past ② toward ③ will also still be equal to the amplitude of waves proceeding past ④ toward ③, since equal amounts of power are being extracted from two equal sets of waves. Therefore, the arc ② ③ ④ still contains a pure standing

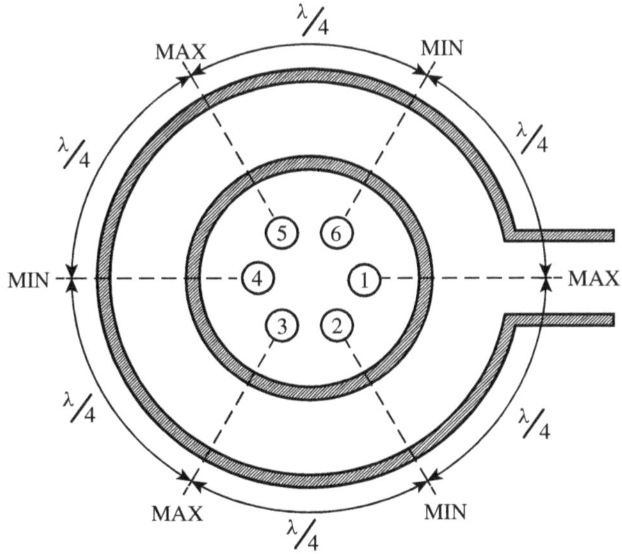

FIGURE 3.1 Cross-sectional view of waveguide ring in the electric plane.

wave. Since identical phase shifts have been introduced at ② and ④ by identical T-junctions, the position of this standing wave is unaltered, and there is still a voltage maximum at ③. A series connection to the ring at ③ in Fig. 3.1 will thus receive no power. Such a connection may be made by introducing a fourth electric-plane junction at ③.

Figure 3.2 shows the modified structure, with series connections to the ring at ①, ②, ③, and ④. It can be concluded that this represents a hybrid circuit with bidirectional characteristics when the ring is driven by a generator in arm ②, with appropriate loads in arms ① and ③. There will therefore be voltage maxima at ②, ④, and ⑥ and additional minima at ①, ③, and ⑤. Arms ① and ③ will thus receive equal amounts of power without disturbing the position of the standing-wave pattern in the arc ① ⑤ ③, and the arm at ④ receives no power. The construction of Fig. 3.2 therefore possesses the essential properties of a hybrid network.

It is necessary to consider the improvement of this circuit by the elimination of undesired reflections within the ring. These arise in two ways. First, there are reflections taking place at each T-junction to the ring due to the abrupt geometry change that characterizes these junctions. It is suggested that these reflections can be reduced or eliminated by introducing appropriate reactive elements in the vicinity of the junctions. If this is done in a sufficiently symmetrical manner, all four junctions of the hybrid ring can be made individually reflectionless. A second source of reflection can be traced to the inherent resistive mismatch that results from using waveguides of approximately the same size for the ring and all the arms. In this case, a generator of impedance Z_0 is connected to two loads essentially in series, which results in resistive mismatch. This mismatch can be

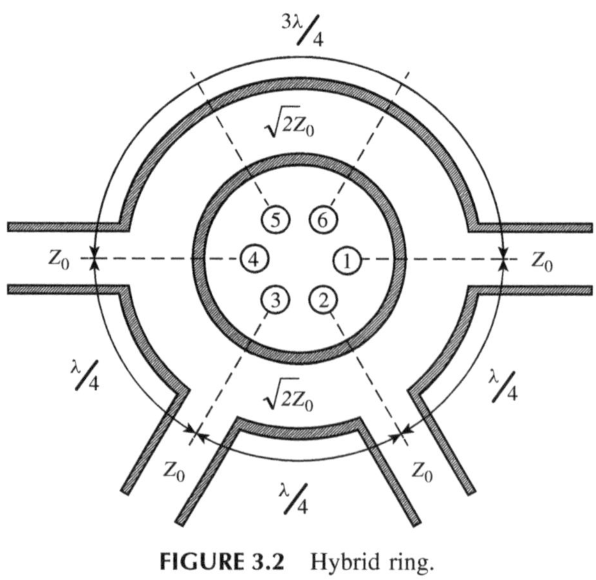

FIGURE 3.2 Hybrid ring.

eliminated readily by altering the characteristic impedance of one or more of the waveguides involved. If the characteristic impedance of all waveguides is $\sqrt{2}\,Z_0$, as shown in Fig. 3.2, the ring will be reflectionless. In this way, the well-known ring hybrid was introduced.

The four series connections to the ring can be replaced, one by one, by shunt connections. This yields five distinctly different hybrid rings in addition to the one discussed in connection with Fig. 3.2. These six circuits are shown in Fig. 3.3. They are indicated schematically to emphasize the perfect generality of these networks. The circuits in Fig. 3.3 may therefore be constructed from any type of transmission line or from any mixture of types. For any of the six circuits, moreover, it is possible to find sets of impedance values that bring about resistive matching in all directions. For the circuits in Fig. 3.3, for one orientation of the polarization, the ring lies in the electric plane; for another, in the magnetic plane; and for any orientation in general, the input waves may be resolved into components parallel with and perpendicular to the plane of the ring, for each of which a hybrid circuit is provided. In this case, the matching of resistive impedances for

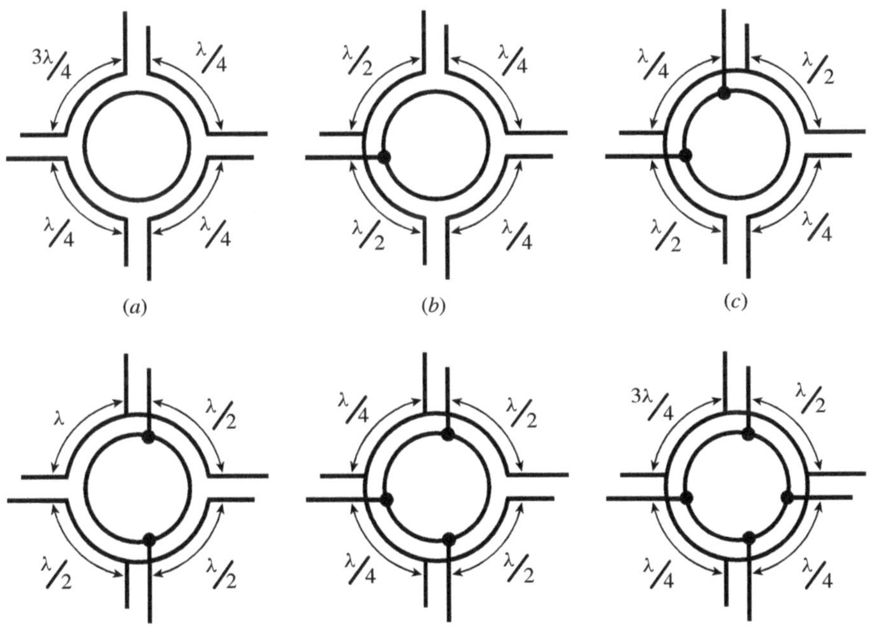

FIGURE 3.3 Six fundamental hybrid rings: (*a*) typical ring hybrid; (*b*) one port of a transmission-line section with a half wavelength is connected with the ground plane; (*c*) two ports of a transmission-line section with a quarter wavelength are connected with the ground plane; (*d*) two ports of two transmission-line sections with a half wavelength and one wavelength are connected with the ground plane; (*e*) two transmission-line sections with a quarter wavelength are connected with the ground plane; (*f*) all the transmission-line sections are connected with the ground plane.

both components becomes somewhat more involved. The circumference dimensions of hybrid rings can be changed in discrete steps without altering any of the circuit characteristics by using either or both of the following rules:

1. An integral number of wavelengths may be added to or subtracted from any arc (i.e., a portion of ring between centers of adjacent connections).
2. A pair of half wavelengths may be added to or subtracted from any two arcs.
3. These rules can be applied when it is desired to simplify the construction or to alter the structure to conform more readily to available space.

As the series connections are progressively replaced by shunt connections with the added quarter-wavelength and three-quarter-wavelength lines, the ring becomes larger unless a wavelength or pair of half wavelengths is judiciously subtracted at each step in the progression. The proportions indicated in Fig. 3.3 conform in each case to the smallest ring in which all connections are separated by finite arcs. From each of the circuits of Fig. 3.3(*b*)–(*e*), an additional pair of half wavelengths may be removed, but this eliminates the separation between certain adjacent connections. In such instances, if the properties of the original hybrid ring are to be preserved, care must be taken to preserve connections that are truly adjacent and not superposed symmetrically to the same point on the ring.

Although such highly condensed circuits appear superficially attractive in general, the transmission constructions are likely to call for an expansion from the dimension in Fig. 3.3. In connection with the transmission-line embodiment of the circuit in Fig. 3.3(*a*), the reactances associated with the junctions do not upset the performance because equal phase shifts are introduced as the waves travel past the two driven arms. It is clear that the same argument can be applied to the highly symmetrical circuits of Fig. 3.3(*d*) and (*f*). Not so obvious, however, is the situation with regard to the rings involving asymmetrically disposed and mixed connections, since different phase shifts may be expected at series and shunt branches. Consider, therefore, the circuit of Fig. 3.3(*b*). When this is driven from the shunt arm or from the opposite series arm, equal phase shifts take place at the two identical adjacent series arms, and balance is secured with the dimensions as given. If the junction reactances are reduced or eliminated by symmetrical tuning, balance and equal power division will still be retained. General considerations based on reciprocity may now be cited to show that this tuning automatically brings about balance, equal division of power, and impedance match when the circuit is driven from either of the opposed series arms, even though the connections adjacent to these arms are of such totally different character. The same argument can be applied to Fig. 3.3(*c*). The completely asymmetrical circuit of Fig. 3.3(*c*) remains as the only one whose dimensions may perhaps require substantial alteration when appreciable phase shifts are involved.

There remains to be discussed the variation of hybrid ring characteristics with frequency. In the case of variable electrical properties, the balance between opposite arms is the most critical. Consider what happens when waves of a frequency different from that considered originally are sent into the ring in Fig. 3.1. There will still be cancellation of fields at point ④, but now the spacing in the standing-wave pattern is changed, and the voltage maximum is displaced from point ③. An arm attached at ③ will thus not be completely uncoupled from the arm at ①.

This reasoning can be applied to the circuits shown in Fig. 3.3(*a*), (*d*), and (*f*) and to those pairs of opposite connections in Fig. 3.3(*b*) and (*e*) that are balanced by virtue of two paths around the ring which differ by a half wavelength, that is, to all cases in which the opposite connections are both parallel or both series. A different situation exists when balance is obtained between a series connection and a shunt connection located at geometrically opposite points across the ring, since the standing-wave pattern at the point diametrically opposite to the feed point does not shift with frequency. This argument applies to Fig. 3.3(*c*) and one pair of opposite arms in Fig. 3.3(*b*) and (*e*). When a high degree of balance is desired between one particular pair of arms over as broad a bandwidth as possible, the circuits of Fig. 3.3(*b*) and (*e*) will be preferred, because they offer driven arms that are connected symmetrically. If the loads attached to the arms of a hybrid ring are maintained nearly reflectionless over a band of frequencies, and if the junction reactances in the ring are effectively canceled throughout this band, the power division or resistive matching of the network will be unaffected by such variables.

A hybrid network for microwaves can also be secured in the form of a compound junction comprising a series connection and a parallel connection made to a guide at the same point on its longitudinal axis. One form of such a junction is shown in Fig. 3.4(*a*) as a cross section in the electric plane. Lines of electric intensity in successive positions of the wavefront are drawn to indicate what happens when power is introduced from the waveguide series arm. Equal intensities appear in the collinear guide arms, with a 180° phase difference, while the voltages induced in the coaxial line mutually cancel. If the two ends of the main waveguide are terminated in their characteristic impedance, the power will be divided equally between the two loads, and no power will appear in the coaxial line. The same junction is shown again in Fig. 3.4(*b*), but the power is introduced from the parallel arm (i.e., from the coaxial line). The two arms of the main guide receive equal intensities in phase, and no net voltage appears across the series branch.

The ring hybrid shown in Fig. 3.5(*b*) is one of the structures that Tyrrel suggested (see Fig. 3.2). When all loads are terminated in equal impedances Z_0, the impedances of the transmission-line sections Z_1, Z_2, Z_3, and Z_4 are equally $\sqrt{2}\,Z_0$. The variation of the scattering parameters versus frequency for a coupling level of -3 dB is plotted in Fig. 3.5, where the power division and isolation performances are shown in Part (*a*) and the matching properties in Part (*b*). It is clear that the ring hybrid is not a wideband structure, a relative bandwidth of 20% being realistic. The ring hybrid also suffers from the fact that with increased frequency the discontinuities of the junctions increasingly degrade the performance.

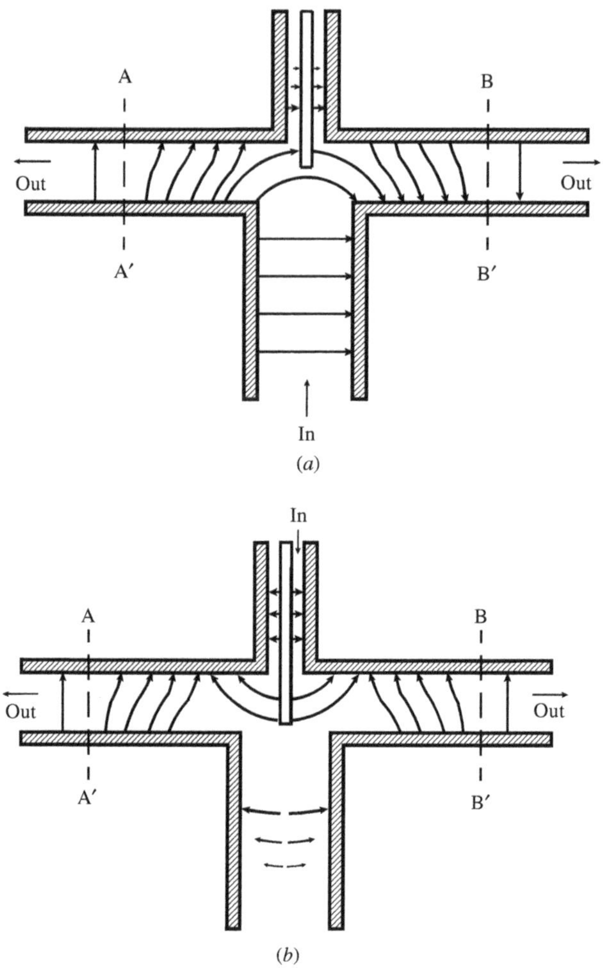

FIGURE 3.4 Spreading of a wavefront into a compound junction from (*a*) the series arm and (*b*) the parallel arm.

3.3 CONVENTIONAL RING HYBRIDS

Ring hybrids have been developed in various forms. Coupled transmission lines, the even- and odd-mode technique, lumped elements, and the uniplanar technique are several methods available to evaluate their performance. In this section, their background and application to ring hybrids are treated.

3.3.1 Coupled Transmission Lines

Following Tyrrel, a number of engineers have discussed ways to increase the bandwidths of ring hybrids. One of them is to use coupled transmission lines,

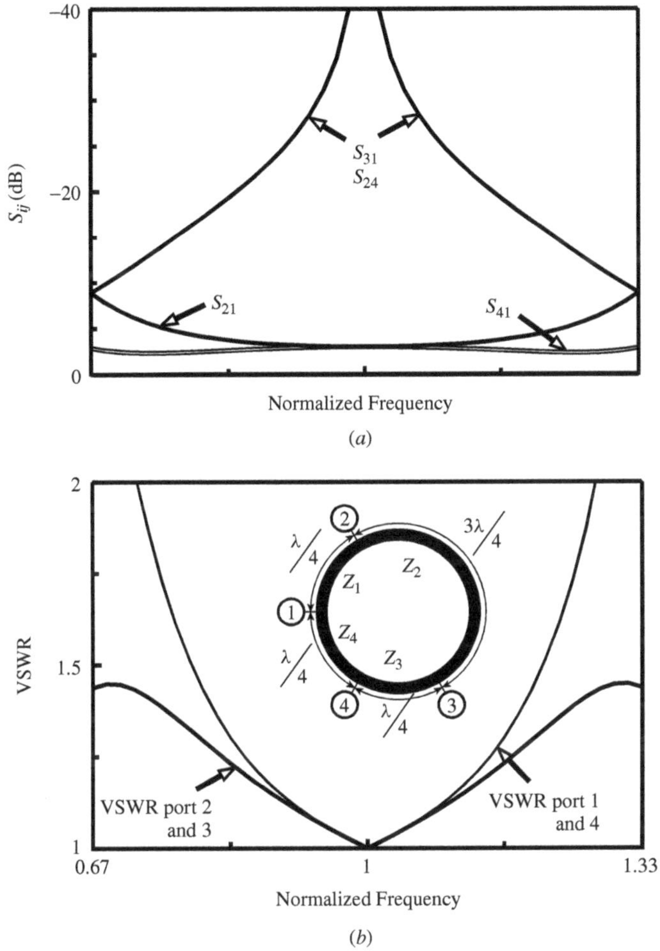

FIGURE 3.5 Responses of the scattering parameters of a 3-dB ring hybrid versus frequencies: (*a*) power divisions and isolations; (*b*) reflection coefficients.

so knowledge of their behavior is indispensable for an understanding of such ring hybrids. Figure 3.6(*a*) shows a pair of coupled transmission lines terminated in equal impedances Z_0, where a voltage source V_g with its internal impedance (reference impedance) Z_0 is fed into port ①. It is symmetric and its two even- and odd-mode equivalent circuits are depicted in Fig. 3.6(*b*) and (*c*). By definition, the characteristic impedances of the corresponding transmission lines are Z_{0e} and Z_{0o} with even- and odd-mode excitations. With reference to (2.103), the vector amplitudes of the signals emerging from the four ports are

$$A_1 = \tfrac{1}{2}S_{11e} + \tfrac{1}{2}S_{11o}, \tag{3.1a}$$

$$A_2 = \tfrac{1}{2}S_{11e} - \tfrac{1}{2}S_{11o}, \tag{3.1b}$$

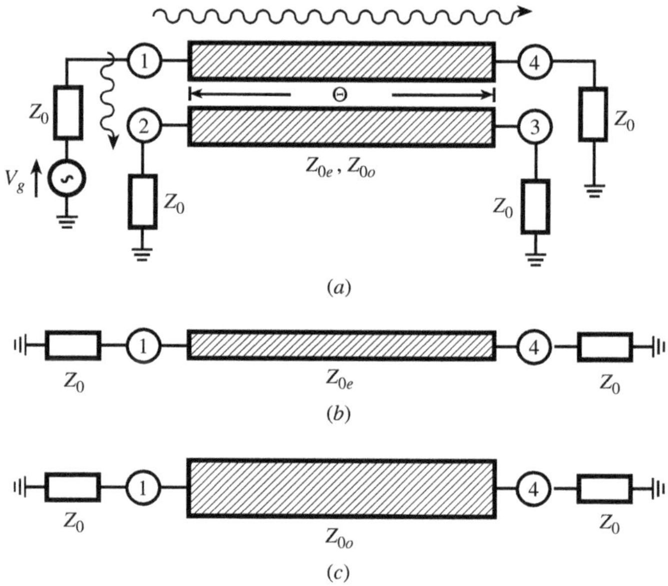

FIGURE 3.6 Pair of coupled transmission lines: (*a*) four-port coupled transmission lines; (*b*) even-mode equivalent circuit; (*c*) odd-mode equivalent circuit.

$$A_3 = \tfrac{1}{2} S_{21e} - \tfrac{1}{2} S_{21o}, \tag{3.1c}$$

$$A_4 = \tfrac{1}{2} S_{21e} + \tfrac{1}{2} S_{21o}, \tag{3.1d}$$

where S_{11e}, S_{11o}, S_{21e}, and S_{21o} are reflected and transmitted scattering parameters for even and odd modes. Referring to (2.72) and (2.75), they are obtained as

$$S_{11e} = \frac{A_e Z_0 + B_e - C_e Z_0^2 - D_e Z_0}{A_e Z_0 + B_e + C_e Z_0^2 + D_e Z_0}, \tag{3.2a}$$

$$S_{11o} = \frac{A_o Z_0 + B_o - C_o Z_0^2 - D_o Z_0}{A_o Z_0 + B_o + C_o Z_0^2 + D_o Z_0}, \tag{3.2b}$$

$$S_{12o} = \frac{2Z_0}{A_o Z_0 + B_o + C_o Z_0^2 + D_o Z_0}, \tag{3.2c}$$

$$S_{21e} = \frac{2Z_0}{A_e Z_0 + B_e + C_e Z_0^2 + D_e Z_0}, \tag{3.2d}$$

where the subscripts e and o denote even and odd modes.

For a matched pair of coupled transmission lines, the magnitude of A_1 in (3.1*a*) should be zero, which gives

$$(A_e + A_o)Z_0 + (B_e + B_o) - (C_e + C_o)Z_0^2 - (D_e + D_o)Z_0 = 0. \tag{3.3}$$

Assuming $\Theta = \pi/2$ for simplicity, the *ABCD* parameters of transmission lines with Z_{0e} and Z_{0o} are given as

$$\begin{bmatrix} A & B \\ C & D \end{bmatrix}_e = \begin{bmatrix} 0 & jZ_{0e} \\ \dfrac{j}{Z_{0e}} & 0 \end{bmatrix}, \tag{3.4a}$$

$$\begin{bmatrix} A & B \\ C & D \end{bmatrix}_o = \begin{bmatrix} 0 & jZ_{0o} \\ \dfrac{j}{Z_{0o}} & 0 \end{bmatrix}, \tag{3.4b}$$

where Z_{0e} is a characteristic impedance of one coupled transmission line, measured with respect to ground with equal currents flowing in the same direction (even mode), and Z_{0o} is a characteristic impedance of one coupled transmission line, measured with respect to ground with equal currents flowing in opposite direction (odd mode).

A condition for perfect matching is found from (3.3) and (3.4) as

$$Z_{0e}Z_{0o} = Z_0^2. \tag{3.5}$$

The magnitude of A_2 is defined as a coupling coefficient C, which gives

$$C = Z_0^2 \frac{B_e C_o - B_o C_e}{(B_e + C_e Z_0^2)(B_o + C_o Z_0^2)}. \tag{3.6}$$

The coupling coefficient is, with a match condition of $Z_{0e}Z_{0o} = Z_0^2$, simplified as

$$C = \frac{Z_{0e} - Z_{0o}}{Z_{0e} + Z_{0o}}. \tag{3.7}$$

The signal at port ③, A_3, is

$$A_3 = Z_0 \frac{(B_o - B_e) + (C_o - C_e)Z_0^2}{(B_e + C_e Z_0^2)(B_o + C_o Z_0^2)}. \tag{3.8}$$

If a pair of coupled lines is $\pi/2$ long and perfectly matched, perfect isolation can always occur at port ③, and the A_3 in (3.8) becomes zero with (3.1) and (3.3)– (3.5).

With the termination impedance Z_0 and the coupling coefficient C specified, the even- and odd-mode impedances required are derived from (3.5) and (3.7) as

$$Z_{0e} = Z_0 \sqrt{\frac{1 + C}{1 - C}}, \tag{3.9a}$$

$$Z_{0o} = Z_0 \sqrt{\frac{1 - C}{1 + C}}. \tag{3.9b}$$

As discussed above, if a voltage is fed into port ① of a pair of coupled transmission lines, part of the excited power is coupled with port ② and part of the power is delivered to port ④, with no power delivered to port ③. In the analyses, it is assumed that the even and odd modes of the coupled transmission lines have the same propagation velocities, so that the transmission lines have the same electrical length for both modes. However, this condition is not generally satisfied for microstrip lines or other non-TEM lines, so the directivity becomes poorer.

Coupled-line filters can be obtained by terminating two of four ports in shorts, opens, or a combination of short and open. Figure 3.7 shows two coupled-line filters that are widely used for building ring hybrids. Terminating ports ② and ④ of the coupled lines in Fig. 3.6(a) in shorts results in the circuit in Fig. 3.7(a). The admittance parameters of the coupled-line filter in Fig. 3.7(a) are obtained from those of the pair of coupled lines in Fig. 3.6(a) and are given [15] as

$$Y_{11} = Y_{22} = Y_{33} = Y_{44} = -j\frac{Y_{0e} + Y_{0o}}{2} \cot \Theta, \tag{3.10a}$$

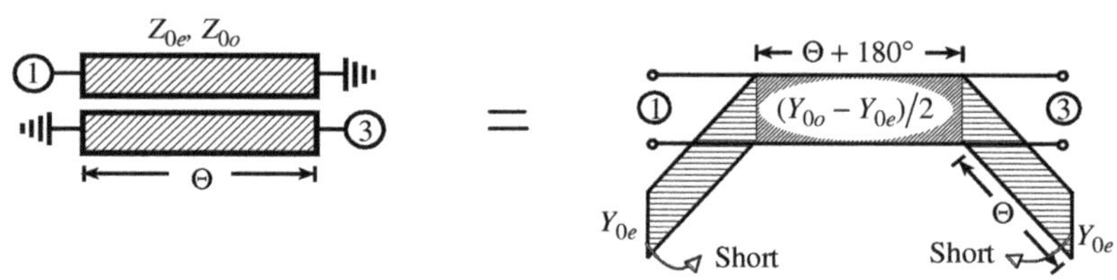

$$Z_{I1} = Z_{I2} \quad \frac{2Z_{0e}Z_{0o}\sin\Theta}{[(Z_{0e} - Z_{0o})^2 - (Z_{0e} + Z_{0o})^2 \cos\Theta]^{1/2}}, \quad \cosh(a + j\beta) = \frac{Z_{0e} + Z_{0o}}{Z_{0e} - Z_{0o}} \cos\Theta$$

when $\Theta = \lambda/4$, $Z_{0e} = \left(\frac{C}{1-C}\right) Z_{I1}$, $Z_{0o} = \left(\frac{C}{1+C}\right) Z_{I1}$

(a)

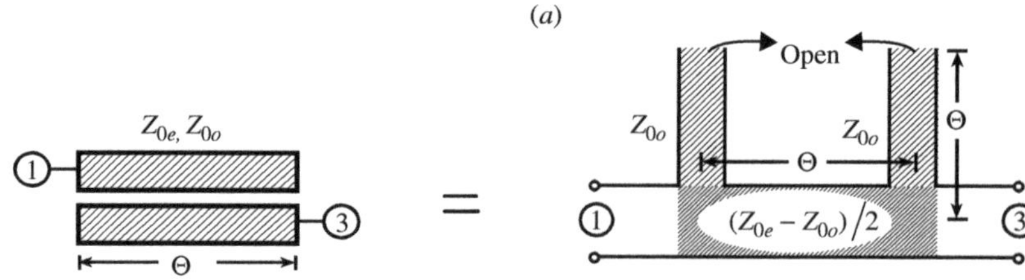

$$Z_{I1} = Z_{I2} \quad \frac{[(Z_{0e} - Z_{0o})^2 - (Z_{0e} + Z_{0o})^2 \cos^2\Theta]^{1/2}}{2Z_{0e}Z_{0o}\sin\Theta}, \quad \cosh(a + j\beta) = \frac{Z_{0e} + Z_{0o}}{Z_{0e} - Z_{0o}} \cos\Theta$$

(b)

FIGURE 3.7 Coupled-line filters (I): (a) two ports are short-circuited; (b) two ports are open-circuited.

$$Y_{12} = Y_{21} = Y_{34} = Y_{43} = -j\frac{Y_{0o} - Y_{0e}}{2}\cot\Theta, \tag{3.10b}$$

$$Y_{13} = Y_{31} = Y_{24} = Y_{42} = -j\frac{Y_{0o} - Y_{0e}}{2}\csc\Theta, \tag{3.10c}$$

$$Y_{14} = Y_{41} = Y_{23} = Y_{32} = -j\frac{Y_{0e} + Y_{0o}}{2}\csc\Theta. \tag{3.10d}$$

Applying the conditions of $V_2 = 0$ and $V_4 = 0$ gives

$$\begin{bmatrix} Y'_{11} & Y'_{12} \\ Y'_{21} & Y'_{22} \end{bmatrix} = \begin{bmatrix} Y_{11} & Y_{13} \\ Y_{31} & Y_{33} \end{bmatrix} = \begin{bmatrix} -j\dfrac{Y_{0e} + Y_{0o}}{2}\cot\Theta & -j\dfrac{Y_{0o} - Y_{0e}}{2}\csc\Theta \\ -j\dfrac{Y_{0o} - Y_{0e}}{2}\csc\Theta & -j\dfrac{Y_{0e} + Y_{0o}}{2}\cot\Theta \end{bmatrix}, \tag{3.11a}$$

$$= -jY_{0e}\cot\Theta\begin{bmatrix} 1 & 0 \\ 0 & 1 \end{bmatrix} + j\frac{(Y_{0o} - Y_{0e})}{2}\begin{bmatrix} -\cot(\pi+\Theta) & \csc(\pi+\Theta) \\ \csc(\pi+\Theta) & -\cot(\pi+\Theta) \end{bmatrix}, \tag{3.11b}$$

where Y'_{11}, Y'_{12}, Y'_{21}, and Y'_{22} are the admittance parameters of the coupled-line filter in Fig. 3.7(*a*) and Y_{11}, Y_{13}, Y_{31}, and Y_{33} are those of the four-port coupled lines in Fig. 3.6(*a*). The first term in (3.11*b*) indicates that two short stubs with each characteristic admittance Y_{0e} are connected in parallel with ports ① and ③. The second term in (3.11*b*) is, referring to Fig. 2.17(*a*), an admittance matrix of a two-port transmission line with a characteristic admittance $(Y_{0o} - Y_{0e})/2$ and a length of $180° + \Theta$. Thus, its equivalent circuit is suggested as shown in Fig. 3.7(*a*). Based on the derived admittance parameters and referring to Table 2.3, the image impedances are given and the A term is obtained from the ratio Y'_{22}/Y'_{12} in Table 2.1. The coupled-line filter in Fig. 3.7(*a*) is symmetric and the image propagation constant is therefore given as

$$\cosh\gamma = A. \tag{3.12}$$

As explained above, the image impedances and propagation constants are given as in Fig. 3.7(*a*).

For the design of a ring-hybrid coupled-line filter, Z_{0e} and Z_{0o} must be determined from a specified coupling coefficient C and image impedances Z_{I1} and Z_{I2}. It is generally designed at a center frequency, and the length of the coupled-line filter is $\lambda/4$. When $\Theta = \lambda/4$, the image impedance in Fig. 3.7(*a*) is

$$Z_{I1} = \frac{2Z_{0e}Z_{0o}}{Z_{0e} - Z_{0o}}. \tag{3.13}$$

From (3.7) and (3.13), the even- and odd-mode impedances are computed as $Z_{0e} = CZ_{I1}/(1 - C)$ and $Z_{0o} = CZ_{I1}/(1 + C)$. The dual circuit in Fig. 3.7(*a*) is shown in Fig. 3.7(*b*) and obtained by terminating two ports of the coupled lines in opens. The impedance parameters of the coupled transmission lines are obtained [15] as

$$Z_{11} = Z_{22} = Z_{33} = Z_{44} = -j\frac{Z_{0e} + Z_{0o}}{2}\cot\Theta, \tag{3.14a}$$

$$Z_{12} = Z_{21} = Z_{34} = Z_{43} = -j\frac{Z_{0e} - Z_{0o}}{2}\cot\Theta, \tag{3.14b}$$

$$Z_{13} = Z_{31} = Z_{24} = Z_{42} = -j\frac{Z_{0e} - Z_{0o}}{2}\csc\Theta, \tag{3.14c}$$

$$Z_{14} = Z_{41} = Z_{23} = Z_{32} = -j\frac{Z_{0e} + Z_{0o}}{2}\csc\Theta. \tag{3.14d}$$

Setting $I_2 = 0$ and $I_4 = 0$ in Fig. 3.6(a) gives

$$\begin{bmatrix} Z'_{11} & Z'_{12} \\ Z'_{21} & Z'_{22} \end{bmatrix} = \begin{bmatrix} Z_{11} & Z_{13} \\ Z_{31} & Z_{33} \end{bmatrix} = \begin{bmatrix} -j\dfrac{Z_{0e}+Z_{0o}}{2}\cot\Theta & -j\dfrac{Z_{0e}-Z_{0o}}{2}\csc\Theta \\ -j\dfrac{Z_{0e}-Z_{0o}}{2}\csc\Theta & -j\dfrac{Z_{0e}+Z_{0o}}{2}\cot\Theta \end{bmatrix}$$
$$= -jZ_{0o}\cot\Theta\begin{bmatrix} 1 & 0 \\ 0 & 1 \end{bmatrix} + j\frac{Z_{0e}-Z_{0o}}{2}\begin{bmatrix} -\cot\Theta & -\csc\Theta \\ -\csc\Theta & -\cot\Theta \end{bmatrix}, \tag{3.15}$$

In a similar way, (3.15) indicates that two open stubs with Z_{0o} are connected in series with ports ① and ③ and that a transmission line with $(Z_{0e} - Z_{0o})/2$ and Θ is between the two open stubs. This suggests the equivalent circuit shown in Fig. 3.7(b). Based on the impedance parameters derived, the image impedances are readily found. Since those in Fig. 3.7(a) and (b) are dual circuits, the image propagation constants are equal and their image impedance product is always $Z_{0e}Z_{0o}$. Using this product, the image impedance of Fig. 3.7(b) is calculated and written in Fig. 3.7(b). In addition to the two circuits discussed above, those in Fig. 3.8 may be used for building ring hybrids. Their image parameters are found similarly and written in Fig. 3.8.

3.3.2 Ring Hybrids with Coupled Transmission Lines

The coupled-line filters discussed in connection with Fig. 3.7 can be used for building ring hybrids. To see how they operate, first consider their scattering parameters when $\Theta = \pi/2$. In a matched coupled-line coupler in Fig. 3.6(a), the scattering matrix with 3-dB coupling becomes

$$\begin{bmatrix} b_1 \\ b_2 \\ b_3 \\ b_4 \end{bmatrix} = \frac{1}{\sqrt{2}}\begin{bmatrix} 0 & 1 & 0 & -j \\ 1 & 0 & -j & 0 \\ 0 & -j & 0 & 1 \\ -j & 0 & 1 & 0 \end{bmatrix}\begin{bmatrix} a_1 \\ a_2 \\ a_3 \\ a_4 \end{bmatrix}, \tag{3.16}$$

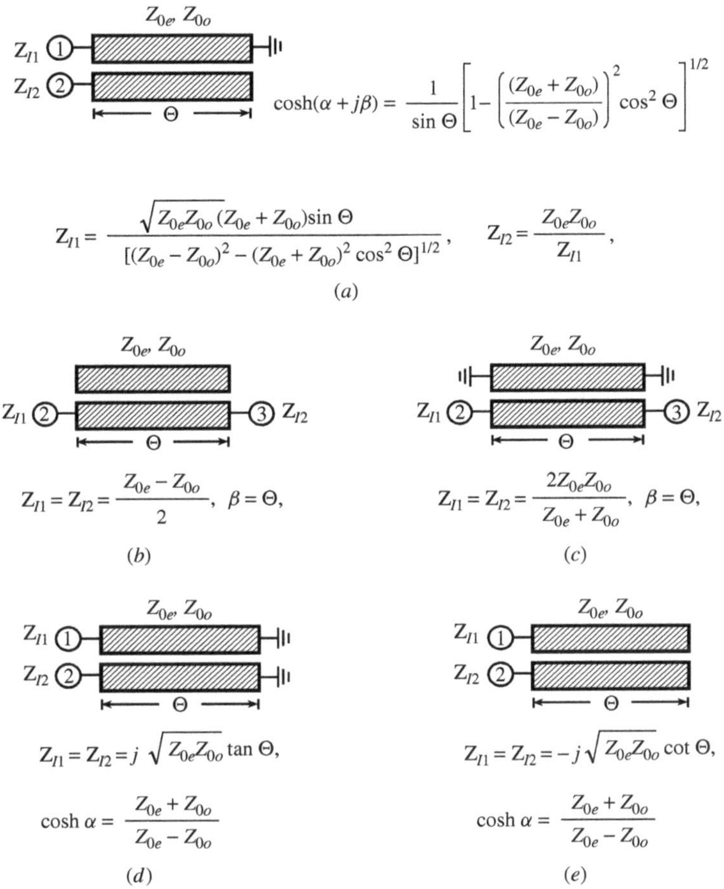

FIGURE 3.8 Coupled-line filters (II): (*a*) two ports are terminated in one short and one open; (*b*) two ports are terminated in two opens; (*c*) two ports are terminated in two shorts; (*d*) two ports are terminated in two shorts; (*e*) two ports are terminated in two opens.

where a_i and b_i are normalized incident and reflected waves at port i. When ports ② and ④ are terminated in short circuits as shown in Fig. 3.7(*a*) and other ports are matched, $a_2 = -b_2$ and $a_4 = -b_4$. When ports ② and ④ are terminated in open circuits as shown in Fig. 3.7(*b*) and other ports are matched, $a_2 = -b_2$ and $a_4 = b_4$. Applying the short or open concepts to the two circuits in Fig. 3.7 gives their relationship as

$$S_s = \begin{bmatrix} 0 & j \\ j & 0 \end{bmatrix}, \tag{3.17a}$$

$$S_o = \begin{bmatrix} 0 & -j \\ -j & 0 \end{bmatrix}, \tag{3.17b}$$

$$S_s = \begin{bmatrix} -1 & 0 \\ 0 & -1 \end{bmatrix} S_o, \qquad (3.17c)$$

where the subscripts s and o correspond to those in Fig. 3.7(a) and (b), respectively.

By the results in (3.17), the coupled-line filter with two shorts in Fig. 3.7(a) can be equivalent to a transmission-line section of 270° electrical length, and that with two opens in Fig. 3.7(b) simulates that with a 90° electrical length. Therefore, the filter with two shorts in Fig. 3.7(a) can also equate to the situation where the filter with two opens in Fig. 3.7(b) is connected in series with a 180° phase shifter. Due to this, the coupled-line filter with two shorts in Fig. 3.7(a) can be replaced by a three-quarter-wavelength transmission-line section in the ring hybrids. Figure 3.9 shows several possible ways in which coupled-line filters are used in ring hybrids.

The ring hybrid in Fig 3.9(a) consists of one coupled-line filter in Fig. 3.7(a) and three transmission-line sections. Since the coupled-line filter in Fig. 3.9(a) and a transmission-line section are out of phase by 180°, if a voltage source is driven at port ①, the power excited at port ① is divided between ports ② and ④, and two waves at the two ports are 180° out of phase and isolated at port ③. A transmission-line section and the coupled-line filter in Fig. 3.9(a) are 180° out

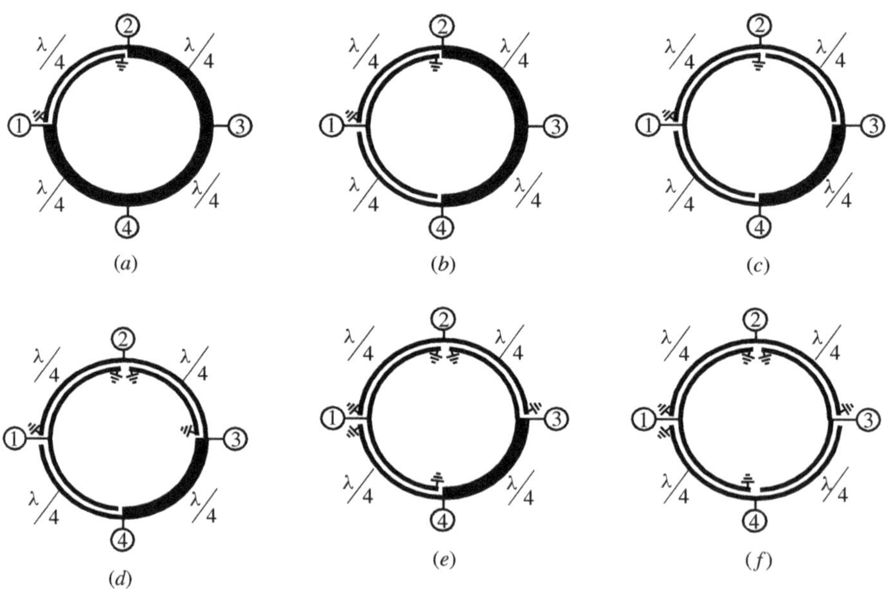

FIGURE 3.9 Ring hybrids consisting of coupled transmission lines: (a) one coupled-line filter and three transmission lines; (b) two coupled-line filters and two transmission lines; (c) three coupled-line filters and one transmission line; (d) three coupled-line filters and one transmission line; (e) three coupled-line filters and one transmission line; (f) four coupled-line filters.

of phase regardless of frequencies, so the resulting ring hybrid in Fig. 3.9(*a*) has wideband performance. This fact was suggested by March [5] and has been used for various applications, so it warrants further discussion. The second ring hybrid, in Fig. 3.9(*b*), consists of two coupled-line filters and two transmission-line sections. As discussed above, since the coupled-line filter in Fig. 3.7(*b*) behaves like a transmission-line section with its image impedance, the resulting performances are similar to those of the ring hybrid in Fig. 3.9(*a*). Figure 3.9(*c*) shows a ring hybrid consisting of three coupled-line filters and one transmission-line section. The coupled-line filter between ports ② and ③ in Fig. 3.9(*c*) also operates like a transmission-line section with its image impedance. The ring hybrid in Fig. 3.9(*d*) consists of three coupled-line filters and one transmission-line section. The coupled-line filters connected between ports ② and ③ in Fig. 3.9(*c*) and (*d*) operate like transmission-line sections with their image impedances.

Figure 3.9(*e*) consists of three coupled-line filters and one transmission-line section. Differently from those in Fig. 3.9(*a*)–(*d*), the power excited at port ① is divided between ports ② and ④ and the two waves at ports ② and ④ are in phase, whereas the two waves at ports ② and ④ are out of phase by 180° when power is excited at port ③. Instead of a transmission-line section, several coupled-line filters in Figs. 3.7(*b*) and 3.8(*b*) and (*c*) can be used. One of them, in Fig. 3.7(*b*), is connected between ports ③ and ④ as shown in Fig. 3.9(*f*) instead of the transmission-line section in Fig. 3.9(*e*). The resulting ring hybrid in Fig. 3.9(*f*) is therefore similar to that in Fig. 3.9(*e*). In addition to the ring hybrids in Fig. 3.9, there are more consisting of coupled-line filters. As can be imagined, the more coupled lines there are, the less convenient fabrication is, so the first one has been widely used.

3.3.3 Wideband Ring Hybrids

The limiting factor in the ring hybrid in Figs. 3.2 and 3.5(*b*) is a with three-quarter-wavelength transmission-line section, which restricts the useful frequency range to $f \simeq 0.23 f_0$, where f_0 is the design center frequency. To solve this problem, a ring hybrid consisting of one coupled-line filter and three transmission-line sections is shown in Fig. 3.10, where the scattering parameter ratio of S_{21} to S_{41} is that of d_1 to d_2, the characteristic and image impedances of transmission-line sections and the coupled-line filter are Z_1, Z_2, Z_3, and Z_4, and all ports are terminated in Z_0. The ring hybrid in Fig. 3.10 has a wide bandwidth and the additional advantages of decreased frequency sensitivity and reduced size.

The design equations of the ring hybrid are given from (7) and (8) in [12], as

$$Z_1 = Z_3 = Z_0 \sqrt{\frac{d_1^2 + d_2^2}{d_1^2}}, \tag{3.18a}$$

$$Z_2 = Z_4 = Z_0 \sqrt{\frac{d_1^2 + d_2^2}{d_2^2}}. \tag{3.18b}$$

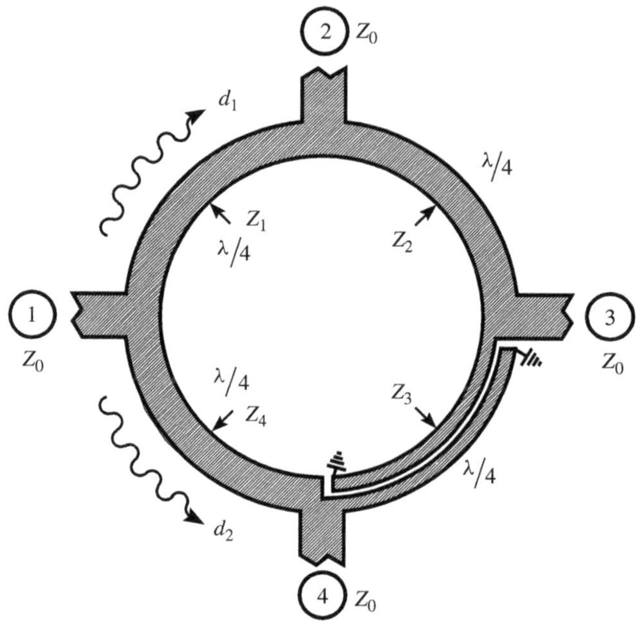

FIGURE 3.10 Ring hybrid with one coupled-line filter.

In the case of $d_1 = d_2$, the characteristic and image impedances are equally $\sqrt{2}\,Z_0$, and for the performance of the ring hybrid, the image impedance of the coupled-line filter between ports ③ and ④ should be equal to Z_1. To see how the impedance level of the coupled-line filter changes, critical angles Θ_{c1} and Θ_{c2} should be found by setting the denominator of Z_{I1} in Fig. 3.7(a) equal to zero. Solving for Θ gives

$$\cos\Theta_{c1} = \frac{Z_{0e} - Z_{0o}}{Z_{0e} + Z_{0o}}, \tag{3.19a}$$

$$\Theta_{c2} = 180^\circ - \Theta_{c1}. \tag{3.19b}$$

The impedance level of the coupled-line filter increases slowly with a change in frequency on either side of f_0, becoming ∞ at Θ_{c1} and Θ_{c2}. For wideband performance, Θ_{c1} should be small so that $\cos\Theta_{c1}$ approaches unity. Therefore, the requirement for maximum bandwidth is $Z_{0e} >> Z_{0o}$. Although Z_3 is equal to Z_1 at midband and the power split varies as a function of frequency, this variation is fairly slow and octave bandwidths are obtained readily with a coupled-line filter.

Two main problems arise in realizing a coupled-line filter with -3.0 ± 0.3 dB of coupling. They are a high level of even-mode impedance with a general measurement system and an extremely small gap. From the relation between coupling and image impedance written in Fig. 3.7(a), taking a slightly lower value of d_2/d_1 in (3.18a) and a lower coupling coefficient in Fig. 3.7(a) reduces the even-mode impedance. How the variation of Z_{0e} and Z_{0o} will affect the performance of the ring hybrid will be simulated next. When the four ports of the ring hybrid are all terminated in 50 Ω and the ratio d_2/d_1 is reduced to 0.85, new values of Z_3, Z_{0e}, and Z_{0o} are 65.65 Ω, 158.6 Ω, and 27.2 Ω, respectively. A reduction in Z_{0e} and Z_{0o} to 0.9 of their values allows the image impedance Z_3 to be within ±10% of its required value over the range prescribed.

Without any change, $Z_1 = Z_3 = 70.7$ Ω, and its correspondent even- and odd-mode impedances are $Z_{0e} = 171.4$ Ω and $Z_{0o} = 29.31$ Ω for a 3-dB coupling. Figure 3.11 compares two cases with other conditions unchanged. Figure 3.11(*a*) and (*b*) are simulation results with the even-mode impedance unchanged, and those with reduced conditions are plotted in Fig. 3.11(*c*) and (*d*). Due to the reduction in the image impedance from 70.71 Ω to 65.66 Ω, perfect matching at ports ③ and ④ does not occur, as shown in Fig. 3.11(*d*). The two cases are an example of a lower even-mode impedance, but as illustrated by Ahn et al. [12], to use looser coupling than 3 dB is also desirable: for example, 4 or 5 dB of coupling. In any case, the output results are about the same as those using 3-dB coupling in the band of interest.

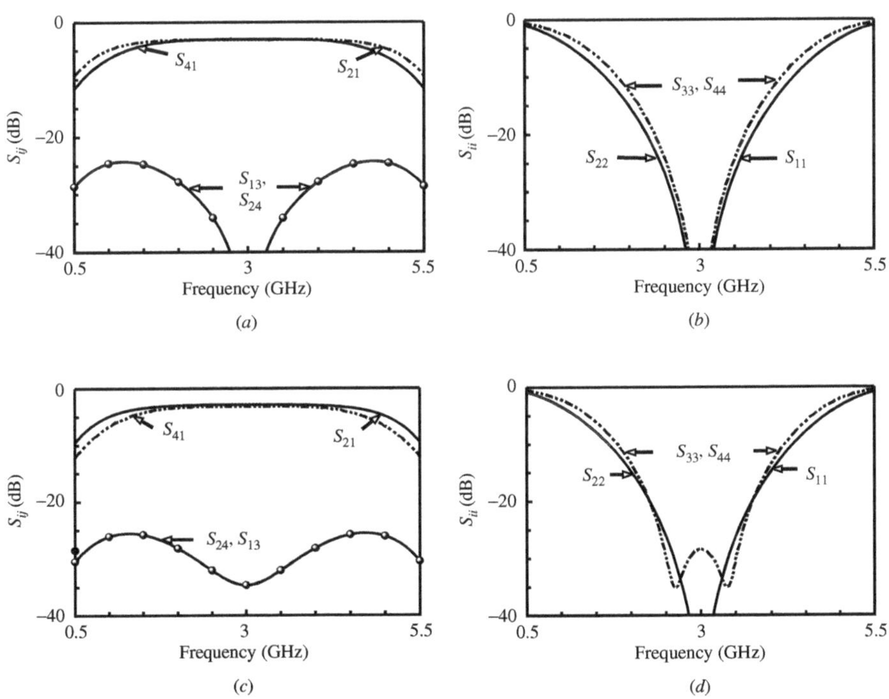

FIGURE 3.11 Simulated results of a ring hybrid with one coupled-line filter: (*a, b*) ring hybrid with $Z_{0e} = 171.4$ Ω and $Z_{0o} = 29.4$ Ω; (*c, d*) ring hybrid with $Z_{0e} = 158.6$ Ω and $Z_{0o} = 27.2$ Ω. Power division and isolation are shown in parts (*a*) and (*c*); all-port matchings are shown in parts (*b*) and (*d*).

3.3.4 Symmetric Ring Hybrids with Arbitrary Power Divisions

The important common properties so far discussed for ring hybrids are that (1) the output arms are isolated from each other; (2) the input impedance is terminated in matching impedances; and (3) the power-split ratio is 0 dB, or power is divided equally. In addition to these properties, a ring hybrid whose power-split ratio is adjusted by varying the impedances between two arms is required. The ring hybrid for arbitrary power division [6] is illustrated in Fig. 3.12, where the power is applied to port ① and the applied power is divided between ports ③ and ④ with the scattering parameter ratio d_1:d_2. The characteristic admittances of the

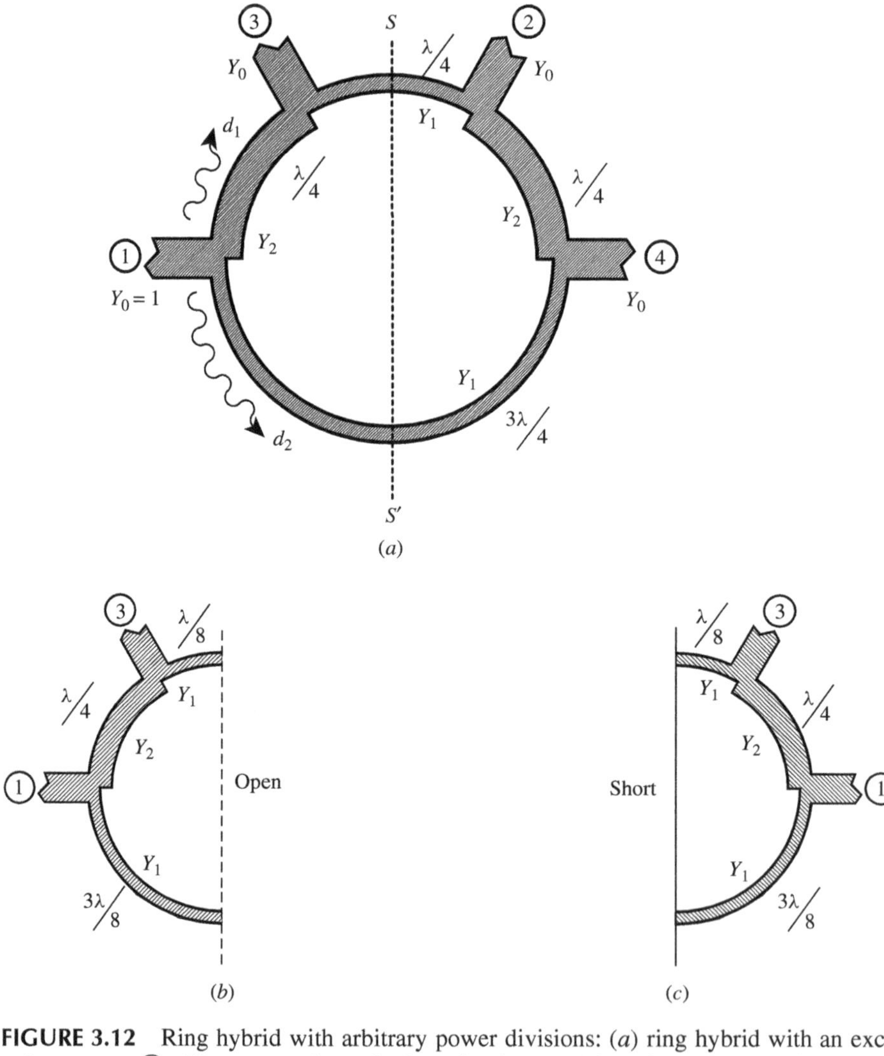

FIGURE 3.12 Ring hybrid with arbitrary power divisions: (*a*) ring hybrid with an excitation at port ①; (*b*) even-mode equivalent circuit; (*c*) odd-mode equivalent circuit.

termination loads are equal and normalized to unity. The variable parameters are the two admittances Y_1 and Y_2, which determine the degree of coupling of the output arms and the match at the input arm.

The design equations of the ring hybrid with arbitrary termination impedances and arbitrary power divisions were discussed sufficiently by Ahn et al. [12], but are derived in this section using symmetry properties. The analysis of this ring hybrid consists of the usual procedure of reducing the four-port network to two two-port networks by taking advantage of the symmetry about plane $S-S'$. When two in-phase waves of the same amplitude are applied to ports ② and ③ or to ports ① and ④, the symmetric plane $S-S'$ is open circuited, and only one half of the circuit (the even-mode equivalent circuit) needs to be analyzed. When two waves with equal amplitude and out of phase by 180° are applied to ports ② and ③ or to ports ① and ④, the voltage is zero at plane $S-S'$. As a result, the ring can be short-circuited at this plane, and again, only one half of the circuit (the odd-mode equivalent circuit) needs to be analyzed. The even- and odd-mode equivalent circuits are depicted in Fig. 3.12(*b*) and (*c*), respectively. Once the scattering matrices for the even and odd modes are known, the resulting signals out of the four ports can be found by superimposing the waves of the two modes, as discussed in Section 2.5.

The even- and odd-mode equivalent networks at arbitrary frequencies are shown in Fig. 3.13(*a*) and (*b*), respectively. The length of the transmission line between ports ① and ③ is Θ and the lengths of the open and short stubs are $3\Theta/2$ and $\Theta/2$. The effective length varies with the frequency change and is expressed as

$$\Theta = \frac{\pi}{2}\frac{f}{f_0}, \tag{3.20}$$

where f is an operating frequency and f_0 is a design center frequency. When $f = f_0$, Θ becomes 90°.

For the even-mode network in Fig. 3.13(*a*), assuming that the currents flowing into the network are I_1 and I_3, those into the transmission line are I_{t1} and I_{t3} and those into the shunt stubs are I_{s1} and I_{s3}. If the voltages at ports ① and ③ are V_1 and V_3, the relation between the voltages and I_{s1} and I_{s3} is

$$I_{s1} = jY_1 \tan\frac{3\Theta}{2} V_1, \tag{3.21a}$$

$$I_{s3} = jY_1 \tan\frac{\Theta}{2} V_3. \tag{3.21b}$$

The relation between I_{t1}, I_{t3} and V_1, V_3 is given from Fig. 2.17(*a*), as

$$I_{t1} = -jY_2 \cot\Theta V_1 + jY_2 \csc\Theta V_3, \tag{3.22a}$$

$$I_{t3} = jY_2 \csc\Theta V_1 - jY_2 \cot\Theta V_3. \tag{3.22b}$$

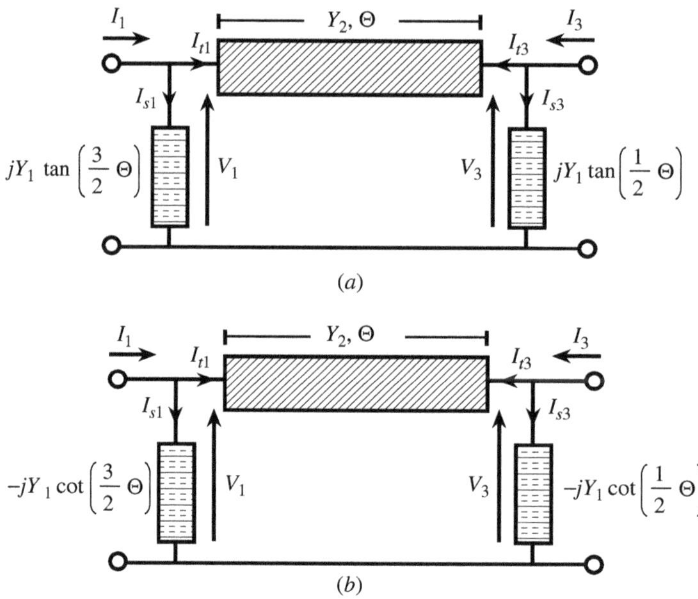

FIGURE 3.13 (*a*) Even- and (*b*) odd-mode networks at arbitrary frequencies.

From (3.21) and (3.22), the admittance parameters of the even-mode network are found as

$$\boldsymbol{Y_e} = j\begin{bmatrix} Y_1 \tan\dfrac{3\Theta}{2} - Y_2 \cot\Theta & Y_2 \csc\Theta \\ Y_2 \csc\Theta_2 & Y_1 \tan\dfrac{\Theta}{2} - Y_2 \cot\Theta \end{bmatrix}. \tag{3.23}$$

Similarly, those of the odd-mode network in Fig. 3.13(*b*) are given as

$$\boldsymbol{Y_o} = j\begin{bmatrix} -Y_1 \cot\dfrac{3\Theta}{2} - Y_2 \cot\Theta & Y_2 \csc\Theta \\ Y_2 \csc\Theta_2 & -Y_1 \cot\dfrac{\Theta}{2} - Y_2 \cot\Theta \end{bmatrix}. \tag{3.24}$$

When $f = f_0$, the scattering parameters of the even- and odd-mode networks are easily obtained from the conversions in Table 2.2 as

$$\boldsymbol{S_e} = \begin{bmatrix} \dfrac{1 - Y_1^2 - Y_2^2 + j2Y_1}{1 + Y_1^2 + Y_2^2} & \dfrac{-j2Y_2}{1 + Y_1^2 + Y_2^2} \\ \dfrac{-j2Y_2}{1 + Y_1^2 + Y_2^2} & \dfrac{1 - Y_1^2 - Y_2^2 - j2Y_1}{1 + Y_1^2 + Y_2^2} \end{bmatrix}, \tag{3.25}$$

$$S_o = \begin{bmatrix} \frac{1 - Y_1^2 - Y_2^2 - j2Y_1}{1 + Y_1^2 + Y_2^2} & \frac{-j2Y_2}{1 + Y_1^2 + Y_2^2} \\ \frac{-j2Y_2}{1 + Y_1^2 + Y_2^2} & \frac{1 - Y_1^2 - Y_2^2 + j2Y_1}{1 + Y_1^2 + Y_2^2} \end{bmatrix}. \tag{3.26}$$

By superposition using (2.103), the vector amplitudes of the signal emerging from the four ports are

$$A_1 = \frac{1}{2}(S_{11e} + S_{11o}) = \frac{1 - Y_1^2 - Y_2^2}{1 + Y_1^2 + Y_2^2}, \tag{3.27a}$$

$$A_2 = \frac{1}{2}(S_{12e} - S_{12o}) = 0, \tag{3.27b}$$

$$A_3 = \frac{1}{2}(S_{12e} + S_{12o}) = \frac{-j2Y_2}{1 + Y_1^2 + Y_2^2}, \tag{3.27c}$$

$$A_4 = \frac{1}{2}(S_{11e} - S_{11o}) = \frac{j2Y_1}{1 + Y_1^2 + Y_2^2}, \tag{3.27d}$$

where the subscript number in the scattering parameters indicates two-port even- and odd- mode networks and that in the vector amplitude is the port number in Fig. 3.12(*a*). For perfect matching at port ①, the following condition should be satisfied:

$$Y_1^2 + Y_2^2 = 1. \tag{3.28}$$

The next factor to be considered is how to satisfy the power division ratio. Since the output voltage ratio at ports ③ and ④ is that of A_3 to A_4, the power division ratio between the two ports is

$$d_1^2 : d_2^2 = Y_2^2 : Y_1^2. \tag{3.29}$$

Based on (3.28) and (3.29), the two unknown variables Y_1 and Y_2 can be determined. By unitary (the principle of power conservation) and reciprocal properties, the final scattering matrix of the ring hybrid is

$$S = \begin{bmatrix} 0 & 0 & -jY_2 & jY_1 \\ 0 & 0 & -jY_1 & -jY_2 \\ -jY_2 & -jY_1 & 0 & 0 \\ jY_1 & -jY_2 & 0 & 0 \end{bmatrix}. \tag{3.30}$$

3.3.5 Conventional Lumped-Element Ring Hybrids

If the ring hybrids are to be realized only with distributed elements, their size becomes quite large at frequencies below about 2 GHz. At these frequencies, they

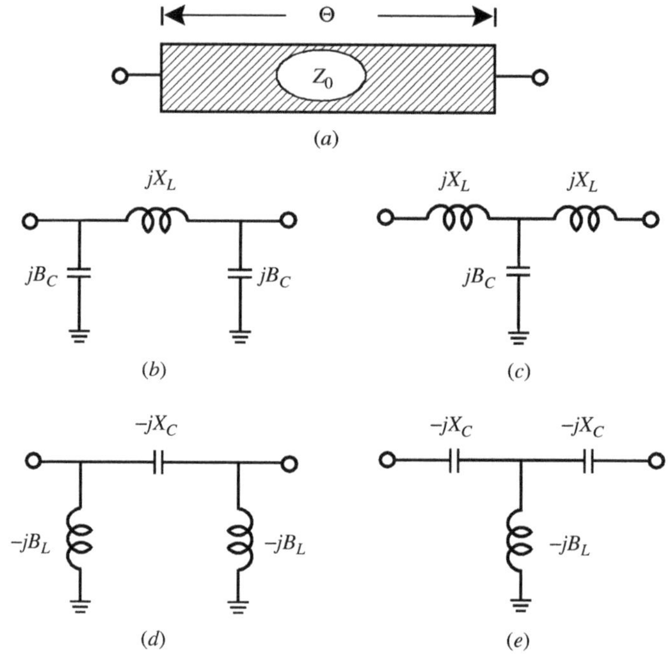

FIGURE 3.14 Lumped-element equivalent circuits of a transmission-line section. (*a*) transmission-line section with electrical length Θ; (*b, c*) Π and T lumped-element equivalent circuits with $\Theta \leq 180°$; (*d, e*) Π and T lumped-element equivalent circuits with $180° < \Theta \leq 360°$.

can be realized with lumped inductors and capacitors to reduce their size. For this, a derivation of a transmission-line model with lumped elements is required [16,17]. Also, how to derive T and Π lumped equivalent circuits and how to apply them to lumped-element ring hybrids are discussed.

Figure 3.14 shows a transmission-line section with an electrical length Θ and its T and Π lumped equivalent circuits. Using the symmetry properties discussed in Section 2.5, the even- and odd-mode admittances and impedances of a transmission-line section in Fig. 3.14(*a*) are found from (2.108) as

$$Y_{O/C} = \frac{j}{Z_0} \tan \frac{\Theta}{2}, \tag{3.31a}$$

$$Z_{O/C} = -jZ_0 \cot \frac{\Theta}{2}, \tag{3.31b}$$

$$Y_{S/C} = \frac{-j}{Z_0} \cot \frac{\Theta}{2}, \tag{3.31c}$$

$$Z_{S/C} = jZ_0 \tan \frac{\Theta}{2}. \tag{3.31d}$$

When the electrical length of a transmission-line section in Fig. 3.14(*a*) is less than 180°, the T and Π lumped equivalent circuits are those in Fig. 3.14(*b*) and (*c*). Using (2.106), those of the Π lumped equivalent circuit in Fig. 3.14(*b*) are

$$Y_{\mathrm{O/C}} = jB_C, \tag{3.32a}$$

$$Y_{\mathrm{S/C}} = jB_C + \frac{2}{jX_L}. \tag{3.32b}$$

Equating the equations in (3.32) with those in (3.31) gives

$$X_L = Z_0 \sin\Theta, \tag{3.33a}$$

$$B_C = \frac{1}{Z_0}\tan\frac{\Theta}{2}. \tag{3.33b}$$

When the electrical length of a transmission line in Fig. 3.14(*a*) is greater than 180° and less than or equal to 360°, its T and Π lumped equivalent circuits can be derived similarly to those in Fig. 3.14(*d*) and (*e*), respectively. The even- and odd-mode admittances of the T equivalent circuit in Fig. 3.14(*e*) are, in a similar way, found as

$$Z_{\mathrm{O/C}} = -jX_C - \frac{2}{jB_L}, \tag{3.34a}$$

$$Z_{\mathrm{S/C}} = -jX_C. \tag{3.34b}$$

Equating the equations in (3.34) with those in (3.31) yields

$$X_C = -\frac{1}{Z_0}\tan\frac{\Theta}{2}, \tag{3.35a}$$

$$B_L = -Z_0 \sin\Theta. \tag{3.35b}$$

Lumped-element modeling closely approximates the transmission-line section at the design center frequency, but it has a narrow bandwidth. To increase the bandwidth, two sections may be cascaded. Figure 3.15 shows two lumped-element models connected in cascade. Since two transmission-line sections are cascaded and each length is Θ/2, the results derived above can be used for the Π lumped-element equivalent circuit in Fig. 3.15(*a*). They are

$$X_{L2} = Z_0 \sin\frac{\Theta}{2}, \tag{3.36a}$$

$$B_{C2} = \frac{1}{Z_0}\tan\frac{\Theta}{4}. \tag{3.36b}$$

Similarly, the X_{C2} and B_{L2} in Fig. 3.15(*d*) are

$$X_{C2} = -\frac{1}{Z_0}\tan\frac{\Theta}{4}, \tag{3.37a}$$

$$B_{L2} = -Z_0 \sin\frac{\Theta}{2}. \tag{3.37b}$$

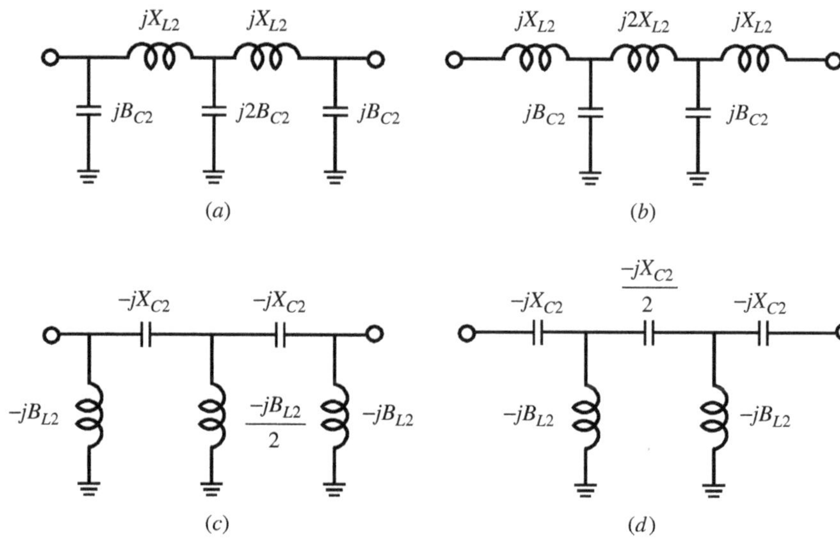

FIGURE 3.15 Lumped-element equivalent circuits of two transmission-line sections in cascade: (*a, b*) Π and T lumped-element equivalent circuits with $\Theta \leq 180^\circ$; (*c, d*) Π and T lumped-element equivalent circuits with $180^\circ < \Theta \leq 360^\circ$.

Using the T and Π lumped-element equivalent circuits in Figs. 3.14 and 3.15, four lumped-element ring hybrids may be introduced in Fig. 3.16, where the electrical length between ports ③ and ④ is 270° and those between the remaining ports are each 90°. The lumped-element ring hybrid in Fig. 3.16(*a*) is produced by three Π and one T lumped equivalent circuits. That in Fig. 3.16(*b*) consists of four π lumped equivalent circuits. The shunt capacitance comes from a transmission-line section with a 90° electrical length and the shunt inductance from that with a 270° electrical length, as shown in Fig. 3.14. The two elements are resonant at the design frequency and therefore removed. That is the reason that no shunt element is connected at ports ③ and ④. Those in Fig. 3.16(*c*) and (*d*) consist of all T- network equivalent circuits. Since lumped-element ring hybrids have small bandwidths, a transmission-line section can consist of two lumped-element circuits in cascade and is connected between ports ③ and ④ in Fig. 3.16(*c*) to increase the bandwidth. The bandwidth of that in Fig. 3.16(*c*) is approximately 35%, which is wider than that of the ring hybrid in Fig. 3.5 (*b*).

3.3.6 Mixed Small Ring Hybrids

As discussed so far, the ring hybrids can divide an input signal into two signals, either in phase or out of phase. They can also be used to combine two signals to obtain the sum and difference of the two signals. Due to their important characteristics, they have been used for power dividers and combiners, mixers, balanced amplifiers, and so on. However, as semiconductor techniques are developed and

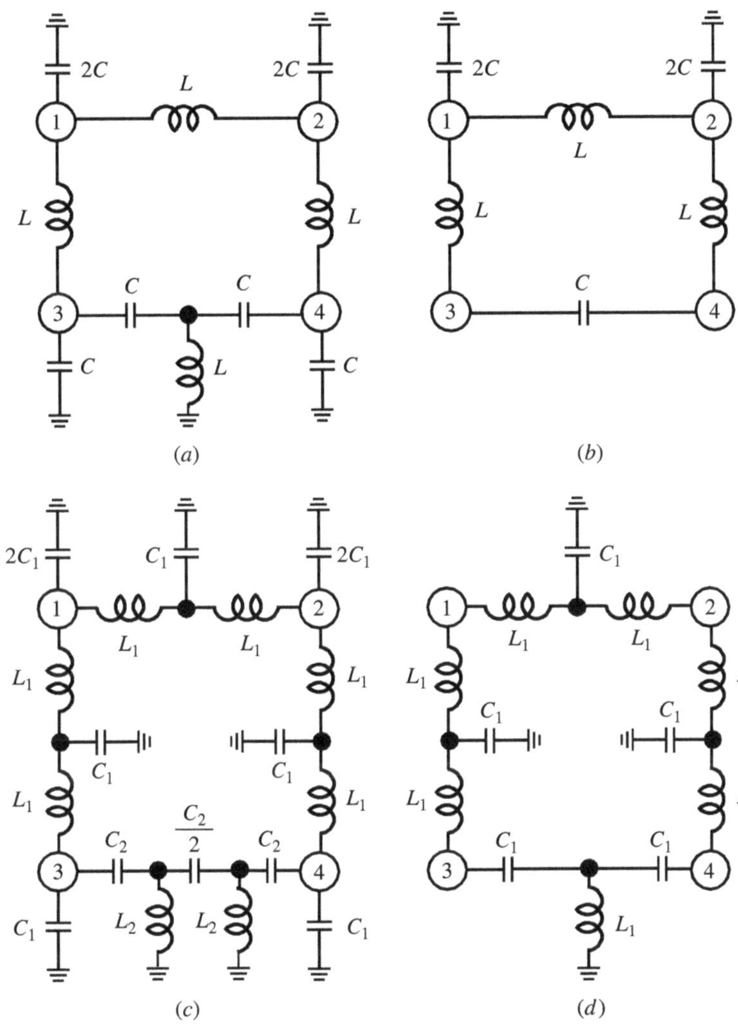

FIGURE 3.16 Lumped-element ring hybrids: (*a*) ring hybrid with three π-circuits and one T-circuit; (*b*) ring hybrid one with four π-circuits; (*c*) ring hybrid one with three T-circuits and one wideband circuit; (*d*) ring hybrid one with four T-circuits.

more compact sizes demanded, their size needs to be reduced. Lumped-element ring hybrids are one choice for small size. However, the bandwidths are quite narrow, and fabrication over 20 GHz is somewhat difficult. To overcome these disadvantages, the mixed small ring hybrids will be treated with distributed elements only [18] or with mixed distributed and lumped elements [19]. For these, the reduction of transmission-line size is important.

A transmission-line section with electrical length Θ and its small equivalent circuit are shown in Fig. 3.17(*a*) and (*b*). The even- and odd-mode equivalent

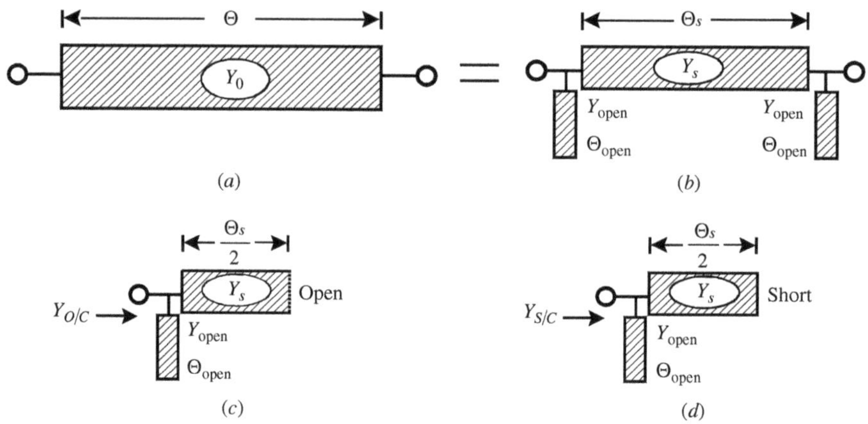

FIGURE 3.17 Transmission-line sections: (*a*) transmission line with electrical length Θ; (*b*) small transmission line with two shunt open stubs: (*c*) even-mode equivalent circuit; (*d*) odd-mode equivalent circuit.

circuits are depicted in Fig. 3.17(*c*) and (*d*), respectively. For the transmission-line section in Fig. 3.17(*a*), its even- and odd-mode admittances are given from (3.31), as

$$Y_{\mathrm{O/C}} = jY_0 \tan\frac{\Theta}{2}, \tag{3.38a}$$

$$Y_{\mathrm{S/C}} = -jY_0 \cot\frac{\Theta}{2}. \tag{3.38b}$$

Those in Fig. 3.17(*c*) and (*d*) are

$$Y_{\mathrm{O/C}} = jY_{\mathrm{open}} \tan\Theta_{\mathrm{open}} + jY_s \tan\frac{\Theta_s}{2}, \tag{3.39a}$$

$$Y_{\mathrm{S/C}} = jY_{\mathrm{open}} \tan\Theta_{\mathrm{open}} - jY_s \cot\frac{\Theta_s}{2}, \tag{3.39b}$$

where Y_s and Θ_s are the characteristic admittance and electrical length of the transmission line, and Y_{open} and Θ_{open} are those of each shunt stub in Fig. 3.17(*b*). Equating the equations in (3.39) to those in (3.38) gives us

$$Y_s = Y_0 \frac{\sin\Theta_s}{\sin\Theta}, \tag{3.40a}$$

$$Y_{\mathrm{open}} = Y_0 \frac{\cos\Theta_s - \cos\Theta}{\sin\Theta \tan\Theta_{\mathrm{open}}}. \tag{3.40b}$$

Figure 3.18 shows conventional distributed and mixed small ring hybrids. The transmission-line section between ports ② and ③ in Fig. 3.18(*a*) is three-quarters of a wavelength long and the others are a quarter-wavelength long. A

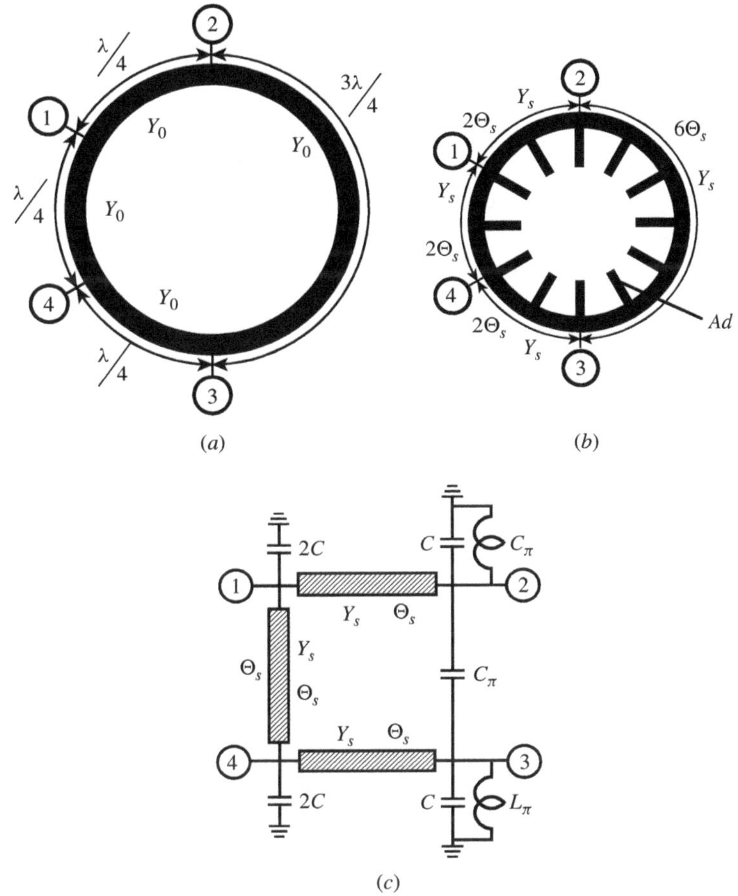

FIGURE 3.18 Ring hybrids: (*a*) distributed ring hybrid; (*b*) small ring hybrid; (*c*) mixed small ring hybrid.

corresponding small ring hybrid is depicted in Fig. 3.18(*b*), and a corresponding mixed small ring hybrid in Fig. 3.18(*c*). One transmission-line section between ports ① and ② of the ring hybrid in Fig. 3.18(*a*) consists of two small transmission lines in Fig. 3.17(*b*) and that between ports ② and ③ is composed of six small lines. The resulting configuration becomes that in Fig. 3.18(*b*), where the admittance of Ad is

$$\mathrm{Ad} = j2Y_{\mathrm{open}} \tan \Theta_{\mathrm{open}}. \tag{3.41}$$

To reduce the size more, the open stubs in Fig. 3.17(*b*) can be realized with capacitances for MMICs. The capacitance C_{ap} for the Ad is

$$C_{ap} = 2\frac{Y_{\mathrm{open}} \tan \Theta_{\mathrm{open}}}{\omega}. \tag{3.42}$$

As an additional way to reduce the size of ring hybrids, the three-quarter-wavelength transmission-line section between ports ② and ③ in Fig. 3.18(*a*) can be replaced with a Π lumped-element circuit, as in Fig. 3.14(*d*), and the remaining transmission-line sections can be replaced with one or two small transmission lines, as in Fig. 3.17(*b*). The resulting circuit is shown in Fig. 3.18(*c*), where the resonant elements C and L_π may be removed at a design center frequency. The mixed ring hybrid is fabricated without any inductor and the size can be reduced by more than 80% compared to a ring hybrid with all distributed elements. The size reduction and easy availability of shunt connections on the uniplanar structure make this structure suitable for MMICs. However, the problem with the narrow bandwidth remains.

3.4 CONVENTIONAL 3-dB UNIPLANAR RING HYBRIDS

One of the main requirements for a suitable transmission-line structure is that the structure be *planar* in configuration. A planar configuration implies that the characteristics of the element can be determined by the dimensions in a single plane. For example, the width of a microstrip line on a dielectric substrate can be adjusted to control its impedance. This results in convenient circuit fabrication through the techniques of photolithography and photoetching of thin films. Use of these techniques at microwave frequencies has led to the development of MICs and MMICs.

There are several transmission-line structures that satisfy the requirement of being planar, the most useful and common being microstrips, slotlines, coplanar waveguides (CPWs), coplanar strips, and coupled slotlines. Their cross-sectional views are shown in Fig. 3.19. The structure of coupled slotlines is very similar to that of the CPW. However, they have two orthogonal modes, and the odd-mode field distribution is different from that of the CPW. As can be shown in Fig. 3.19, the slotline, coplanar waveguide (CPW), coplanar strip, and coupled slotlines do not use the back side of the substrate and are therefore called *uniplanar* structures. They allow easy series and shunt connection of passive and active solid-state

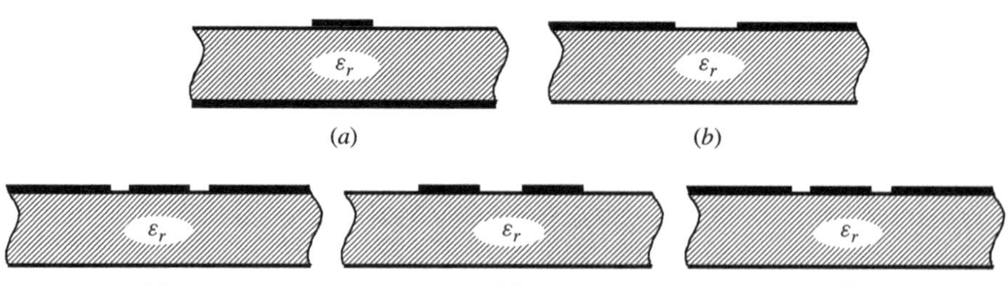

FIGURE 3.19 Planar transmission lines used in microwave integrated circuits: (*a*) microstrip; (*b*) slotline; (*c*) coplanar waveguide; (*d*) coplanar strip; (*e*) coupled slotlines.

devices and do not need via holes for ground connections. Bonding wires are needed for the difference between ground and other planes, so the performance of uniplanar circuits can be adjusted by changing the position of the bonding wires. The deterioration of the insertion loss and return loss due to the bonding wires is very small and does not affect circuit performance. Thus, with a simple fabrication process, uniplanar structures are being used more as MMIC techniques are developed.

3.4.1 Uniplanar T-Junctions

The T-junction is a simple three-port network that can be used for power division or power combining and can be implemented in virtually any type of transmission-line media. The T-junction is, in the absence of transmission-line loss, lossless, so it cannot be matched simultaneously at all ports. Figure 3.20 classifies uniplanar T-junctions according to the combination of transmission lines and CPWs; slotlines and coupled slotlines are used as the input transmission lines. When the power is excited at the CPW line in Fig. 3.20(*a*), an E-field (electric)

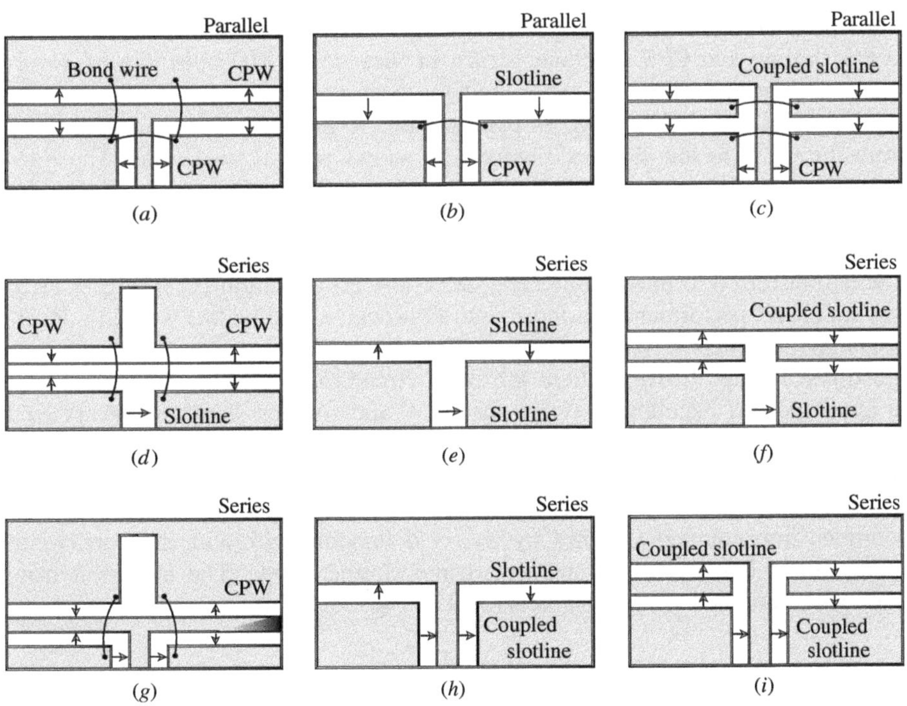

FIGURE 3.20 Uniplanar T-junctions: (*a*) CPW–CPW; (*b*) CPW–slotlines; (*c*) CPW–coupled slotlines; (*d*) slotline–CPWs; (*e*) slotline–slotlines; (*f*) slotline–coupled slotlines; (*g*) coupled slotline–CPWs; (*h*) coupled slotlines–slotline; (*i*) coupled slotlines–coupled slotlines. (From Ref. 8 with permission from IEEE.)

pattern is produced, and the bonding wire is needed to make ground planes of the input and output CPW lines. The arrows in these figures show the electric field in the CPWs and slotlines. (Only the near field is considered.) Therefore, the same field patterns are generated in both output lines, which are connected in parallel.

In a similar way, when the CPWs are the input lines, the connections of the T-junction are parallel, as shown in Fig. 3.20(*b*) and (*c*). In Fig. 3.20(*d*)–(*f*), the input lines are slotlines. For the case in Fig. 3.20(*d*), the plane on the right side of the slotline is grounded, and the ground plane is common with that of the CPW on the right side. The plane on the left side of the input line is, on the other hand, not grounded and is common with a plane of the CPW on the left side. Therefore, the E-field patterns of the outputs are generated as shown in Fig. 3.20(*d*), and the two output arms are connected in series. In a similar way, the two output arms in Fig. 3.20(*e*) and (*f*) are connected in series. The same principle can be applied for coupled slotlines, so those in Fig. 3.20(*g*)–(*i*) are connected in series. In conclusion, CPW input lines result in parallel T-junctions, whereas slotline and coupled-slotline input lines produce series T-junctions.

3.4.2 Transitions

Other basic components for uniplanar structures are the transitions from the CPW to the slotline and CPS (coplanar strip), or vice versa. The objective of developing the various transitions is to find the best combination of CPW short or open circuits and slotline short or open circuits to generate broadband matching from the CPW to the slotline. Figure 3.21 shows several transitions. The transition from CPW to slotline is depicted in Fig. 3.21(*a*). The angle of the slot radial stub in Fig. 3.21(*a*) is normally 90° and the slot radial stub is a broadband open. Due to the radial open stub, almost all the power excited at the CPW line is transferred to the slotline. The CPW-to-CPS transition [9] in Fig. 3.21(*b*) has wideband performance (a more than two-octave bandwidth) with low insertion loss (less than −1 dB). The transition from CPW to coupled slotlines is illustrated in Fig. 3.21(*c*), where the open circuits and bonding wires are used to create perfect matching between the CPW and coupled slotlines. The ground plane on the right side of the CPW in Fig. 3.21(*c*) is common with that of the coupled slotlines, so the electric field of the slotlines is generated as shown. In this transition process, the electromagnetic field distributed in the CPW is concentrated into coupled slotlines by use of a bonding wire and an open circuit so that the CPW mode is transformed into slotline mode. The transition from slotlines to coupled slotlines is described in Fig. 3.21(*d*).

3.4.3 Wideband Uniplanar Baluns

For wider bandwidths, Fig. 3.22 shows T-junctions using wideband radial stubs; a CPW–CPW balun and a CPW–slotline balun [11] are depicted in Fig. 3.22(*a*) and (*b*), respectively. The circuit of Fig. 3.22(*a*) consists of one CPW–slotline T-junction and two slotline–CPW transitions. The E-field of the input CPW

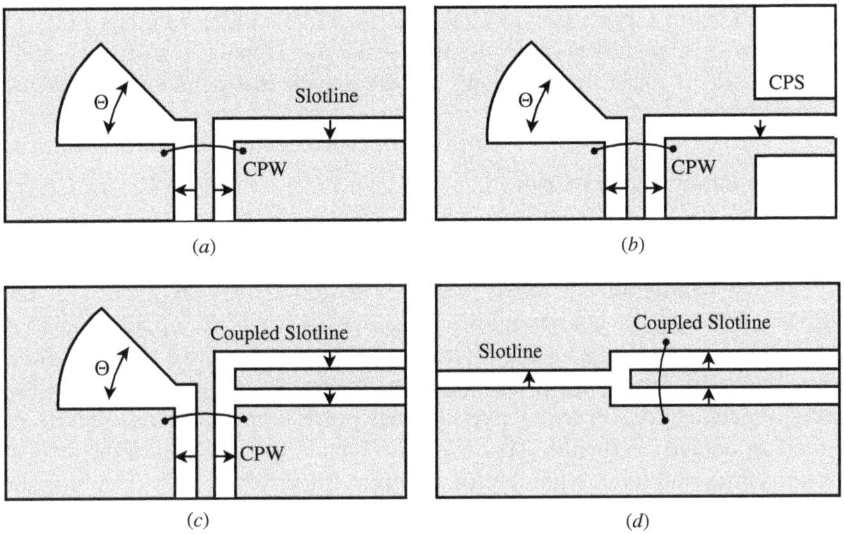

FIGURE 3.21 Transitions: (*a*) from CPW into slotline; (*b*) from CPW into CPS; (*c*) from CPW into coupled slotlines; (*d*) from slotline into coupled slotlines. [(*a, b*) From Ref. 9 with permission from IEEE.]

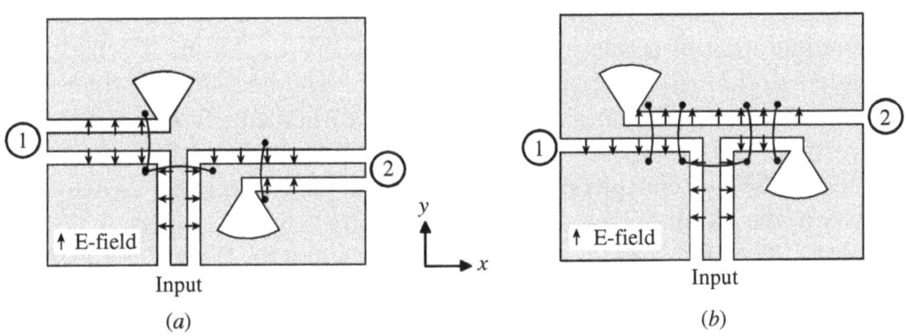

FIGURE 3.22 Circuit configurations and schematic diagrams of E-field distribution for uniplanar baluns (T-junctions): (*a*) CPW–CPW balun; (*b*) CPW–slotline balun. (From Ref. 11 with permission from IEEE.)

(near the CPW–slotline T-junction) is directed from the center conductor to the CPW ground planes and produces two slotline waves with negative signs in the y-direction. The negative E-field of the slotline at port ① produces a positive E-field in the CPW, whereas that at port ② produces a negative E-field in the CPW. Thus, Fig. 3.22(*a*) is truly a T-junction with a 180° phase shift.

The circuit of Fig. 3.22(*b*) consists of one CPW T-junction and two CPW–slotline transitions. The E-field of the input CPW is directed from the center conductor toward the ground planes and produces the two positive E-fields of CPWs as explained for Fig. 3.20(*a*). One of the two slotlines of each CPW is connected with a port; port ① takes a negative value, whereas port ② is connected with a positive value. In this way, the signals at ports ① and ② are always out of phase by 180° and are frequency independent (i.e., of broad bandwidth).

3.4.4 Uniplanar Ring Hybrids

The bandwidth of a conventional ring hybrid is about 20%, so several design techniques have been developed to extend the bandwidth (Figs. 3.9 and 3.10). Although the bandwidth has been increased to more than one octave of bandwidth, the difficulty in constructing a ring hybrid limits its application to low frequencies since the coupled-line filter in Fig. 3.10 requires two shorts at the end, and gaps, which are very small for microstrip technology, must be etched away. Other approaches [20,21] using hypothetical ports with matching circuits have achieved an octave bandwidth. However, the matching circuits require very wide microstrip lines and a large number of different impedances [20]. The broadband design technique [21] is also useful in the sum mode (in-phase mode) operation only. Both cases demand intensive optimization to obtain desirable performances.

Therefore, the uniplanar ring hybrids with frequency-independent elements may be good choices [10,11,22]. Several examples are shown in Fig. 3.23, where three fundamental uniplanar ring hybrids and two with broadband baluns are represented. The uniplanar ring hybrid shown in Fig. 3.23(*a*) consists of four CPW-to-slotline T-junctions and four slotline sections. Figure 3.23(*b*) shows the circuit configuration of a ring hybrid with four CPW-to-slotline T-junctions and one coupled-slotline filter section. Although the coupled-slotline filter section requires no via holes for the short circuit, it is difficult to fabricate because of the very small gap between the two coupled slotlines with a 3-dB coupling. Figure 3.23(*c*) shows the physical configuration of a uniplanar crossover ring hybrid where the slotline T-junction between ports ① and ② is used as a phase inverter [10]. The ring hybrids with broadband baluns in Fig. 3.22 are depicted in Fig. 3.23(*d*) and (*e*). Figure 3.23(*d*) consists of four CPW feeds, three slotline sections, and one broadband balun, and Fig. 3.23(*e*) consists of four CPW feeds, three CPW sections, and one broadband balun. All the ring hybrids except the one in Fig. 3.23(*a*) have broadband performances.

Figure 3.24 shows the simulated results of the ring hybrids in Fig. 3.23(*d*) and (*e*), where a divided power of −3.01 dB, matching properties of less than −124 dB, and isolation of less than −130 dB at a center frequency of 3 GHz are obtained. As expected, the simulated results show broadband performances. The two ring hybrids in Fig. 3.23(*d*) and (*e*) were fabricated, and their photos are shown in Fig. 3.25(*a*) and (*b*), respectively.

The uniplanar circuits are normally measured by an on-wafer measurement, but the sizes of the ring hybrids in Fig. 3.25 are too large for that. Thus, special packages are needed. The packages shown in Fig. 3.25 are for measuring ring hybrids with microstrip connectors. To connect the ground of CPW feeding lines with the package wall, "(ii) For ground" in Fig. 3.25 is required, and "(iii) For fixing" keeps the ring hybrid stable during measurement.

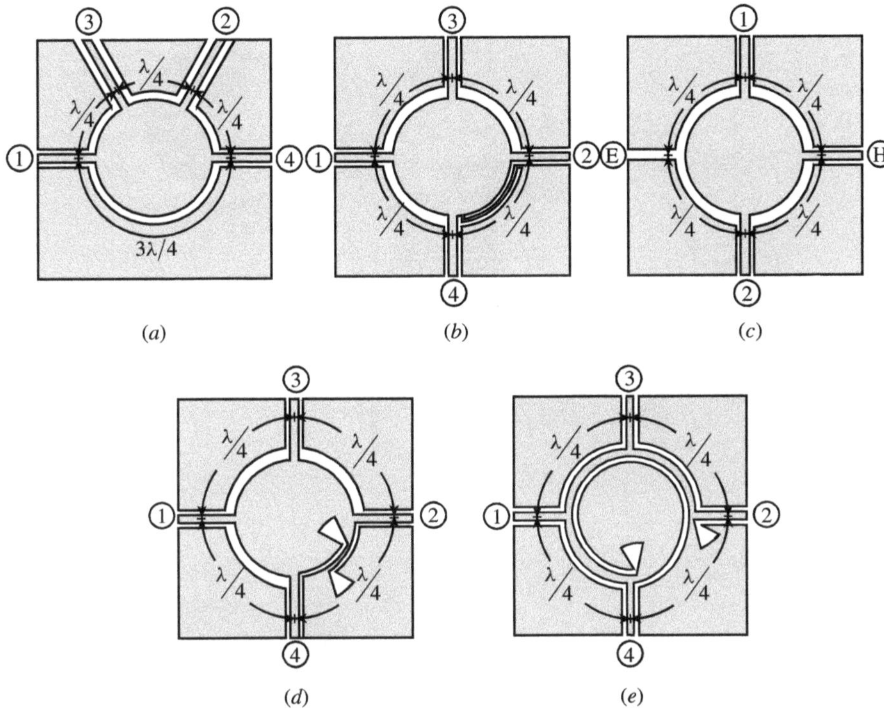

FIGURE 3.23 Circuit configuration of uniplanar ring hybrids: (*a*) conventional ring hybrid; (*b*) ring hybrid with a coupled-slotline filter; (*c*) crossover ring hybrid; (*d, e*) ring hybrids with broadband baluns. (From Refs. 9 and 11 with permission from IEEE.)

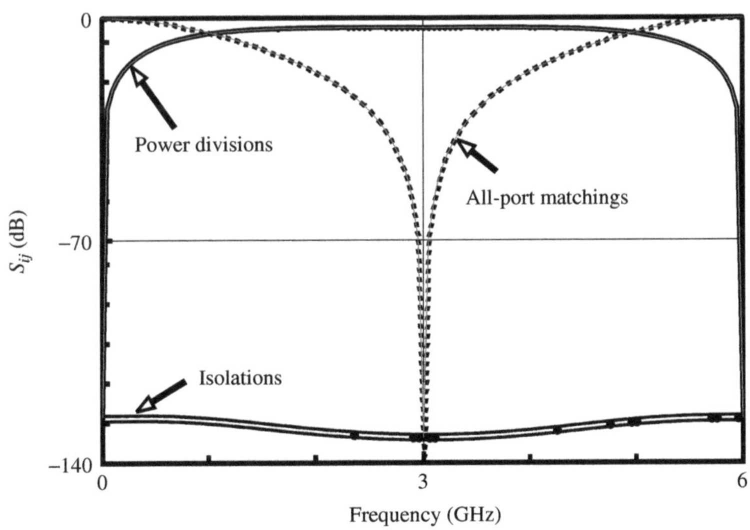

FIGURE 3.24 Simulation results of a uniplanar ring hybrid with equal power division and equal termination impedances.

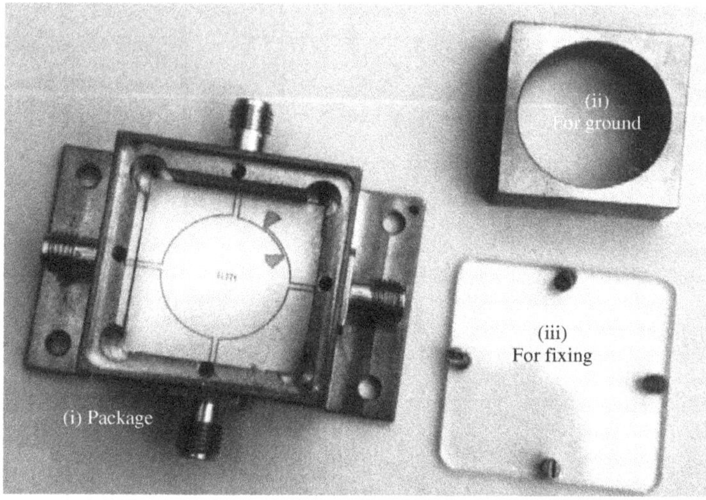

(*a*)

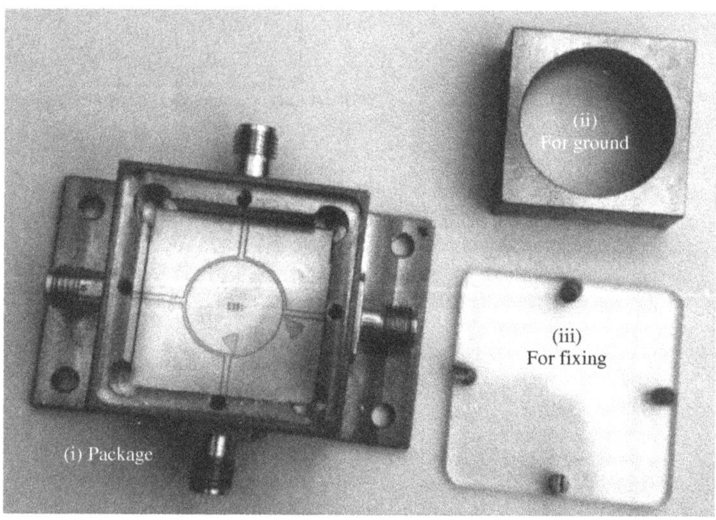

(*b*)

FIGURE 3.25 Fabrication of ring hybrids: (*a*) with a CPW–CPW balun; (*b*) with a CPW–slotline balun.

EXERCISES

3.1 Design the parallel coupled transmission lines in Fig. 3.6 to satisfy the following conditions:

(a) $C = -3\ dB,\ Z_0 = 30\ \Omega$

(b) $C = -6\ dB,\ Z_0 = 50\ \Omega$

3.2 Derive the image impedances in Figs. 3.7 and 3.8.

3.3 Design a ring hybrid like Fig. 3.9(*b*) when all ports are terminated in equal impedances 50 Ω and the power-split ratio is −4 dB.

3.4 For the ring hybrid in Fig. 3.10:

(a) Determine the characteristic impedances of the transmission-line sections for $S_{21}/S_{31} = 2$.

(b) Design a ring hybrid with $S_{21}/S_{31} = 2$ and a 5-dB coupling coefficient for the coupled transmission lines.

3.5 Design the ring hybrid in Fig. 3.12 with a 3-dB power-split ratio.

3.6 Design the ring hybrid in Fig. 3.18(*b*) to provide $S_{21}/S_{31} = 2$.

3.7 Design the ring hybrid in Fig. 3.18(*c*) with $S_{21}/S_{31} = 2$ at a design frequency of 3 GHz.

3.8 Design the ring hybrids in Fig. 3.23(*d*) and (*e*) for an equal power division for $\varepsilon_r = 10$ and $h = 0.1$ mm.

REFERENCES

1. W. A. Tyrrel, Hybrid Circuits for Microwaves, *Proc. IRE*, Vol. 35, November 1947, pp. 1294–1306.
2. T. Morita and L. S. Sheingold, A Coaxial Magic-T, *IRE Trans. Microwave Theory Tech.*, Vol. 1, November 1953, pp. 17–23.
3. V. I. Albanese and W. P. Peyser, An Analysis of a Broad-Band Coaxial Hybrid Ring, *IRE Trans. Microwave Theory Tech.*, Vol. 6, October 1958, pp. 369–373.
4. W. V. Tyminski and A. E. Hylas, A Wide-Band Hybrid Ring for UHF, *Proc. IRE*, Vol. 41, January 1953, pp. 81–87.
5. S. March, Wideband Stripline Hybrid Ring, *IEEE Trans. Microwave Theory Tech.*, Vol. 16, June 1968, pp. 361–362.
6. C. Y. Pon, Hybrid-Ring Directional Coupler for Arbitrary Power Divisions, *IRE Trans. Microwave Theory Tech.*, Vol. 9, November 1961, pp. 529–535.
7. T. Hirota, Y. Tarusawa, and H. Ogawa, Uniplanar MMIC Hybrids: A Proposed New MMIC Structure, *IEEE Trans. Microwave Theory Tech.*, Vol. 35, June 1987, pp. 576–581.
8. H. Ogawa and A. Minagawa, Uniplanar MIC Balanced Multiplier: A Proposed New Structure for MIC's, *IEEE Trans. Microwave Theory Tech.*, Vol. 35, December 1987, pp. 1363–1367.
9. C.-H. Ho, L. Fan, and K. Chang, Broad-Band Uniplanar Hybrid-Ring and Branch-Line Couplers, *IEEE Trans. Microwave Theory Tech.*, Vol. 41, December 1993, pp. 2116–2124.
10. C.-H. Ho, L. Fan, and K. Chang, Slotline Annular Ring Elements and Their Applications to Resonator, Filter and Coupler Design, *IEEE Trans. Microwave Theory Tech.*, Vol. 41, September 1993, pp. 1648–1650.

11. C.-H. Ho, L. Fan, and K. Chang, New Uniplanar Coplanar Waveguide Hybrid-Ring Couplers and Magic-T's, *IEEE Trans. Microwave Theory Tech.*, Vol. 42, December 1994, pp. 2440–2448.
12. H.-R. Ahn, I. Wolff, and I.-S. Chang, Arbitrary Termination Impedances, Arbitrary Power Division and Small-Sized Ring Hybrids, *IEEE Trans. Microwave Theory Tech.*, Vol. 44, December 1997, pp. 2241–2247.
13. T. Q. Ho and S. M. Hart, Broad-Band Coplanar Waveguide to Slotline Transition, *IEEE Microwave Guided Wave Lett.*, Vol. 2, No. 10, October 1992, pp. 415–416.
14. D. M. Pozar, Microwave Engineering, *Addison-Wesley*, Reading, MA, 1990, p. 391.
15. E. M. T. Jones and J. T. Bolljahn, Coupled-Strip Transmission-Line Filters and Directional Couplers, *IRE Trans. Microwave Theory Tech.*, Vol. 4, April 1956, pp. 75–81.
16. S. J. Parisi, 180° Lumped Element Hybrid, *IEEE MTT-S Dig.*, 1989, pp. 1243–1246.
17. R. K. Gupta and W. J. Gestinger, Quasi-Lumped-Element 3- and 4-Port Networks for MIC an MMIC Applications, *IEEE MTT-S Dig.*, 1984, pp. 409–411.
18. M.-L. Chung, Miniaturized Ring Coupler of Arbitrary Reduced Size, *IEEE Microwave Component Lett.*, Vol. 15, No. 1, January 2005, pp. 16–18.
19. T. Hirota, A. Minakawa, and M. Muraguchi, Reduced-Size Branch-Line and Rat-Race Hybrids for Uniplanar MMIC's, *IEEE Trans. Microwave Theory Tech.*, Vol. 38, March 1990, pp. 270–275.
20. D. Kim and Y. Natio, Broad Band Design of Improved Hybrid-Ring 3-dB Directional Coupler, *IEEE Trans. Microwave Theory Tech.*, Vol. 30, November 1982, pp. 2040–2046.
21. G. F. Mikucki and A. K. Agrawal, A Broad-Band Printed Circuit Hybrid Ring Power Divider, *IEEE Trans. Microwave Theory Tech.*, Vol. 37, January 1989, pp. 112–117.
22. D. Chana, A New Coplanar Waveguide/Slotline Double Balance Mixer, *IEEE MTT-S Dig.*, 1989, pp. 967–968.

CHAPTER FOUR

Asymmetric Ring Hybrids

4.1 INTRODUCTION

Although considered narrowband devices, ring hybrids have found extensive use in microwave integrated circuits. Ring hybrids have a definite advantage over branch-line hybrids in applications in antenna arrays, mixers, and balancing amplifiers, because they have wider bandwidths than those of branch-line hybrids and do not require a phase-compensation element. In practice, ring hybrids are used together with other active and/or passive devices. Thus, to obtain a desired performance, additional matching networks are necessary for conventional ring hybrids. Therefore, ring hybrids terminated in arbitrary impedances can significantly reduce the size of MMICs or MICs [13–18]. Since 1947 [1], ring hybrid studies and applications have focused on just symmetric structures [1–12]; in 1994 the first asymmetric-structure ring hybrids, which can be used for equal power division, were investigated by Ahn et al. [13]. Since Pon [6], several authors have dealt with ring hybrids with arbitrary power divisions [11,12]. However, these analyses are also applied only to symmetric structures, and the procedures to find the correct design equations are somewhat complex [6]. In 1997, the design equations of asymmetric ring hybrids were first derived for arbitrary power divisions, arbitrary termination impedances, and small sizes. Based on the design method suggested, in this chapter we focus on how to derive the design equations and how to reduce the size of asymmetric ring hybrids.

4.2 DERIVATION OF DESIGN EQUATIONS OF ASYMMETRIC RING HYBRIDS

The asymmetric ring hybrid is shown in Fig. 4.1, where $\Theta_1 = \Theta_2 = \Theta_3 = \lambda/4$ and $\Theta_4 = 3\lambda/4$ at a design center frequency. It is terminated in arbitrary

Asymmetric Passive Components in Microwave Integrated Circuits, by Hee-Ran Ahn

admittances Y_a, Y_b, Y_c, and Y_d. This structure is therefore no longer symmetric and may be called an asymmetric ring hybrid. In the case of conventional ring hybrids in Chapter 3, the termination admittances were equal; the characteristic admittances of transmission-line sections Y_1, Y_2, Y_3, and Y_4 were the same or two of them were equal. Furthermore, as there were still symmetry planes in the ring structure, even- and odd-mode analyses were possible, although finding design equations was somewhat complicated [6]. In the case of an asymmetric ring hybrid terminated in arbitrary impedances, the termination admittances and characteristic admittances of transmission-line sections differ from each other; as there is no symmetric plane, even- and odd-mode analyses are not possible. Because of the conditions, a new design method is necessary.

For the ring hybrid in Fig. 4.1, it is well understood that the excited power at port ① is split between ports ② and ④, while port ③ is isolated. This operation can be regarded as a three-port power divider [15–17] with two outputs being out of phase. For the excitation at port ③, similar arguments lead to operation as a three-port power divider.

An asymmetric ring hybrid may be described by a set of scattering parameters given by

$$\begin{bmatrix} b_1 \\ b_2 \\ b_3 \\ b_4 \end{bmatrix} = \begin{bmatrix} S_{11} & S_{12} & S_{13} & S_{14} \\ S_{21} & S_{22} & S_{23} & S_{24} \\ S_{31} & S_{32} & S_{33} & S_{34} \\ S_{41} & S_{42} & S_{43} & S_{44} \end{bmatrix} \begin{bmatrix} a_1 \\ a_2 \\ a_3 \\ a_4 \end{bmatrix}, \tag{4.1}$$

where $\boldsymbol{a}$ and $\boldsymbol{b}$ are reflected and incident normalized wave vectors. If port ③ is terminated in a reflection coefficient Γ_3, (4.1) may be expressed by

$$\begin{bmatrix} \boldsymbol{B}_1 \\ \boldsymbol{B}_2 \end{bmatrix} = \begin{bmatrix} \boldsymbol{P}_{11} & \boldsymbol{P}_{12} \\ \boldsymbol{P}_{21} & \boldsymbol{P}_{22} \end{bmatrix} \begin{bmatrix} \boldsymbol{A}_1 \\ \boldsymbol{A}_2 \end{bmatrix}, \tag{4.2}$$

where

$$\boldsymbol{B}_1 = \begin{bmatrix} b_1 \\ b_2 \\ b_4 \end{bmatrix}, \qquad \boldsymbol{A}_1 = \begin{bmatrix} a_1 \\ a_2 \\ a_4 \end{bmatrix}, \qquad \boldsymbol{B}_2 = b_3, \qquad \boldsymbol{A}_2 = a_3$$

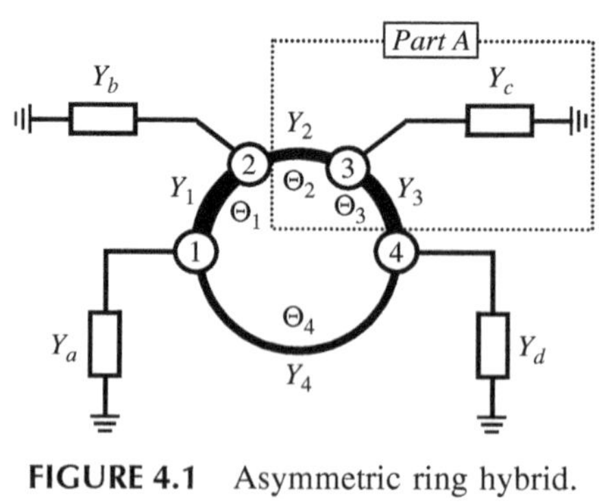

FIGURE 4.1 Asymmetric ring hybrid.

$$P_{11} = \begin{bmatrix} S_{11} & S_{12} & S_{14} \\ S_{21} & S_{22} & S_{24} \\ S_{41} & S_{42} & S_{44} \end{bmatrix}, \qquad P_{12} = \begin{bmatrix} S_{13} \\ S_{23} \\ S_{43} \end{bmatrix},$$

$$P_{21} = \begin{bmatrix} S_{31} & S_{32} & S_{34} \end{bmatrix}, \qquad P_{22} = [S_{33}],$$

$$B_2 = \Gamma_L^{-1} A_2, \qquad \Gamma_L = \Gamma_3$$

A straightforward calculation derives

$$B_1 = P_{11}A_1 + P_{12}\Gamma_L(U - P_{22}\Gamma_L)^{-1}P_{21}A_1, \tag{4.3}$$

which can be rewritten as

$$\begin{bmatrix} b_1 \\ b_2 \\ b_4 \end{bmatrix} = \begin{bmatrix} S_{11} & S_{12} & S_{14} \\ S_{21} & S_{22} & S_{24} \\ S_{41} & S_{42} & S_{44} \end{bmatrix} \begin{bmatrix} a_1 \\ 0 \\ 0 \end{bmatrix} + \frac{S_{31}a_1\Gamma_3}{1 - \Gamma_3 S_{33}} \begin{bmatrix} S_{13} \\ S_{23} \\ S_{43} \end{bmatrix}. \tag{4.4}$$

Assuming that $|S_{31}| = 0$ in (4.4) indicates that there is no correlation with Γ_3 in terms of port ① excitation. So under the conditions of $|S_{31}| = 0$, the excitation at port ① can be interpreted as that in Fig. 4.2. If the ring hybrid in Fig. 4.1 is lossless and passive, scattering parameters should satisfy a unitary condition

$$SS^{*T} = U_n, \tag{4.5}$$

where U_n is an identity matrix of order n. Therefore, if the output arms are isolated from each other and the input impedances are matched looking into any arm (when the other arms are terminated in matched impedances), (4.5) gives, in detail,

$$|S_{21}|^2 + |S_{41}|^2 = 1, \tag{4.6a}$$

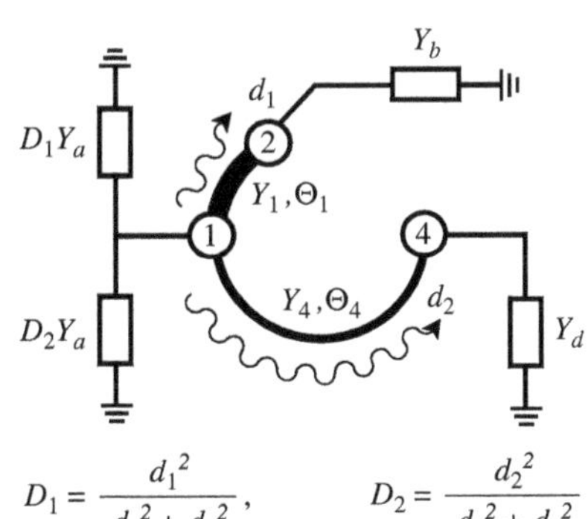

FIGURE 4.2 Excitation at port ①.

$$|S_{12}|^2 + |S_{32}|^2 = 1, \tag{4.6b}$$

$$|S_{23}|^2 + |S_{43}|^2 = 1, \tag{4.6c}$$

$$|S_{14}|^2 + |S_{34}|^2 = 1. \tag{4.6d}$$

The power-split ratio is proportional to the square of the admittance ratio of the two variable admittances in the ring [6,11,12]. Therefore, if the ratio of $|S_{21}|$ to $|S_{41}|$ is that of d_1 to d_2 in Fig. 4.2, the termination admittance Y_a at port ① is divided into two parts, D_1Y_a and D_2Y_a, depending on the scattering parameter ratio, where D_1 and D_2 are $d_1^2/(d_1^2 + d_2^2)$ and $d_2^2/(d_1^2 + d_2^2)$, respectively. If $\Theta_1 = \lambda/4$ and $\Theta_4 = 3\lambda/4$ in Fig. 4.2, the characteristic admittances of the transmission-line sections, Y_1 and Y_4, are

$$Y_1 = \sqrt{D_1 Y_a Y_b} = \sqrt{\frac{d_1^2}{d_1^2 + d_2^2} Y_a Y_b}, \tag{4.7a}$$

$$Y_4 = \sqrt{D_2 Y_a Y_d} = \sqrt{\frac{d_2^2}{d_1^2 + d_2^2} Y_a Y_d}, \tag{4.7b}$$

where d_1 and d_2 are determined by the needed power division ratio. Under the assumption of $|S_{31}| = 0$, the characteristic admittances Y_1 and Y_4 are derived.

The configuration in Fig. 4.2 is true in terms of excitation at port ①, so perfect matching appears at port ① but not at the output ports, ② and ④. For the perfect matching at ports ② and ④, Part A in Fig. 4.1, which was removed in Fig. 4.2, must be considered. For a Y- or T-junction such as a Wilkinson power divider, an isolation resistor is needed for perfect matching at the output ports [15–20]. In the same way, the circuit of Part A needs to be connected as shown in Fig. 4.1 for perfect matching at ports ② and ④.

When power is fed into port ① in Fig. 4.1, the power is divided and delivered to port ②. Most of the power at port ② is transferred to the load Y_b, but an extremely small amount of the power flows forward to port ③. Similarly, almost all the power reaching port ④ is delivered to the load Y_d, and only an extremely small amount of power flows forward to port ③. This is shown in Fig. 4.3. For these two waves at port ③ to be isolated from port ①, two conditions should be satisfied: the two waves should be out of phase by 180°, and the scattering parameter ratio of the two waves should be $d_2 : d_1$, as shown in Fig. 4.3. If the lengths of the transmission-line sections with Y_1 and Y_4 are assumed to be $\lambda/4$ and $3\lambda/4$, respectively, the transmission-line sections with Y_2 and Y_3 should both be $\lambda/4$ long to satisfy the first condition. The second condition mentioned above is satisfied by (4.5) and (4.6).

The circuit in Fig. 4.3 is passive and therefore has reciprocity. Thus, it can be considered as being excited at port ③. In a similar way, the characteristic

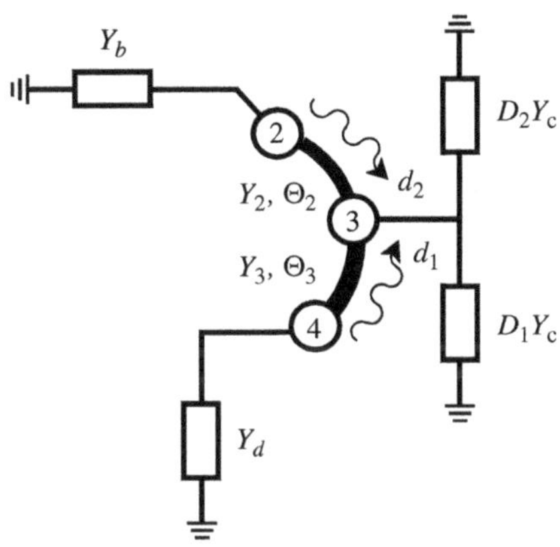

$$D_1 = \frac{d_1^2}{d_1^2 + d_2^2}, \qquad D_2 = \frac{d_2^2}{d_1^2 + d_2^2}$$

FIGURE 4.3 Excitation at port ③.

admittances of Y_2 and Y_3 are

$$Y_2 = \sqrt{D_2 Y_b Y_c} = \sqrt{\frac{d_2^2}{d_1^2 + d_2^2} Y_b Y_c}, \tag{4.8a}$$

$$Y_3 = \sqrt{D_1 Y_c Y_d} = \sqrt{\frac{d_1^2}{d_1^2 + d_2^2} Y_c Y_d}. \tag{4.8b}$$

From (4.7) and (4.8), for $Y_a = Y_b = Y_c = Y_d$, the results are the same as those given by Pon [6]. If $d_1 : d_2 = 1 : 1$, the results are in agreement with those of Ahn et al. [13]. For $Y_a = Y_b = Y_c = Y_d$ and $d_1 : d_2 = 1 : 1$, the design equations are those of the well-known 3-dB ring hybrid.

Figure 4.4 shows the results simulated on the basis of derived design equations (4.7) and (4.8). The frequency responses of the power-split ratio are plotted in Fig. 4.4(*a*), the isolation characteristics in Fig. 4.4(*b*), the reflection coefficients in Fig. 4.4(*c*), and the phase responses in Fig. 4.4(*d*) and (*e*). The ring hybrid for the simulation is terminated in 50 Ω, 45.45 Ω, 71.4 Ω, and 62.5 Ω, and its power-split ratio is 2 dB [i.e., $20\log(d_1/d_2) = 2$ dB]. An Advanced Design System (ADS) circuit simulator was used for these simulations of the ring hybrid. The simulated results in Fig. 4.4(*a*) show that the power division ratio is really 2 dB with $|S_{21}| = -2.124$ dB, $|S_{41}| = -4.124$ dB, $|S_{43}| = -2.124$ dB, and $|S_{23}| = -4.124$ dB. The simulation results in Fig. 4.4(*b*) show that the powers at the isolated ports ($|S_{31}| = -158.656$ dB and $|S_{42}| = -160.656$ dB) are theoretically zero at the center frequency. The value $|S_{21}| = -2.124$ dB is obtained from $10\log[d_1^2/(d_1^2 + d_2^2)]$.

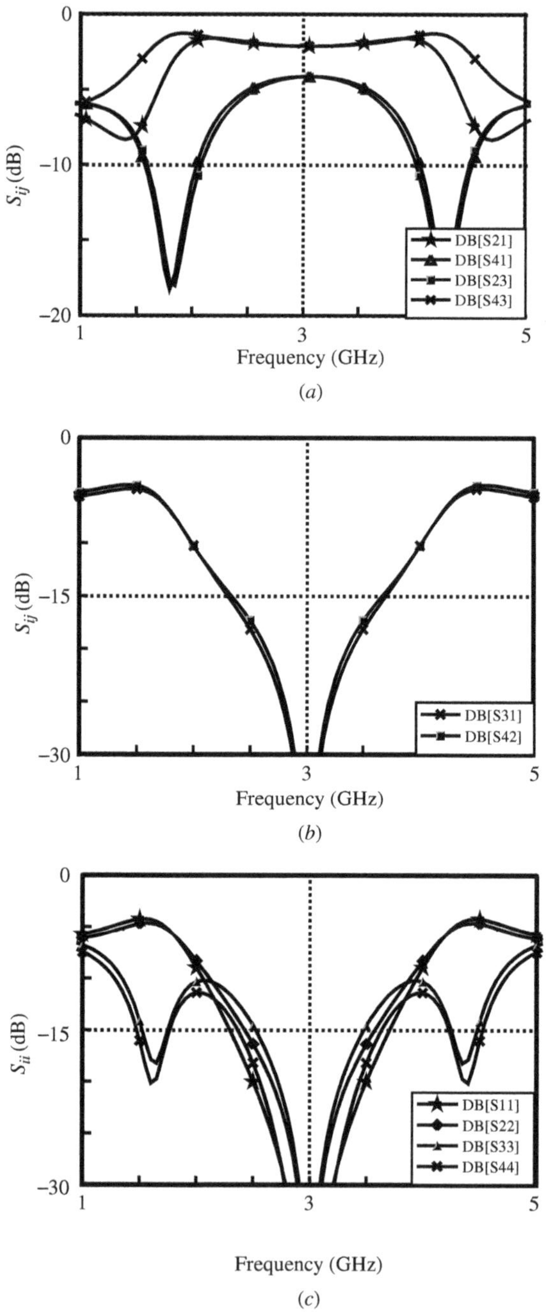

FIGURE 4.4 Simulation results of a ring hybrid terminated in $Z_a = 50\ \Omega$, $Z_b = 45.45\ \Omega$, $Z_c = 71.4\ \Omega$, and $Z_d = 62.5\ \Omega$; power-split ratio, 2 dB. (*a*) [*S*]-parameters for power division; (*b*) [*S*]-parameters for isolation; (*c*) [*S*]-parameters for matching; (*d*) phases of S_{23} and S_{43}; (*e*) phases of S_{21} and S_{41}.

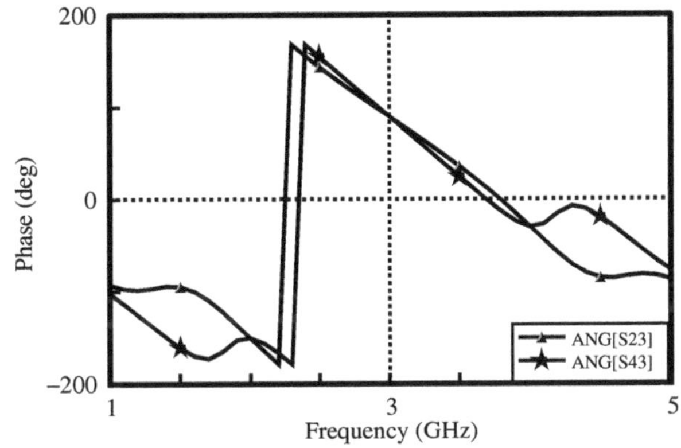

(*d*)

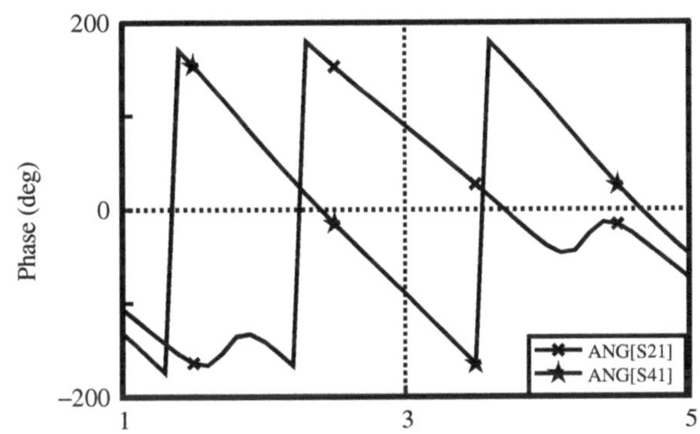

(*e*)

FIGURE 4.4 (*continued*)

4.3 SMALL ASYMMETRIC RING HYBRIDS

The voltages in the two output arms are either in phase or out of phase, depending on the input arms chosen, and the lengths of the transmission-line sections are $\lambda/4$ or $3\lambda/4$. To reduce the size of the ring hybrid, transmission-line sections whose electrical lengths are less than $\lambda/4$ can be used; lengths of $\lambda/8$, $\lambda/6$ [19], and $\lambda/5$ [20,21] are examples. In any case, three arcs of the ring are of equal length, and the fourth should provide a $\pm 180^\circ$ phase shift compared to the others. If the lengths of the transmission-line sections are not $\lambda/4$ or $3\lambda/4$, the characteristic impedances are changed in proportion to the lengths. The characteristic admittances may be derived as

$$Y_1 = \sqrt{\frac{D_1 Y_a Y_b}{1 - \cot^2 \Theta_1}} = \sqrt{\frac{d_1^2}{(d_1^2 + d_2^2)(1 - \cot^2 \Theta_1)} Y_a Y_b}, \tag{4.9a}$$

TABLE 4.1 Simulation Results (dB) for a Small Asymmetric Ring Hybrid[a]

Power-Split Ratio	Matching	Isolation	Power Division
0	$S_{11} = -49.37$	$S_{31} = -40.4$	$S_{21} = -3.013$
	$S_{22} = -32.59$	$S_{24} = -38.4$	$S_{41} = -3.009$
	$S_{33} = -31.69$		$S_{23} = -3.014$
	$S_{44} = -44.82$		$S_{43} = -3.013$
2	$S_{11} = -44.14$	$S_{31} = -41.9$	$S_{21} = -2.126$
	$S_{22} = -34.03$	$S_{24} = -39.6$	$S_{41} = -4.123$
	$S_{33} = -32.83$		$S_{23} = -4.127$
	$S_{44} = -41.10$		$S_{43} = -2.127$
4	$S_{11} = -41.30$	$S_{31} = -44.0$	$S_{21} = -1.457$
	$S_{22} = -35.37$	$S_{24} = -41.5$	$S_{41} = -5.454$
	$S_{33} = -33.97$		$S_{23} = -5.458$
	$S_{44} = -39.15$		$S_{43} = -1.457$

[a] $Z_a = 1/Y_a = 50\ \Omega$, $Z_b = 1/Y_b = 45.45\ \Omega$, $Z_c = 1/Y_c = 71.4\ \Omega$, and $Z_d = 1/Y_d = 62.5\ \Omega$; $\Theta_1 = \Theta_2 = \Theta_3 = 80^\circ$, $\Theta_4 = 180^\circ + \Theta_1$.

$$Y_2 = \sqrt{\frac{D_2 Y_b Y_c}{1 - \cot^2 \Theta_2}} = \sqrt{\frac{d_2^2}{(d_1^2 + d_2^2)(1 - \cot^2 \Theta_2)} Y_b Y_c}, \tag{4.9b}$$

$$Y_3 = \sqrt{\frac{D_1 Y_c Y_d}{1 - \cot^2 \Theta_3}} = \sqrt{\frac{d_1^2}{(d_1^2 + d_2^2)(1 - \cot^2 \Theta_3)} Y_c Y_d}, \tag{4.9c}$$

$$Y_4 = \sqrt{\frac{D_2 Y_a Y_d}{1 - \cot^2 \Theta_4}} = \sqrt{\frac{d_2^2}{(d_1^2 + d_2^2)(1 - \cot^2 \Theta_4)} Y_a Y_d}. \tag{4.9d}$$

From (4.9), if $Y_a = Y_b = Y_c = Y_d$, $\Theta_1 = \Theta_2 = \Theta_3 = \lambda/6$, $\Theta_4 = 4\lambda/6$, and $d_1 : d_2 = 1 : 1$, the results are the same as those suggested by Kim and Yang [19]. If $Y_a = Y_b = Y_c = Y_d$, $d_1 : d_2 = 1 : 1$ and for general arc lengths, the results are those presented by Murgulescu et al. [20] and Fan et al. [21].

On the basis of (4.9), a ring hybrid with termination impedances of 50 Ω, 45.45 Ω, 71.4 Ω, and 62.5 Ω and an electrical length of 80° was simulated at a design center frequency of 3 GHz. The simulation results of the scattering parameters given in Table 4.1 show that the performances are the same as those of conventional ring hybrids with 90° and 270° electrical lengths.

4.4 WIDEBAND OR SMALL ASYMMETRIC RING HYBRIDS

4.4.1 Microstrip Asymmetric Ring Hybrids

The limiting factor in conventional ring hybrids is that a transmission-line section with $180^\circ + \Theta$, which provides a 180° phase shift compared to other sections, is effective only in the vicinity of the center frequency. To overcome this problem, a wideband ring hybrid using a coupled-line filter [5] may be realized in stripline technology, but it requires complicated fabrication. Here, an easy way to realize such ring hybrids with microstrip technology is introduced. They can be used for arbitrary power divisions, termination impedances, and lengths of transmission-line sections, whereas conventional ring hybrids operate only with equal power splitting, equal termination impedances, and an equal wavelength of 90°.

To realize a coupled-line filter section, a set of Z_{0e} (even-mode impedance) and Z_{0o} (odd-mode impedance) are needed. With reference to Fig. 3.7(a), the image impedance is given as

$$Z_I = \frac{2Z_{0e}Z_{0o}\sin\Theta}{[(Z_{0e}-Z_{0o})^2-(Z_{0e}-Z_{0o})^2\cos^2\Theta]^{1/2}}. \tag{4.10}$$

The coupling coefficient of the coupled transmission lines [22] is known as

$$C = \left|\frac{j[(Z_{0e}/Z_{0o})^{0.5}-(Z_{0e}/Z_{0o})^{-0.5}]\sin\Theta}{D_{\text{eff}}}\right|, \tag{4.11}$$

where $D_{\text{eff}} = 2\cos\Theta + j[(Z_{0e}/Z_{0o})^{0.5}+(Z_{0e}/Z_{0o})^{-0.5}]\sin\Theta$.

For coupling coefficients of −3 dB, −3.5 dB, −4 dB, and −5 dB, the even- and odd-mode impedances were, using (4.10) and (4.11), calculated and given in Table 4.2, where one set of values, shown in boldface type, can easily be realized. Any transmission-line section of the ring hybrid in Fig. 4.1 can be substituted for by a coupled-line filter section, and the impedances $Z_1 = 54\ \Omega$, $Z_2 = 78.5\ \Omega$, $Z_3 = 57.2\ \Omega$, and $Z_4 = 98.9\ \Omega$ in Table 4.2 were produced from the ring hybrid with termination impedances of 50 Ω, 45 Ω, 42 Ω, and 60 Ω, a power split ratio of 4 dB, and $\Theta_1 = \Theta_2 = \Theta_3 = \Theta_4 = 75°$ in Fig. 4.1.

Figure 4.5 shows a microstrip ARH (asymmetric ring hybrid) in which a coupled-line filter is adopted between ports ② and ③. The main problem here is how to fabricate the coupled lines with −3 dB coupling [8]. Figure 4.6 shows the simulated performances of the ring hybrid in Fig. 4.5 based on the coupling coefficients. For the simulations, all other elements except the coupled-line filter had physically realizable values. The results in Fig. 4.6 show wideband performances

TABLE 4.2 Even- and Odd-Mode Coefficients (Ω) for an Asymmetric Ring Hybrid[a]

	Z_{0e}	Z_{0o}	Z_{0e}	Z_{0o}	Z_{0e}	Z_{0o}	Z_{0e}	Z_{0o}
C:	−3 dB		−3.5 dB		−4 dB		−5 dB	
Z_1	134.3	21.8	111.5	21	92.7	20.1	67.8	18.3
Z_2	195.1	31.7	160.5	30.45	**134.7**	**29.2**	98.5	26.5
Z_3	142.1	23.1	117	22.2	98.1	21.2	71.8	19.3
Z_4	254	40	202	38.4	169.7	36.7	124	33.4

[a] $R_a = 50\ \Omega$, $R_b = 45\ \Omega$, $R_c = 42\ \Omega$, and $R_d = 60\ \Omega$; $d_1/d_2 = 4$ dB; $Z_1 = 54\ \Omega$, $Z_2 = 78.5\ \Omega$, $Z_3 = 57.2\ \Omega$, and $Z_4 = 98.9\ \Omega$ with $\Theta_1 = \Theta_2 = \Theta_3 = \Theta_4 = 75°$; C, coupling coefficient. Boldface values denote the case referred to in the text.

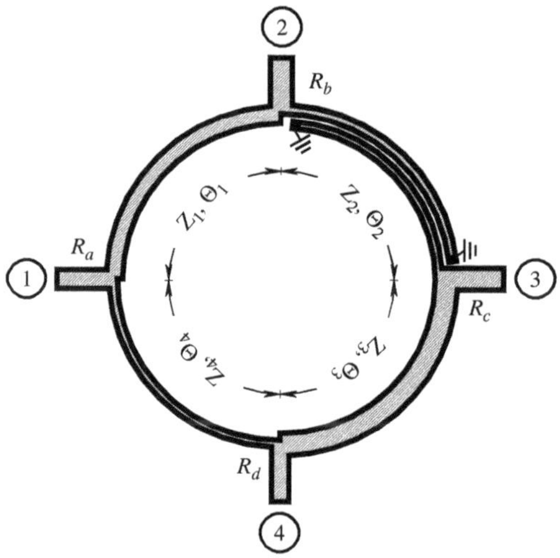

FIGURE 4.5 Configuration of a microstrip ring hybrid. (From Ref. 14 with permission from IEEE.)

TABLE 4.3 Ring Hybrid of Table 4.2 with a Coupling Coefficient of 4 dB[a]

Termination Impedance	MS Feeding Transformer Lines (μm)	MS Ring Transmission Lines (μm)
Port ①; 50 Ω	Z_{01}; $w = 609.7$, $l = 7935$	Z_1; $w = 520.4$, $l = 8071$
Port ②; 45 Ω	Z_{02}; $w = 677.0$, $l = 7895$	**Z_2; w=297.7, s=32, l=8048**
Port ③; 42 Ω	Z_{03}; $w = 724.0$, $l = 7868$	Z_3; $w = 432.9$, $l = 8067$
Port ④; 60 Ω	Z_{04}; $w = 503.0$, $l = 8005$	Z_4; $w = 88.36$, $l = 8371$

[a] $Z_{01} = 50\ \Omega$, $Z_{02} = 47.43\ \Omega$, $Z_{03} = 45.8\ \Omega$, and $Z_{03} = 54.8\ \Omega$. w, Microstrip line width; s, gap width; l, line length. Boldface values denote the case referred to in the text.

regardless of the coupling coefficients, which; indicates that a coupled-line filter with −4 or −5 dB coupling can be used instead of that with −3 dB coupling [14]. One microstrip ARH was, based on Tables 4.2 and 4.3, fabricated on Al_2O_3 substrate ($\varepsilon_r = 10$, $h = 635$ μm). The fabrication data are given in Table 4.3, where the coupled-line filter section with boldface values in Table 4.2 is included.

4.4.2 Uniplanar Asymmetric Ring Hybrids

Another approach to realizing wideband ring hybrids was demonstrated in coplanar technology [8,9,20,21]. However, all these are limited to equal power division and equal termination impedances. Here, a uniplanar ring hybrid proposed by

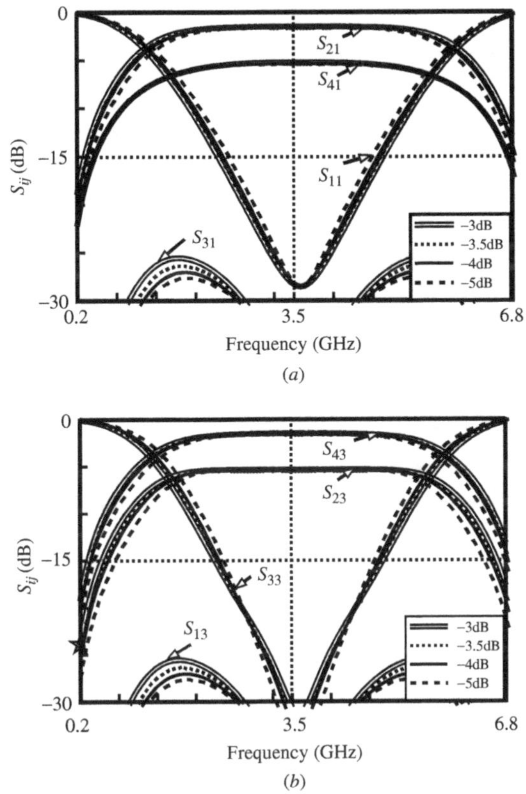

FIGURE 4.6 Simulation results for a microstrip ring hybrid terminated in 50 Ω, 45 Ω, 42 Ω, and 60 Ω; power-split ratio, 4 dB; lengths of the four arcs, 75°: (*a*) excitation at port ①; (*b*) excitation at port ③.

TABLE 4.4 Fabrication Data for the Ring Hybrid of Fig. 4.7[a]

Termination Impedance	Transmission Line Admittance (S)	CPS Physical Size (μm)
Port ①; 50 Ω	Y_1; 0.0184	Y_1; $w = 700$, $s = 47$
Port ②; 45.45 Ω	Y_2; 0.0132	Y_2; $w = 300$, $s = 94$
Port ③; 38.46 Ω	Y_3; 0.0187	Y_3; $w = 1000$, $s = 52$
Port ④; 55.56 Ω	Y_4; 0.0105	Y_4; $w = 120$, $s = 89$

[a] $R_a = 50$ Ω, $R_b = 45$ Ω, $R_c = 38.4$ Ω, and $R_d = 55.56$ Ω; $d_1/d_2 = 4$ dB; $\Theta_1 = \Theta_2 = \Theta_3 = \Theta_4 = 75°$. w, Strip line width; s, gap width; l, line length.

Murgulescu et al. [20] will be introduced on the basis of the derived design equations (4.9). The circuit configuration is described in more detail in Fig. 4.7, and the fabrication data are given in Table 4.4, where the power-split ratio of the ring hybrid is 4 dB, the termination impedances are 50 Ω, 45.45 Ω, 38.46 Ω,

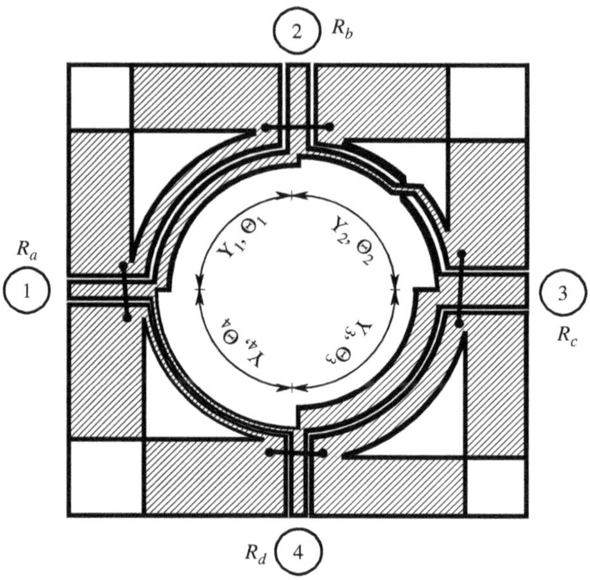

FIGURE 4.7 Coplanar ring hybrid. (From Ref. 14 with permission from IEEE.)

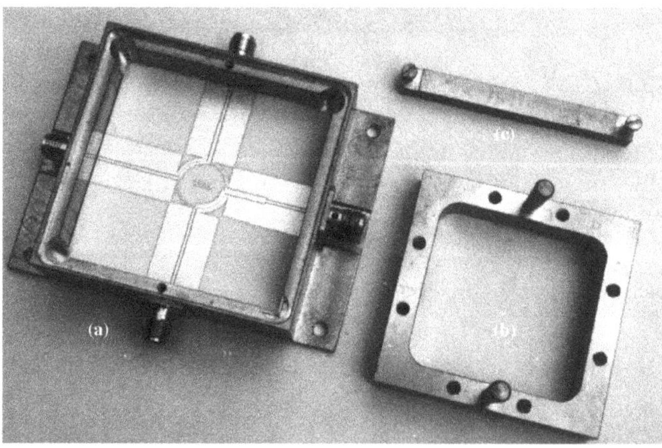

FIGURE 4.8 Packages used for measuring a coplanar ring hybrid: (*a*) coplanar ring hybrid; (*b*) for making grounds of CPW inputs; (*c*) for fixing. (From Ref. 14 with permission from IEEE.)

and 55.56 Ω, and the center frequency is 3 GHz. The ring hybrid was fabricated on Al_2O_3 substrate ($\varepsilon_r = 10$, $h = 635$ μm) with uniplanar technology.

This ring hybrid is not terminated in 50 Ω, so additional transformers were necessary for a general measurement system. A uniplanar circuit can be measured with on-wafer measurements, but it could not be done here, due to the

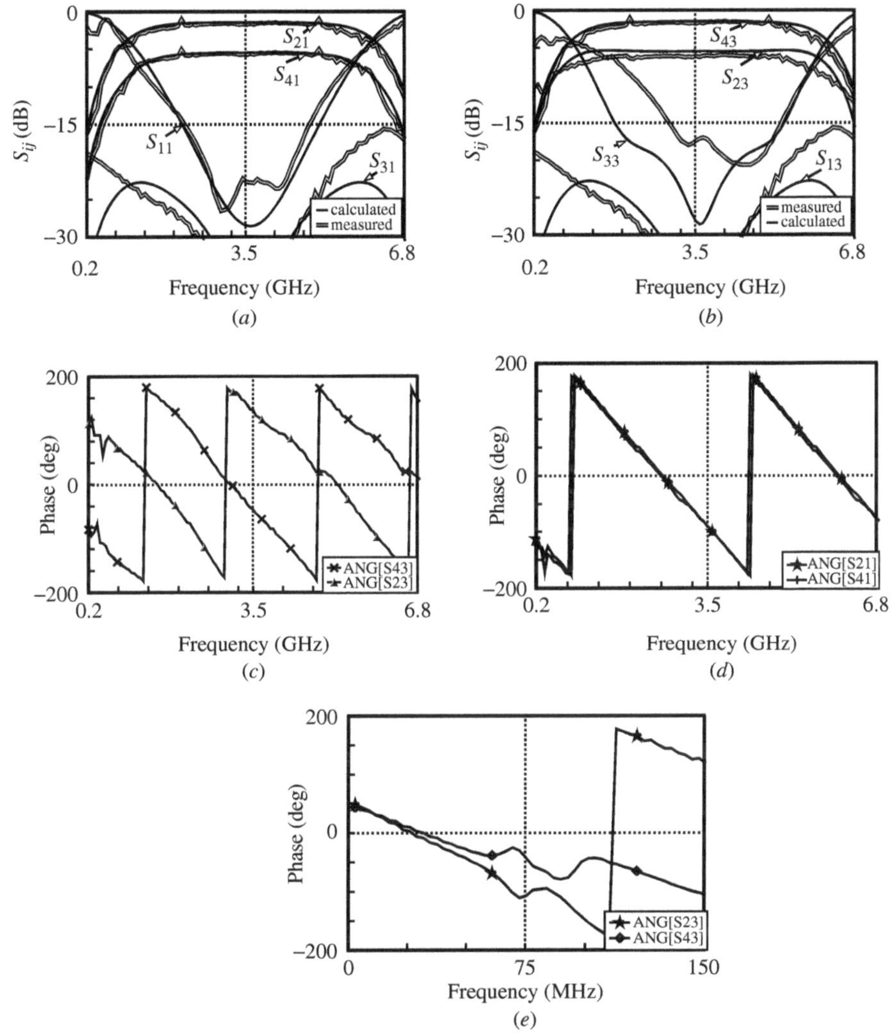

FIGURE 4.9 Measured and calculated responses of a coplanar ring hybrid terminated in arbitrary impedances of 50 Ω, 45.45 Ω, 55.55 Ω, and 38.46 Ω. The power-split ratio is 4 dB, and the electrical length of each arc is 75°. (*a*) Measured and calculated [*S*]-parameters (port ① excitation); (*b*) measured and calculated [*S*]-parameters (port ③ excitation); (*c*) measured phase responses of S_{23} and S_{43}; (*d*) measured phase responses of S_{21} and S_{41}; (*e*) measured phase responses of S_{23} and S_{43} (in the low-frequency range).

large circuit size. There are special connectors for large coplanar circuits, but they are available only for two-port circuits. Therefore, the special packaging shown in Fig. 4.8 was needed for the general measurement system. Since the coplanar circuits require no back-side metallization, the space between the substrate and the bottom of package (*a*) is filled with air. Also, package (*b*) is needed to provide good contact between the walls of package (*a*) and the ground planes of the CPW. The additional bar (*c*) is necessary to make the circuit stable during measurement.

The results are compared with the predictions in Fig. 4.9. When power is excited at port ① or ③, the frequency responses of the scattering parameters are plotted in Fig. 4.9(*a*) and (*b*), respectively. The frequency dependencies of the phases are depicted in Fig. 4.9(*c*)–(*e*). The deviations between the simulated and measured results for $|S_{33}|$ in Fig. 4.9(*b*) are due to unexpected stray inductances in the course of the line crossover between ports ② and ③ in Fig. 4.7. Nevertheless, the results with excitation at port ① show good performance in Fig. 4.9(*a*). This indicates that the derived equations (4.4) are correct. Figure 4.9(*e*) shows the phase changes of S_{23} and S_{43} with increasing frequency. As the frequency approaches 150 MHz, the phase difference between S_{23} and S_{43} goes to 180°, as shown in Fig. 4.9(*e*).

4.5 MINIATURIZED RING HYBRIDS TERMINATED IN ARBITRARY IMPEDANCES

If a ring hybrid for the UHF or VHF band is realized with only distributed elements, it will be quite large, so miniaturization through MMIC techniques is indispensable. For that, there have been many trials, including that of Parisi [23], where three spiral inductors with relatively high resistive loss were used to realize a lumped-element ring hybrid on a thin substrate with a high dielectric constant. However, the ring hybrid produces definitely high loss. To reduce the loss from the inductors, a new type of lumped-element ring hybrid was suggested by Ahn et al. [13] and is described in this section.

A lumped-element ring hybrid will be obtained from a modified ring hybrid, which consists of three $3\lambda/4$ transmission-line sections and one with $\lambda/4$. The modified ring hybrid is derived using dual properties, from which the lumped-element ring hybrid can be introduced. It will have three inductors, and one of the coupled-line filters in Figs. 3.7 and 3.8 may be substituted to reduce the loss from the three inductors. However, since the realization of coupled transmission lines is limited by microstrip technology, a coupling scale factor, d, will be introduced in the process of replacing the three inductors by the coupled microstrip filter section. In addition to having low loss, the ring hybrid may be terminated in arbitrary impedances and used for arbitrary power divisions. Therefore, it is called an *asymmetric lumped-element ring hybrid* (ALERH).

4.5.1 Asymmetric Lumped-Element Ring Hybrids

Figure 4.10 shows two different types of ring hybrids, a conventional ring hybrid in Fig. 4.10(*a*) and a modified ring hybrid in Fig. 4.10(*b*). The modified ring hybrid consists of three $3\lambda/4$ transmission-line sections and one with $\lambda/4$ at the center frequency. To reduce the size of the ring hybrid, miniaturization of the

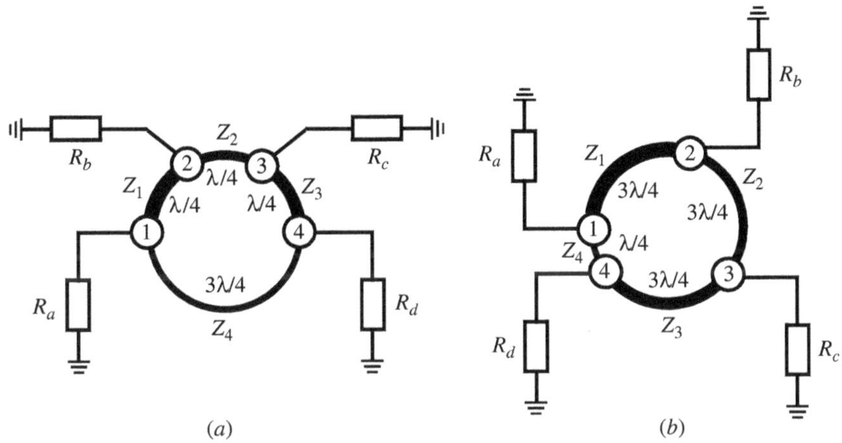

FIGURE 4.10 Ring hybrids terminated in arbitrary impedances: (*a*) conventional ring hybrid; (*b*) modified ring hybrid.

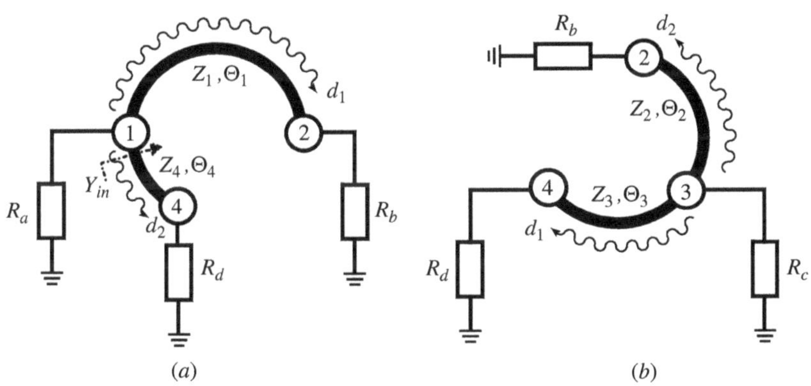

FIGURE 4.11 Two equivalent circuits of the modified ring hybrid with power excitations: (*a*) equivalent circuit (port ① excitation); (*b*) equivalent circuit (port ③ excitation).

transmission lines is necessary. For the lumped equivalent circuits of a transmission line in Fig. 3.14(*b*) and (*d*), one inductor is placed between two output ports when its electrical length is less than 180°, and one capacitance is needed when its electrical length is more than 180°. Due to that fact, to reduce loss introduced by the inductors, the modified ring hybrid is a good candidate for a lumped-element ring hybrid [13]. On the basis of (4.3), Fig. 4.11 shows two equivalent circuits of a modified ring hybrid. When the power is excited at port ①, its equivalent circuit is depicted in Fig. 4.11(*a*), and that with the excitation at port ③ is shown in Fig. 4.11(*b*). The ratio of d_1 to d_2 in Fig. 4.11 is that of S_{21} to S_{41}. In Fig. 4.11(*a*), the *ABCD* parameters contributed by ports ① and ② are

$$\begin{bmatrix} A & B \\ C & D \end{bmatrix} = \begin{bmatrix} 1 & 0 \\ Y_{in} & 1 \end{bmatrix} \begin{bmatrix} \cos\Theta_1 & j\dfrac{1}{Y_1}\sin\Theta_1 \\ jY_1\sin\Theta_1 & \cos\Theta_1 \end{bmatrix}$$

$$= \begin{bmatrix} \cos\Theta_1 & j\dfrac{1}{Y_1}\sin\Theta_1 \\ Y_{\text{in}}\cos\Theta_1 + jY_1\sin\Theta_1 & j\dfrac{Y_{\text{in}}}{Y_1}\sin\Theta_1 + \cos\Theta_1 \end{bmatrix}, \quad (4.12)$$

where

$$Z_1 = \frac{1}{Y_1}, \qquad Y_{\text{in}} = \frac{1}{Z_{\text{in}}}, \qquad Z_{\text{in}} = Z_4\frac{R_d + jZ_4\tan\Theta_4}{Z_4 + jR_d\tan\Theta_4}.$$

From the relations between the *ABCD* matrix and the admittance matrix Y in Table 2.1, the admittance matrix of (4.12) and a reference admittance matrix $\boldsymbol{y}$ between ports ① and ② in Fig. 4.11(a) are given by

$$\boldsymbol{Y} = \frac{Y_1}{j\sin\Theta}\begin{bmatrix} j\dfrac{Y_{\text{in}}}{Y_1}\sin\Theta_1 + \cos\Theta_1 & -1 \\ -1 & \cos\Theta_1 \end{bmatrix}, \quad (4.13a)$$

$$\boldsymbol{y} = \begin{bmatrix} \dfrac{1}{R_a} & 0 \\ 0 & \dfrac{1}{R_b} \end{bmatrix}. \quad (4.13b)$$

Applying $\Theta_1 = 3\pi/2$ and $\Theta_4 = \pi/2$ to (4.13a) and performing additional calculations, the scattering parameters between ports ① and ② in Fig. 4.11(a) are derived as

$$\boldsymbol{S} = \boldsymbol{k}\boldsymbol{S}^V\boldsymbol{k}_*^{-1} = \begin{bmatrix} 0 & j\sqrt{\dfrac{d_1^2}{d_1^2 + d_2^2}} \\ j\sqrt{\dfrac{d_1^2}{d_1^2 + d_2^2}} & \dfrac{d_2^2}{d_1^2 + d_2^2} \end{bmatrix}, \quad (4.14)$$

where $\boldsymbol{kk}_* = \frac{1}{2}(\boldsymbol{y} + \boldsymbol{y}_*)$ and $\boldsymbol{S}^V = -(\boldsymbol{Y} + \boldsymbol{y})^{-1}(\boldsymbol{Y} - \boldsymbol{y}^*)$.

Based on (4.14), all-port scattering parameters are easily obtained. For example, by the unitary property $\boldsymbol{SS}^{*T} = \boldsymbol{U}$, S_{41} is calculated from the relation $|S_{21}|^2 + |S_{41}|^2 = 1$. Using the unitary and reciprocal properties, all scattering matrices of the hybrid in Fig. 4.10(b) are given as

$$S_{\text{ring}} = \frac{1}{\sqrt{d_1^2 + d_2^2}} \begin{bmatrix} 0 & jd_1 & 0 & -jd_2 \\ jd_1 & 0 & jd_2 & 0 \\ 0 & jd_2 & 0 & jd_1 \\ -jd_2 & 0 & jd_1 & 0 \end{bmatrix}, \tag{4.15}$$

where $d_1^2 + d_2^2 = 1$ when the unitary property is satisfied.

The derived scattering matrix (4.15) shows that phase balancing and magnitude in the scattering parameters of the modified ring hybrid are the same as those of the conventional ring hybrid in Fig. 4.10(*a*). Figure 4.12 shows a lumped-element ring hybrid using the lumped-element transmission-line modeling in Fig. 3.14(*b*) and (*d*). The ALERH is terminated in arbitrary impedances and also allows arbitrary power divisions.

The design equations of the ring hybrid in Fig. 4.10(*b*) are given by

$$Z_1 = \sqrt{\frac{d_1^2 + d_2^2}{d_1^2}}\sqrt{R_a R_b}, \tag{4.16a}$$

$$Z_2 = \sqrt{\frac{d_1^2 + d_2^2}{d_2^2}}\sqrt{R_b R_c}, \tag{4.16b}$$

$$Z_3 = \sqrt{\frac{d_1^2 + d_2^2}{d_1^2}}\sqrt{R_c R_d}, \tag{4.16c}$$

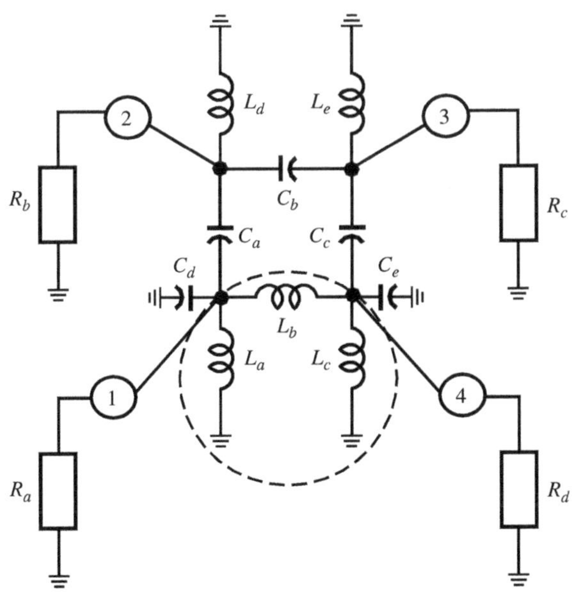

FIGURE 4.12 ALERH terminated in arbitrary impedances. (From Ref. 13 with permission from IEEE.)

$$Z_4 = \sqrt{\frac{d_1^2 + d_2^2}{d_2^2}}\sqrt{R_d R_a}. \tag{4.16d}$$

By the relations (3.33), (3.35), and (4.16), all lumped elements in Fig. 4.12 are given as

$$C_a = \frac{d_1}{\omega\sqrt{R_a R_b}\sqrt{d_1^2 + d_2^2}}, \tag{4.17a}$$

$$C_b = \frac{d_2}{\omega\sqrt{R_b R_c}\sqrt{d_1^2 + d_2^2}}, \tag{4.17b}$$

$$C_c = \frac{d_1}{\omega\sqrt{R_c R_d}\sqrt{d_1^2 + d_2^2}}, \tag{4.17c}$$

$$C_d = C_e = \frac{d_2}{\omega\sqrt{R_a R_d}\sqrt{d_1^2 + d_2^2}}, \tag{4.17d}$$

$$L_a = \frac{\sqrt{d_1^2 + d_2^2}\sqrt{R_a R_b}}{\omega d_1}, \tag{4.17e}$$

$$L_b = \frac{\sqrt{d_1^2 + d_2^2}\sqrt{R_d R_a}}{\omega d_2}, \tag{4.17f}$$

$$L_c = \frac{\sqrt{d_1^2 + d_2^2}\sqrt{R_c R_d}}{\omega d_1}, \tag{4.17g}$$

$$L_d = \frac{\sqrt{d_1^2 + d_2^2}\sqrt{R_a R_b R_c}}{\omega(d_2\sqrt{R_a} + d_1\sqrt{R_c})}, \tag{4.17h}$$

$$L_e = \frac{\sqrt{d_1^2 + d_2^2}\sqrt{R_b R_c R_d}}{\omega(d_2\sqrt{R_d} + d_1\sqrt{R_b})}, \tag{4.17i}$$

where ω is the angular frequency.

A lumped inductor L_b in the dashed circle in Fig. 4.12 is a high-loss element. To reduce the loss introduced by the inductor L_b, a coupled-line filter section can be replaced by the inductors L_a and L_c, together with L_b.

Figure 4.13 shows lumped-element circuits and various coupled transmission filter sections. The four-port coupled transmission lines in Fig. 4.13(a) have two conductors placed in parallel. The length of the conductors is Θ and their even- and odd-mode impedances are Z_{0e}^a, Z_{0o}^a, Z_{0e}^b, and Z_{0o}^b, as indicated. When two of the four ports are terminated in shorts or opens, we get several coupled transmission-line two-port circuits. They are useful for the design of common microwave integrated circuits. The representative configurations in Fig. 4.13(*b*) and (*c*) are symmetric; the rest are asymmetric. Since the lumped-element circuit in Fig. 4.13(*b*) is symmetric, its even- and odd-mode admittances are obtained

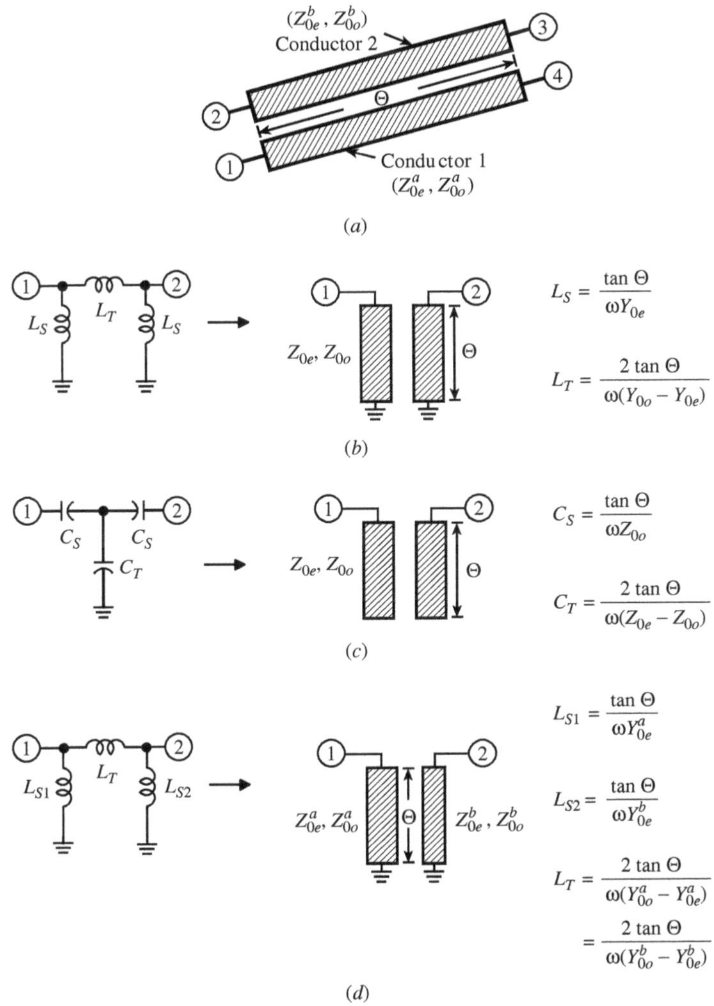

FIGURE 4.13 Lumped-element circuits and their equivalent two-port coupled-line filter sections. (*a*) four-port coupled lines; (*b*) symmetric Π-network lumped-element circuit and its equivalent circuit; (*c*) T lumped-element circuit and its equivalent circuit; (*d*) asymmetric Π-network lumped-element circuit and its equivalent circuit; (*e*) lumped-element circuit with two inductors and a transmission-line section and its equivalent two-port coupled-line filter section; (*f*) lumped-element circuit with two capacitors and a transmission-line section and its equivalent coupled-line filter section.

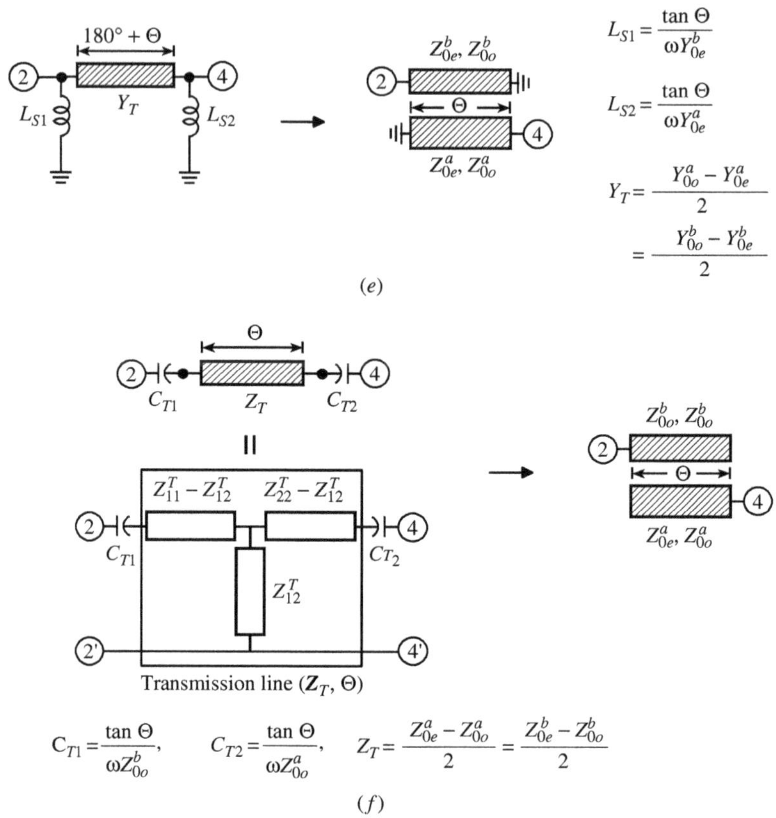

FIGURE 4.13 *(continued)*

from (2.106) as

$$Y_{O/C} = \frac{1}{j\omega L_S}, \tag{4.18a}$$

$$Y_{S/C} = \frac{1}{j\omega L_S} + \frac{2}{j\omega L_T}. \tag{4.18b}$$

Since the coupled transmission lines are symmetric, $Z^a_{0e} = Z^a_{0e} = Z_{0e}$ and $Z^b_{0o} = Z^b_{0o} = Z_{0o}$ and the even- and odd-mode admittances are

$$Y_{O/C} = -jY_{0e}\cot\Theta, \tag{4.19a}$$

$$Y_{S/C} = -jY_{0o}\cot\Theta, \tag{4.19b}$$

where $Y_{0e} = Z_{0e}^{-1}$ and $Y_{0o} = Z_{0o}^{-1}$.

Equating (4.18) to (4.19), we find L_S and L_T as we found those shown in Fig. 4.13(*b*). For the lumped-element circuit in Fig. 4.13(*c*), the even- and odd-mode impedances are

$$Z_{\mathrm{O/C}} = \frac{1}{j\omega C_S} + \frac{2}{j\omega C_T}, \tag{4.20a}$$

$$Z_{\mathrm{S/C}} = \frac{1}{j\omega C_S}, \tag{4.20b}$$

and those of its equivalent two-port coupled transmission lines are

$$Z_{\mathrm{O/C}} = -jZ_{0e}\cot\Theta, \tag{4.21a}$$

$$Z_{\mathrm{S/C}} = -jZ_{0o}\cot\Theta. \tag{4.21b}$$

From (4.20) and (4.21), C_S and C_T are calculated similarly to those shown in Fig. 4.13(*c*).

The coupled transmission lines in Fig. 4.13(*d*) are not symmetric but a plane of shorts with even-mode excitation and a plane of opens with odd-mode excitation existing between the two conductors and $Z_{0e}^a - Z_{0o}^a$ becoming equal to $Z_{0o}^b - Z_{0o}^b$ [24]. The following relation holds:

$$\frac{1}{j\omega L_{S1}} = -jY_{0e}^a\cot\Theta, \tag{4.22a}$$

$$\frac{1}{j\omega L_{S2}} = -jY_{0e}^b\cot\Theta, \tag{4.22b}$$

$$\frac{1}{j\omega L_{S1}} + \frac{2}{j\omega L_T} = \frac{1}{j\omega L_{S2}} + \frac{2}{j\omega L_T} = -jY_{0o}\cot\Theta, \tag{4.22c}$$

where $Y_{0o}^a = Y_{0o}^b = Y_{0o}$.

From (4.22), L_{S1}, L_{S2}, and L_T are found as those shown in Fig. 4.13(*d*). Figure 4.13(*e*) shows a lumped-element circuit and its equivalent two-port coupled-line filter section. The lumped-element circuit consists of two inductors and one transmission line with $180^\circ + \Theta$ and Y_T, and its equivalent two-port coupled-line filter section is obtained by terminating two ports of the coupled transmission lines in Fig. 4.13(*a*) in shorts. Referring to Fig. 2.17(*a*), the admittance parameters of the lumped-element circuit are given as

$$\mathbf{Y}_{\mathbf{lump-(e)}} = j\begin{bmatrix} -\dfrac{1}{\omega L_{S1}} & 0 \\ 0 & -\dfrac{1}{\omega L_{S2}} \end{bmatrix} + jY_T\begin{bmatrix} -\cot(\pi+\Theta) & \csc(\pi+\Theta) \\ \csc(\pi+\Theta) & -\cot(\pi+\Theta) \end{bmatrix}. \tag{4.23}$$

The admittance parameters of the four-port coupled transmission lines in Fig. 4.13(*a*) are given as

$$Y_{11} = Y_{44} = -j\frac{Y_{0e}^a + Y_{0o}^a}{2}\cot\Theta, \tag{4.24a}$$

$$Y_{12} = Y_{21} = Y_{34} = Y_{43} = -j\frac{Y_{0o}^a - Y_{0e}^a}{2}\cot\Theta$$

$$= -j\frac{Y_{0o}^b - Y_{0e}^b}{2}\cot\Theta, \tag{4.24b}$$

$$Y_{13} = Y_{31} = Y_{24} = Y_{42} = -j\frac{Y_{0o}^a - Y_{0e}^a}{2}\csc\Theta$$

$$= -j\frac{Y_{0o}^b - Y_{0e}^b}{2}\csc\Theta, \tag{4.24c}$$

$$Y_{14} = Y_{41} = -j\frac{Y_{0e}^a + Y_{0o}^a}{2}\csc\Theta, \tag{4.24d}$$

$$Y_{23} = Y_{32} = -j\frac{Y_{0e}^b + Y_{0o}^b}{2}\csc\Theta, \tag{4.24e}$$

$$Y_{22} = Y_{33} = -j\frac{Y_{0e}^b + Y_{0o}^b}{2}\cot\Theta. \tag{4.24f}$$

Applying the conditions of $V_1 = 0$ and $V_3 = 0$ to Fig. 4.13(*a*) gives the configuration shown in Fig. 4.13(*e*). The admittance parameters Y'_{11}, Y'_{12}, Y'_{21}, and Y'_{22} of the equivalent coupled transmission lines in Fig. 4.13(*e*) are obtained such that Y'_{11} is the ratio of I_2 to V_2 with $V_4 = 0$, Y'_{12} is that of I_2 to V_4 with $V_3 = 0$, and so on. They are found as

$$\begin{bmatrix} Y'_{11} & Y'_{12} \\ Y'_{21} & Y'_{22} \end{bmatrix} = \begin{bmatrix} Y_{22} & Y_{24} \\ Y_{42} & Y_{44} \end{bmatrix} = \begin{bmatrix} -j\dfrac{Y_{0e}^b + Y_{0o}^b}{2}\cot\Theta & -j\dfrac{Y_{0o}^a - Y_{0e}^a}{2}\csc\Theta \\ -j\dfrac{Y_{0o}^a - Y_{0e}^a}{2}\csc\Theta & -j\dfrac{Y_{0e}^a + Y_{0o}^a}{2}\cot\Theta \end{bmatrix}$$

$$= -j\begin{bmatrix} Y_{0e}^b\cot\Theta & 0 \\ 0 & Y_{0e}^a\cot\Theta \end{bmatrix} + j\frac{(Y_{0o}^a - Y_{0e}^a)}{2}\begin{bmatrix} -\cot(\pi+\Theta) & \csc(\pi+\Theta) \\ \csc(\pi+\Theta) & -\cot(\pi+\Theta) \end{bmatrix}. \tag{4.25}$$

Therefore, L_{S1}, L_{S2}, and Y_T are obtained by equating (4.23) to (4.25); they are given in Fig. 4.13(*e*).

Figure 4.13(*f*) shows a two-port lumped-element circuit and its equivalent coupled-line filter section. The lumped-element circuit consists of two capacitances and a transmission-line section whose characteristic impedance and electrical length are Z_T and Θ. Since the transmission-line section may be expressed with a T-network as shown in Fig. 4.13(*f*), the impedance parameters of the lumped-element circuit are obtained as

$$\mathbf{Z}_{\text{lump-}(f)} = \begin{bmatrix} -\dfrac{j}{\omega C_{T1}} + Z_{11}^{\text{T}} & Z_{12}^{\text{T}} \\ Z_{12}^{\text{T}} & -\dfrac{j}{\omega C_{T1}} + Z_{22}^{\text{T}} \end{bmatrix}, \tag{4.26}$$

where $Z_{11}^{\text{T}} = Z_{22}^{\text{T}} = -jZ_T \cot\Theta$ and $Z_{12}^{\text{T}} = Z_{21}^{\text{T}} = -jZ_T \csc\Theta$, which are obtained from Fig. 2.17(a).

The impedance parameters [24] of four-port coupled transmission lines in Fig. 4.13(a) are given by

$$Z_{11} = Z_{44} = -j\frac{Z_{0e}^a + Z_{0o}^a}{2}\cot\Theta \tag{4.27a}$$

$$Z_{12} = Z_{21} = Z_{34} = Z_{43} = -j\frac{Z_{0e}^a - Z_{0o}^a}{2}\cot\Theta$$

$$= -j\frac{Z_{0e}^b - Z_{0o}^b}{2}\cot\Theta, \tag{4.27b}$$

$$Z_{13} = Z_{31} = Z_{24} = Z_{42} = -j\frac{Z_{0e}^a - Z_{0o}^a}{2}\csc\Theta$$

$$= -j\frac{Z_{0e}^b - Z_{0o}^b}{2}\csc\Theta, \tag{4.27c}$$

$$Z_{14} = Z_{41} = -j\frac{Z_{0e}^a + Z_{0o}^a}{2}\csc\Theta, \tag{4.27d}$$

$$Z_{23} = Z_{32} = -j\frac{Z_{0e}^b + Z_{0o}^b}{2}\csc\Theta, \tag{4.27e}$$

$$Z_{22} = Z_{33} = -j\frac{Z_{0e}^b + Z_{0o}^b}{2}\cot\Theta. \tag{4.27f}$$

By imposing $I_1 = I_3 = 0$ of the four-port coupled transmission lines in Fig. 4.13(a), the impedance matrix of the equivalent coupled-line filter section in Fig. 4.13(f) is found as

$$\begin{bmatrix} Z'_{11} & Z'_{12} \\ Z'_{21} & Z'_{22} \end{bmatrix} = \begin{bmatrix} Z_{22} & Z_{24} \\ Z_{42} & Z_{44} \end{bmatrix}$$

$$= \begin{bmatrix} -j\dfrac{Z_{0e}^b + Z_{0o}^b}{2}\cot\Theta & -j\dfrac{Z_{0e}^a - Z_{0o}^a}{2}\csc\Theta \\ -j\dfrac{Z_{0e}^a - Z_{0o}^a}{2}\csc\Theta & -j\dfrac{Z_{0e}^a + Z_{0o}^a}{2}\cot\Theta \end{bmatrix}$$

$$= -j\begin{bmatrix} Z_{0o}^b\cot\Theta & 0 \\ 0 & Z_{0o}^a\cot\Theta \end{bmatrix} - j\frac{Z_{0e}^a - Z_{0o}^a}{2}\begin{bmatrix} \cot\Theta & \csc\Theta \\ \csc\Theta & \cot\Theta \end{bmatrix}. \tag{4.28}$$

Comparing (4.28) and (4.26) gives the values of C_{T1}, C_{T2}, and Z_T as shown in Fig. 4.13(*f*). As discussed above, the three inductors in Fig. 4.12 are realized with a coupled-line filter section in Fig. 4.13(*b*). Since the two inductors L_a and L_c in (4.17*e*) and (4.17*g*) are dependent on the termination impedances, three cases are possible. One is symmetric ($L_a = L_c$), the other two are asymmetric ($L_a > L_c$ and $L_a < L_c$).

Symmetric Case ($L_a = L_c$) In the symmetric case, $L_a = L_c$, so $R_a R_b = R_c R_d$. Figure 4.14 shows how to convert three inductors into a coupled-line filter section. Since the coupling coefficient required is high, realization is not always possible with microstrip technology. Therefore, the coupling coefficient needs to be reduced. This can be accomplished by introducing a scaling factor d as shown in Fig. 4.14(*b*). C'_d and C'_e are determined so that $C_d L_a = C'_d(L_a/d)$ and $C_e L_c = C'_e(L_c/d)$. From Fig. 4.13(*b*), the even- and odd-mode impedances and needed elements are derived as

$$L'_a = \frac{L_a}{d} = \frac{L_c}{d} = L'_c, \tag{4.29a}$$

$$Z_{0e} = \frac{\omega L'_a}{\tan\Theta}, \tag{4.29b}$$

$$Z_{0o} = \frac{\omega L'_a L_b}{(2L'_a + L_b)\tan\Theta}, \tag{4.29c}$$

$$C'_d = C'_e = d\frac{d_2}{\omega\sqrt{R_a R_d}\sqrt{d_1^2 + d_2^2}}, \tag{4.29d}$$

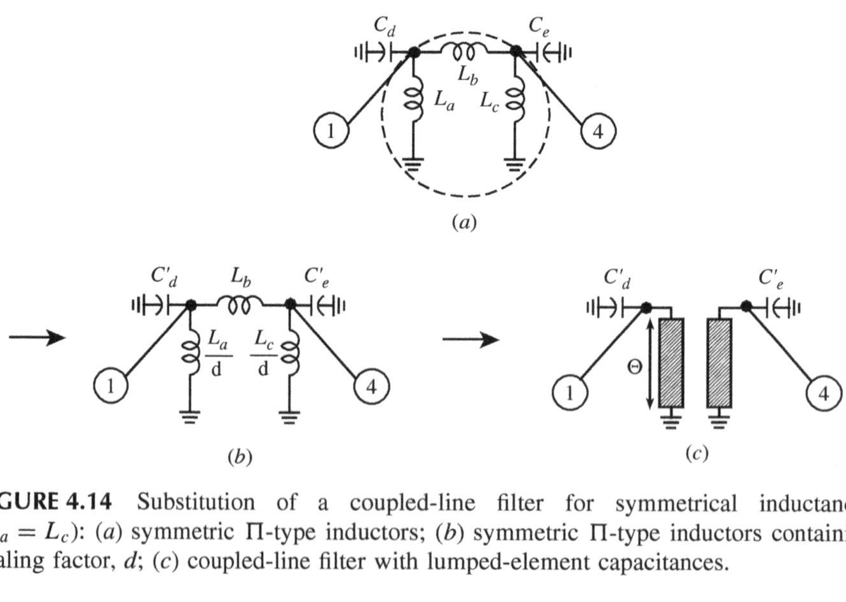

FIGURE 4.14 Substitution of a coupled-line filter for symmetrical inductances ($L_a = L_c$): (*a*) symmetric Π-type inductors; (*b*) symmetric Π-type inductors containing scaling factor, *d*; (*c*) coupled-line filter with lumped-element capacitances.

Asymmetric Cases ($L_a < L_c$) For $L_a < L_c$, replacement by the asymmetric coupled-line filter section in Fig. 4.13(*d*) is possible, but coupling of the symmetric coupled lines is stronger than that of the asymmetric lines [24,25]. Thus, a symmetric coupled-line filter section is used to replace the Π-network in the dashed circle in Fig. 4.12. In order to use symmetric coupled transmission lines, the circuit in Fig. 4.15(*a*) is modified to that in Fig. 4.15(*b*), where L_a is composed of two inductances, L_c and L_{as}, where L_{as} may be determined so as to satisfy the relation $L_a = L_c \| L_{as}$ ($\|$ indicates a parallel connection). In Fig. 4.15(*c*), the inductance L_{as} is removed together with the capacitance C_{ds}, due to the resonance condition of $L_{as}C_{ds} = 1/\omega^2$ at the design center frequency. Therefore, the sum of the two capacitances C_{d1} and C_{ds} is C_d. After the symmetric Π-network is completed in Fig. 4.15(*c*), the symmetric case procedure can be used. The even- and odd-mode impedances, length Θ, and other lumped elements are derived as

$$L_c' = \frac{L_c}{d}, \tag{4.30a}$$

$$Z_{0e} = \frac{\omega L_c'}{\tan \Theta}, \tag{4.30b}$$

$$Z_{0o} = \frac{\omega L_c' L_b}{(2L'_c + L_b)\tan \Theta}, \tag{4.30c}$$

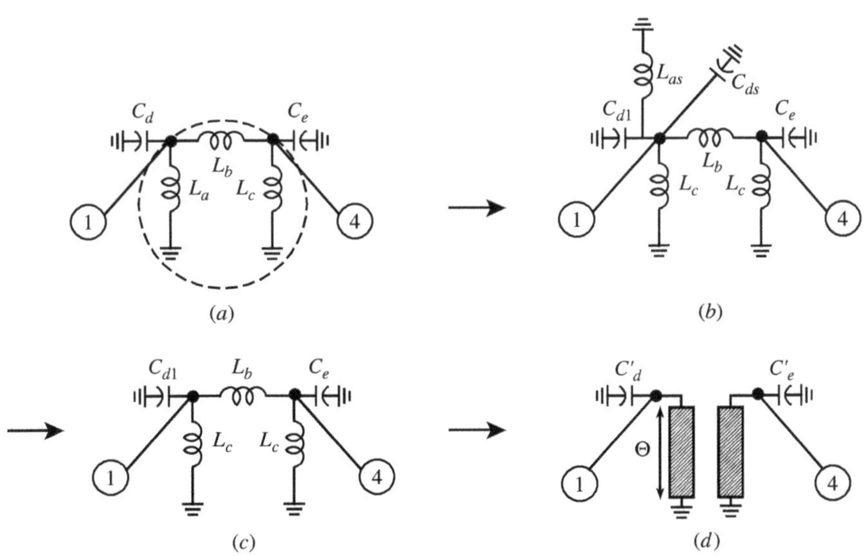

FIGURE 4.15 Substitution of a coupled-line filter for asymmetrical inductances ($L_a < L_c$): (*a*) asymmetric Π-type inductors; (*b*) change from asymmetric Π-type inductors to symmetric ones; (*c*) symmetric Π-type inductors; (*d*) coupled-line filter with lumped-element capacitances.

$$L_{as} = \frac{L_a L_c}{L_c - L_a}, \tag{4.30d}$$

$$C_{ds} = \frac{1}{L_{as}\omega^2}, \tag{4.30e}$$

$$C'_d = d(C_d - C_{ds}), \tag{4.30f}$$

$$C'_e = d\frac{d_2}{\omega\sqrt{R_a R_d}\sqrt{d_1^2 + d_2^2}}. \tag{4.30g}$$

The case of $L_a > L_c$ is similar to that of $L_a < L_c$, and the even- and odd-mode impedances, length Θ, and other lumped elements are derived as

$$L'_a = \frac{L_a}{d}, \tag{4.31a}$$

$$Z_{0e} = \frac{\omega L'_a}{\tan\Theta}, \tag{4.31b}$$

$$Z_{0o} = \frac{\omega L'_a L_b}{(2L'_a + L_b)\tan\Theta}, \tag{4.31c}$$

$$L_{cs} = \frac{L_a L_c}{L_a - L_c}, \tag{4.31d}$$

$$C_{es} = \frac{1}{L_{cs}\omega^2}, \tag{4.31e}$$

$$C'_d = d\frac{d_2}{\omega\sqrt{R_a R_d}\sqrt{d_1^2 + d_2^2}}, \tag{4.31f}$$

$$C'_e = d(C_e - C_{es}). \tag{4.31g}$$

Simulation Results and Experimental Data Three design tables for equal power divisions were generated on the basis of (4.16)–(4.17), and (4.29)–(4.31). They are effective for a Teflon substrate ($\varepsilon_r = 2.54$, $h = 0.77$ mm) and an operating frequency of 900 MHz. For the symmetric case, $R_a R_b$ is equal to $R_c R_d$, and Table 4.5 provides one case of a ring hybrid terminated in equal impedances of 50 Ω. Several choices about the coupled-line filter section are given in Table 4.5, and one of them is shown in boldface type. For that case, $C_a = 2.5$ pF, $C_b = 2.5$ pF, $C_c = 2.5$ pF, $C'_d = 6.25$ pF, $C'_e = 6.25$ pF, $L_e = 6.25$ nH, $L_d = 6.35$ nH, $Z_{0e} = 77.71$ Ω, $Z_{0o} = 43.17$ Ω, and $\Theta = 20^\circ$ with reference to Figs. 4.12 and 4.14.

Table 4.6 is a design table for an asymmetric case ($L_a < L_c$) where the ring hybrid is terminated in 25 Ω, 50 Ω, 50 Ω, and 33.5 Ω. One of several design choices is shown in boldface. For that case, $C_a = 3.4$ pF, $C_b = 2.5$ pF, $C_c = 3.06$ pF, $C_d = 12.55$ pF, $C_e = 13.13$ pF, $L_e = 5.18$ nH, $L_d = 5.62$ nH,

TABLE 4.5 Symmetric Case[a]

	d:	2.5	3.0	3.5	4.0
$\Theta = 16°$	$Z_{0e}(\Omega)$:	98.64	82.20	70.64	61.65
	C:	0.29	0.25	0.22	0.20
			$w = 1.65$ mm	$w = 2.44$ mm	$w = 2.74$ mm
			$s = 0.33$ mm	$s = 0.31$ mm	$s = 0.36$ mm
			$l = 11.6$ mm	$l = 12.9$ mm	$l = 12.4$ mm
$\Theta = 18°$	$Z_{0e}(\Omega)$:	87.05	72.54	62.18	54.41
	C:	0.29	0.25	0.22	0.20
		$w = 1.87$ mm	$w = 1.67$ mm	$w = 2.46$ mm	$w = 2.8$ mm
		$s = 0.2$ mm	$s = 0.33$ mm	$s = 0.32$ mm	$s = 0.34$ mm
		$l = 14.6$ mm	$l = 11.78$ mm	$l = 13.0$ mm	$l = 12.6$ mm
$\Theta = 20°$	$Z_{0e}(\Omega)$:	**77.71**	64.76	55.51	48.57
	C:	**0.29**	0.25	0.22	0.20
		w= 1.88 mm	$w = 1.87$ mm	$w = 1.97$ mm	$w = 2.39$ mm
		s= 0.2 mm	$s = 0.30$ mm	$s = 0.38$ mm	$s = 0.40$ mm
		l= 14.6 mm	$l = 12.6$ mm	$l = 11.4$ mm	$l = 11.46$ mm

[a] All termination impedances are 50 Ω. w, Microstrip coupled line width; s, gap width; l, line length. Boldface values denote the case referred to in the text.

TABLE 4.6 Asymmetric Case $L_a < L_c$[a]

	d:	3.4	3.8	4.2	4.6
$\Theta = 12°$	$Z_{0e}(\Omega)$:	79.87	71.48	64.67	59.05
	C (dB):	0.29	0.27	0.25	0.24
			$w = 2.00$ mm	$w = 2.60$ mm	$w = 2.63$ mm
			$s = 0.22$ mm	$s = 0.19$ mm	$s = 0.23$ mm
			$l = 8.56$ mm	$l = 9.3$ mm	$l = 8.66$ mm
$\Theta = 16°$	$Z_{0e}(\Omega)$:	68.11	60.94	55.13	50.34
	C (dB):	0.29	0.27	0.25	0.24
			$w = 2.05$ mm	$w = 2.73$ mm	$w = 2.65$ mm
			$s = 0.2$ mm	$s = 0.18$ mm	$s = 0.23$ mm
			$l = 8.74$ mm	$l = 9.65$ mm	$l = 8.71$ mm
$\Theta = 18°$	$Z_{0e}(\Omega)$:	63.37	**56.70**	51.30	46.84
	C (dB):	0.29	**0.27**	0.25	0.20
			w= 2.1 mm	$w = 2.73$ mm	$w = 2.58$ mm
			s= 0.2 mm	$s = 0.18$ mm	$s = 0.24$ mm
			l= 8.89 mm	$l = 9.65$ mm	$l = 8.57$ mm

[a] Termination impedances are 25 Ω, 50 Ω, 50 Ω, and 33.5 Ω. w, Microstrip coupled line width; s, gap width; l, line length. Boldface values denote the case referred to in the text.

$Z_{0e} = 56.70\ \Omega$, $Z_{0o} = 32.51\ \Omega$, and $\Theta = 15°$. A schematic layout based on the data is provided in Fig. 4.16, and the simulation results of scattering parameters are plotted in Fig. 4.17, where the scattering parameters with excitation at port ① are plotted in Fig. 4.17(*a*), those with excitation at port ③ are shown in Fig. 4.17(*b*), and the phase responses are shown in Fig. 4.17(*c*).

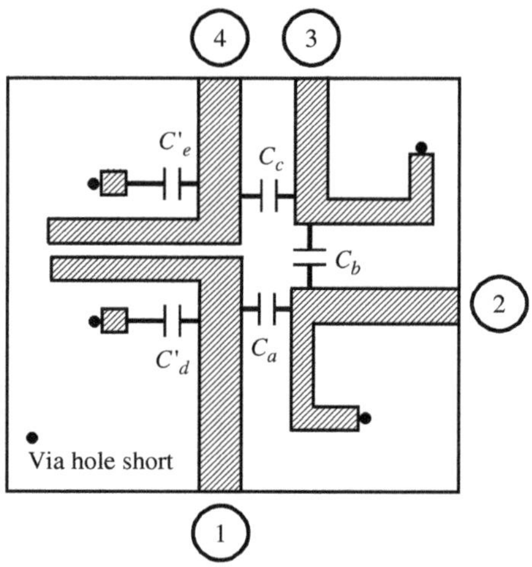

FIGURE 4.16 Miniaturized asymmetric ring hybrid. (From Ref. 13 with permission from IEEE.)

TABLE 4.7 Asymmetric Case $L_a > L_c$[a]

	d:	3.4	3.8	4.2	4.6
$\Theta = 14°$	$Z_{0e}(\Omega)$:	93.26	83.44	75.50	68.93
	C (dB):	0.29	0.24	0.23	0.21
		$w = 1.61$ mm	$w = 2.03$ mm	$w = 2.1$ mm	$w = 2.19$ mm
		$s = 0.28$ mm	$s = 0.28$ mm	$s = 0.34$ mm	$s = 0.38$ mm
		$l = 11.33$ mm	$l = 11.77$ mm	$l = 11.04$ mm	$l = 10.49$ mm
$\Theta = 16°$	$Z_{0e}(\Omega)$:	81.09	72.55	65.64	59.94
	C (dB):	0.29	0.24	0.23	0.21
		$w = 1.62$ mm	$w = 2.13$ mm	$w = 2.32$ mm	$w = 2.42$ mm
		$s = 0.29$ mm	$s = 0.27$ mm	$s = 0.3$ mm	$s = 0.35$ mm
		$l = 11.36$ mm	$l = 12.14$ mm	$l = 11.7$ mm	$l = 11.15$ mm
$\Theta = 18°$	$Z_{0e}(\Omega)$:	**71.56**	64.03	57.93	52.89
	C (dB):	**0.26**	0.24	0.23	0.21
		w= 1.7 mm	$w = 2.59$ mm	$w = 2.53$ mm	$w = 2.53$ mm
		s= 0.27 mm	$s = 0.21$ mm	$s = 0.27$ mm	$s = 0.34$ mm
		l= 11.67 mm	$l = 13.68$ mm	$l = 12.4$ mm	$l = 11.56$ mm

[a]Termination impedances are 50 Ω, 62.5 Ω, 55.5 Ω, and 41.6 Ω. w, Microstrip coupled line width; s, gap width; l, line length. Boldface values denote the case referred to in the text.

Table 4.7 gives a design table for the asymmetric case ($L_a > L_c$), where several possible choices are given. For the design shown in boldface, $C_a = 2.24$ pF, $C_b = 2.21$ pF, $C_c = 2.60$ pF, $C_d' = 8$ pF, $C_e' = 7.63$ pF, $L_e = 7.17$ nH, $L_d = 6.62$ nH, $Z_{0e} = 71.56\ \Omega$, $Z_{0o} = 41.60\ \Omega$, and $\Theta = 18°$. The asymmetric ring hybrid is terminated in 50 Ω, 62.5 Ω, 55.5 Ω, and 41.6 Ω.

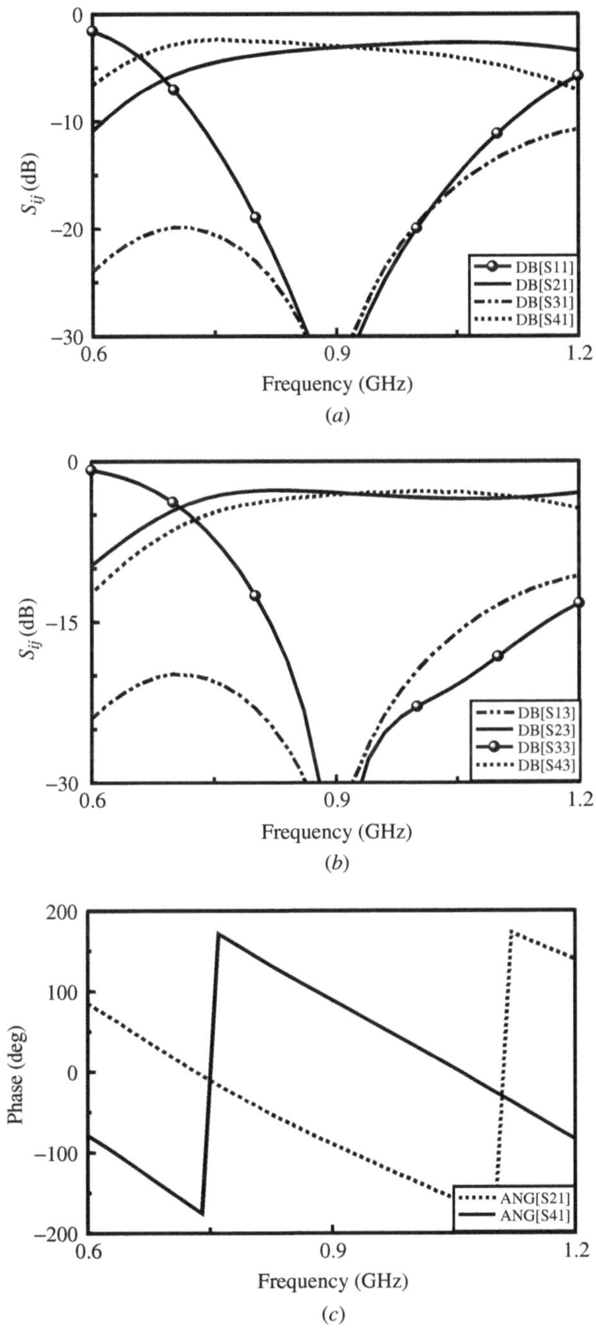

FIGURE 4.17 Simulation results of the scattering parameters in an asymmetric case ($L_a < L_c$): (*a*) in the case of port ① excitation; (*b*) in the case of port ③ excitation; (*c*) phase responses of S_{21} and S_{41}.

In conclusion, new design equations for ring hybrids terminated in arbitrary impedances have been presented that can be applied for arbitrary power divisions, arbitrary arc lengths, arbitrary termination impedances, and miniaturized lumped- and distributed-element ring hybrids. Using this design method, big advantages in total size reduction can be gained.

EXERCISES

4.1 Derive the *ABCD* parameters of Part A connected between ports ② and ④ in Fig. 4.1 when the ring hybrid is terminated in equal impedances of 50 Ω and the power is divided equally.

4.2 Derive (4.4).

4.3 Design the ring hybrid in Fig. 4.5 when all the ports are terminated in $R_a = 50\ \Omega$, $R_b = 40\ \Omega$, $R_c = 60\ \Omega$, and $R_d = 70\ \Omega$, the power-split ratio is 3 dB, and the coupling coefficients of the coupled-line filter section between ports ② and ③ are 3 dB, 4 dB, and 5 dB.

4.4 Simulate ring hybrids with each coupling coefficient given in Exercise 4.3 and compare them.

4.5 Derive (4.14).

4.6 Design and simulate the ring hybrid in Fig. 4.12 when $L_a < L_c$.

4.7 Derive the *ABCD* parameters of the circuit of Part A in Fig. 4.1 and find its equivalent circuit.

4.8 Derive the admittance or impedance matrix of the following coupled-line filter sections in Fig. E4.8.

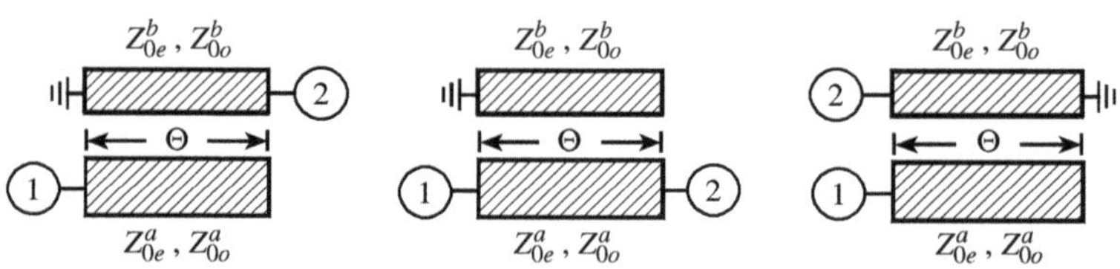

FIGURE E4.8 Three two-port parallel coupled lines: (*a*) short opposite port ② and an open opposite port ①; (*b*) short adjacent to port ① and an open adjacent to port ②; (*c*) open opposite port ① and a short opposite port ②.

REFERENCES

1. W. A. Tyrrel, Hybrid Circuits for Microwaves, *Proc. IRE*, Vol. 35, November 1947, pp. 1294–1306.

2. T. Morita and L. S. Sheingold, A Coaxial Magic-T, *IRE Trans. Microwave Theory Tech.*, Vol. 1, November 1953, pp. 17–23.
3. V. I. Albanese and W. P. Peyser, An Analysis of a Broad-Band Coaxial Hybrid Ring, *IRE Trans. Microwave Theory Tech.*, Vol. 6, October 1958, pp. 369–373.
4. W. V. Tyminski and A. E. Hylas, A Wide-Band Hybrid Ring for UHF, *Proc. IRE*, Vol. 41, January 1953, pp. 81–87.
5. S. March, Wideband Stripline Hybrid Ring, *IEEE Trans. Microwave Theory Tech.*, Vol. 16, June 1968, pp. 361–362.
6. C. Y. Pon, Hybrid-Ring Directional Coupler for Arbitrary Power Divisions, *IRE Trans. Microwave Theory Tech.*, Vol. 9, November 1961, pp. 529–535.
7. T. Hirota, Y. Tarusawa, and H. Ogawa, Uniplanar MMIC Hybrids: A Proposed New MMIC Structure, *IEEE Trans. Microwave Theory Tech.*, Vol. 35, June 1987, pp. 576–581.
8. C.-H. Ho, L. Fan, and K. Chang, Broad-Band Uniplanar Hybrid-Ring and Branch-Line Couplers, *IEEE Trans. Microwave Theory Tech.*, Vol. 41, December 1993, pp. 2116–2124.
9. C.-H. Ho, L. Fan, and K. Chang, New Uniplanar Coplanar Waveguide Hybrid-Ring Couplers and Magic-T's, *IEEE Trans. Microwave Theory Tech.*, Vol. 42, December 1994, pp. 2440–2448.
10. D. Kim and Y. Natio, Broad Band Design of Improved Hybrid-Ring 3-dB Directional Coupler, *IEEE Trans. Microwave Theory Tech.*, Vol. 30, November 1982, pp. 2040–2046.
11. G. F. Mikucki and A. K. Agrawal, A Broad-Band Printed Circuit Hybrid Ring Power Divider, *IEEE Trans. Microwave Theory Tech.*, Vol. 37, January 1989, pp. 112–117.
12. A. K. Agrawal and G. F. Mikucki, A Printed Circuit Hybrid Ring Directional Coupler for Arbitrary Power Division, *IEEE Trans. Microwave Theory Tech.*, Vol. 34, December 1986, pp. 1401–1407.
13. H.-R. Ahn, I.-S. Chang, and S.-W. Yun, Miniaturized 3-dB Ring Hybrid Terminated by Arbitrary Impedances, *IEEE Trans. Microwave Theory Tech.*, Vol. 42, December 1994, pp. 2216–2221.
14. H.-R. Ahn, I. Wolff, and I.-S. Chang, Arbitrary Termination Impedances, Arbitrary Power Division and Small-Sized Ring Hybrids, *IEEE Trans. Microwave Theory Tech.*, Vol. 44, December 1997, pp. 2241–2247.
15. H.-R. Ahn and I. Wolff, Three-Port 3-dB Power Divider Terminated by Different Impedances and Its Application to MMIC's, *IEEE Trans. Microwave Theory Tech.*, Vol. 47, June 1999, pp. 786–794.
16. H.-R. Ahn and I. Wolff, Three-Port 3-dB Power Divider Terminated in Arbitrary Impedances, *IEEE MTT-S Dig.*, Baltimore, June 1998, pp. 781–784.
17. H.-R. Ahn and I. Wolff, General Design Equations of Three-Port Power Dividers, *IEEE MTT-S Dig.*, Boston, June 2000, pp. 1137–1140.
18. H.-R. Ahn and I. Wolff, General Design Equations of Three-Port Power Dividers, Small-Sized Impedance Transformers, and Their Applications to Small-Sized Three-Port 3-dB Power Divider, *IEEE Trans. Microwave Theory Tech.*, Vol. 49, July 2001, pp. 1277–1288.

19. D. I. Kim and G. S. Yang, Design of a New Hybrid-Ring Directional Coupler Using $\lambda/8$ or $\lambda/6$ Sections, *IEEE Trans. Microwave Theory Tech.*, Vol. 39, October 1991, pp. 1179–1783.
20. B.-H. Murgulescu, E. Moisan, P. Legaud, E. Penard, and I. Zaquine, New Wideband, 0.67 λg Circumference 180° Hybrid Ring Coupler, *Electron. Lett.*, Vol. 30, No. 4, February 1994, pp. 299–300.
21. L. Fan, C.-H. Ho, S. Kanamaluru, and K.-Chang, Wide-Band Reduced Sized Uniplanar Magic-T, Hybrid Ring, and De Rondé CPW–Slot Couplers, *IEEE Trans. Microwave Theory Tech.*, Vol. 43, December 1995, pp. 2749–2758.
22. P. Bhartia and I. J. Bhal, *Millimeter Wave Engineering and Applications*, Wiley, New York, 1984, pp. 373–374.
23. S. J. Parisi, 180° Lumped Element Hybrid, *IEEE MTT-S Dig.*, 1989, pp. 1243–1246.
24. V. K. Tripathi, Asymmetric Coupled Transmission Lines in an Inhomogeneous Medium, *IEEE Trans. Microwave Theory Tech.*, Vol. 23, No. 9, September 1975, pp. 734–739.
25. R. H. Jansen, Fast Accurate Hybrid Mode Computation of Nonsymmetrical Coupled Microstrip Characteristics, *Proc. EUMC*, 1977, pp. 135–129.

CHAPTER FIVE

Asymmetric Branch-Line Hybrids

5.1 INTRODUCTION

After the first ring hybrid was suggested in 1947 by Tyrrel [3], numerous ring hybrids [3–11] have been developed. The first branch-line hybrid was invented in the same year, and various branch-line hybrids have been studied separately from ring hybrids for a long time [1,12–18]. However, these studies have focused on symmetric four-port hybrids where the method of even- and odd-mode excitation analyses can be used. Additionally, the relation between ring hybrids and branch-line hybrids had never been investigated. The four-port hybrids are used with active and/or other passive components, and additional matching circuits are necessary to obtain the output performances desired. If they can be terminated in arbitrary impedances, no matching circuit is needed and the total size of microwave integrated circuits can be reduced. For this purpose, the first asymmetric four-port ring hybrids and branch-line hybrids were suggested by Ahn et al. [19–22]. In this chapter the design equations of asymmetric branch-line hybrids terminated in arbitrary impedances are derived along with those of symmetric branch-line hybrids.

5.2 ORIGIN OF BRANCH-LINE HYBRIDS

Whenever a transmission line is used to convey radio-frequency power from one point in a system to another, a knowledge of the power contained in the transmitted and reflected waves is of fundamental importance. It is possible to obtain this knowledge indirectly by employing a calibrated traveling-wave

---

Asymmetric Passive Components in Microwave Integrated Circuits, by Hee-Ran Ahn

detector to measure the voltage standing-wave ratio, from which the transmitted and reflected powers may be calculated. A more straightforward and direct measurement can be accomplished by utilizing a device that samples these waves separately and causes output powers proportional to each of these power parts to appear in two separate power meters. This device has been called by various names: stationary standing-wave detector, directive tap, directional tap, wave selector, directive pickup, and finally, directional coupler. The last name in this list has been generally agreed upon and accepted by representatives of major research and development groups.

The directional coupler may be called a branch-line hybrid, depending on how the coupling is realized [1,2]. It usually takes the form of two adjacent transmission lines with one or more coupling elements between them. One of these lines is the main, or primary, line, and the other is the secondary line. A small fraction of the energy in the main line is transferred, through the coupling elements, to the secondary line. When the two lines are otherwise thoroughly shielded, the ratio of the power that flows in the two lines depends on the physical size and the number and spacing of coupling elements. Since all of these factors can be controlled and held constant, a high degree of stability is ensured. This property is especially important in applications that require high values of attenuation. Because of its remarkable properties, the directional coupler is a useful tool for laboratory and production testing and is an invaluable aid to operators in the field.

Figure 5.1 shows two adjacent transmission lines coupled together weakly in two places by means of simple link circuits. The primary line is driven by a generator and terminated in an arbitrary load impedance Z_L. There exist a transmitted wave traveling from left to right and a reflected wave traveling from

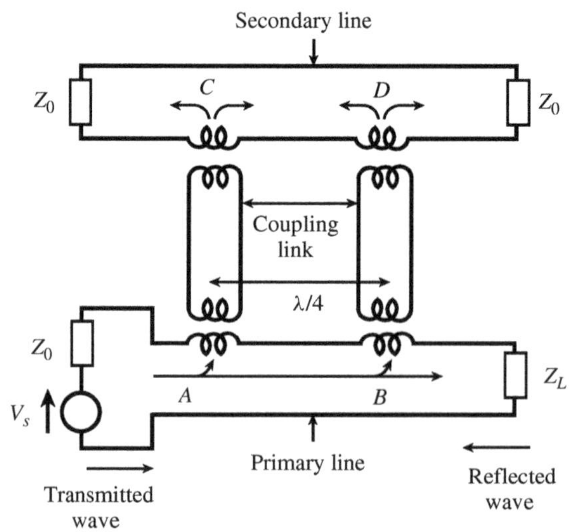

FIGURE 5.1 Directional coupler.

right to left. If only the transmitted wave is considered, at point A a small portion of the wave passes from the primary line into the link circuit and into the secondary line at point C. Here it sets up two waves traveling in opposite directions toward the ends of the secondary line, which are terminated in matching impedances. Similarly, at point B, a small portion of the transmitted wave in the primary line is fed into the secondary line through coupling link BD. At point D, therefore, two waves are also set up in the secondary line. One wave travels to the left and the other to the right. Hence, two waves traveling in each direction exist in the secondary line. The two waves traveling to the right will be in phase and will therefore reinforce each other, since paths ACD and ABD are of the same length. The two waves traveling to the left will cancel and give zero current in the left-hand termination if path $ABDC$ is a half wavelength longer than path AC (i.e., if the spacing between the coupling links is a quarter wavelength and the coupling elements do not introduce a phase shift).

As for the reflected waves, it can be shown by a similar argument that the current flowing in the secondary line will cancel to the right and reinforce to the left if the distance AB or CD is a quarter wavelength. Thus, a directional coupler has the property that the power dissipated at the left termination in the secondary line is proportional to the reflected wave. Since the effective bandwidth for this useful property is only about 10%, multisection branch-line couplers have been utilized to increase bandwidths.

5.3 MULTISECTION BRANCH-LINE COUPLERS

Figure 5.2 shows a five-branch-line coupler terminated in equal admittances $Y_0 = 1$, and the admittance of the main line is Y_0. The three center connecting branch lines have the same width, and each characteristic admittance is Y_c. The two end branch lines share a different width, and the characteristic admittance is Y_a. The spacing between the center lines of adjacent branch lines is assumed to be identical, and the length of all these lines is assumed to be same. First, for simplicity,

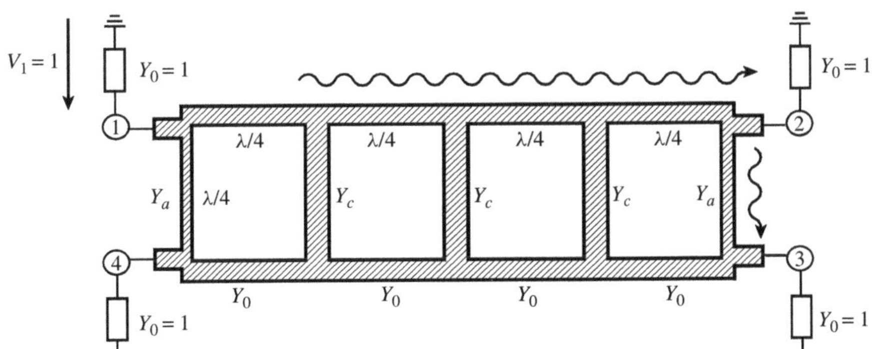

FIGURE 5.2 Typical coupler with five branches.

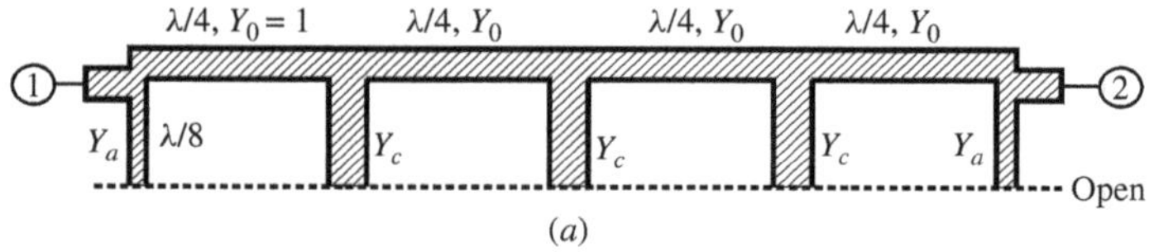

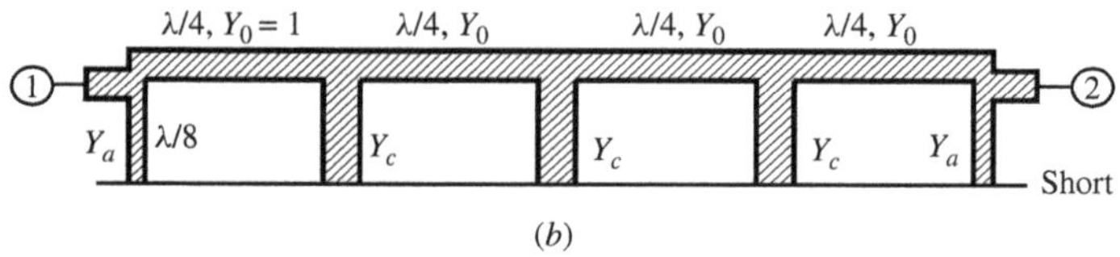

FIGURE 5.3 (*a*) Even- and (*b*) odd-mode equivalent circuits.

each of the two distances is considered to be a quarter wavelength and the junction effect is ignored. A signal of amplitude 1 is applied to port ① as shown in Fig. 5.2. To determine the vector amplitudes out of the other ports, assuming that all ports are matched, the even- and odd-mode concept is used. As explained in Fig. 2.18, a voltage maximum and current null at every point on the plane of symmetry occur for the even-mode excitation, and a voltage null occurs at every point on the plane of symmetry for the odd-mode excitation [23]. The even- and odd-mode equivalent circuits are depicted in Fig. 5.3(*a*) and (*b*), respectively.

In each case the symmetry of the modes will not be disturbed by passing through the circuit. When the power is excited at port ①, the amplitude out of port ② will be the sum of the amplitudes transmitted for the even and odd modes, and the amplitude out from ③ will be the difference of these amplitudes. Similarly, the amplitudes out of ports ① and ④ are the sum and difference of the reflected amplitude. As expressed in symbols in (2.103), the following results are obtained:

$$A_1 = \tfrac{1}{2}S_{11e} + \tfrac{1}{2}S_{11o}, \tag{5.1a}$$

$$A_2 = \tfrac{1}{2}S_{21e} + \tfrac{1}{2}S_{21o}, \tag{5.1b}$$

$$A_3 = \tfrac{1}{2}S_{21e} - \tfrac{1}{2}S_{21o}, \tag{5.1c}$$

$$A_4 = \tfrac{1}{2}S_{11e} - \tfrac{1}{2}S_{11o}, \tag{5.1d}$$

where the subscript attached to the amplitudes is the port number.

Thus, the five-branch-line coupler shown in Fig. 5.2 can be analyzed in terms of the results from the two two-port equivalent circuits shown in Fig. 5.3. The *ABCD* parameters of the two circuits are determined for both even- and odd-mode excitations, from which the reflected and transmitted amplitudes can be found. If the five-branch-line coupler in Fig. 5.2 is assumed to be lossless, the *A* and *D* terms are purely real quantities and the *B* and *C* terms are purely imaginary quantities. The values of the reflected and transmitted amplitudes of each mode are given in terms of the elements of the *ABCD* matrix. Setting $Z_{01} = Z_{02} = 1$

in (2.72) and (2.75), they are found as

$$\frac{\Gamma}{2} = \frac{A + B - C - D}{2(A + B + C + D)}, \tag{5.2a}$$

$$\frac{T}{2} = \frac{1}{A + B + C + D}. \tag{5.2b}$$

The quantities T and Γ are the elements S_{21} and S_{11} of the scattering matrix for a two-port circuit as explained earlier. For the first application of the method described above, consider the three-branch-line coupler in Fig. 5.4(a) and its even- and odd-mode equivalent circuits in Fig. 5.4(b) and (c). The characteristic admittance of the main line is Y_0, and those of the center line and two end branch lines are Y_c and Y_a. The three connecting branch lines are assumed to be a quarter wavelength long, and the branch lines are spaced at quarter wavelengths. The $ABCD$ matrix of the even-mode circuit is

$$\begin{bmatrix} A & B \\ C & D \end{bmatrix}_{e3} = \begin{bmatrix} 1 & 0 \\ jY_a & 1 \end{bmatrix} \begin{bmatrix} 0 & j \\ j & 0 \end{bmatrix} \begin{bmatrix} 1 & 0 \\ jY_c & 1 \end{bmatrix} \begin{bmatrix} 0 & j \\ j & 0 \end{bmatrix} \begin{bmatrix} 1 & 0 \\ jY_a & 1 \end{bmatrix}$$

$$= \begin{bmatrix} Y_aY_c - 1 & j(Y_a^2Y_c - 2Y_a) \\ -jY_c & Y_aY_c - 1 \end{bmatrix}. \tag{5.3}$$

This matrix is the product of five matrices: the first, third, and fifth being those of open-circuit stubs with an eighth wavelength and of characteristic admittances Y_a, Y_c, and Y_a, respectively. The second and fourth matrices are those of quarter-wavelength lines with the characteristic admittance unit. The matrix for the odd

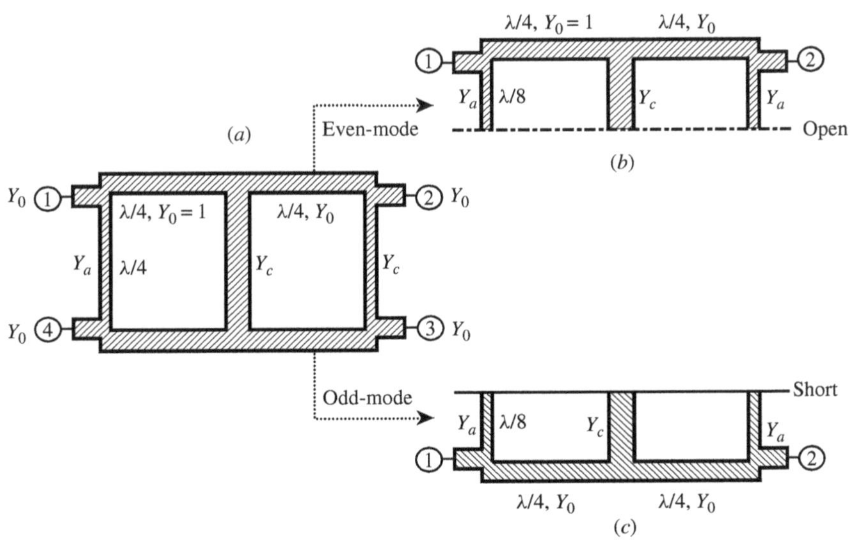

FIGURE 5.4 (a) Branch-line coupler with three branch lines; (b) even-mode equivalent circuit; (c) odd-mode equivalent circuit.

mode is similar except that short-circuited eighth-wavelength stubs are used. The matrix product is found quickly by replacing Y_a and Y_c with $-Y_a$ and $-Y_c$.

$$\begin{bmatrix} A & B \\ C & D \end{bmatrix}_{o3} = \begin{bmatrix} Y_a Y_c - 1 & -j(Y_a^2 Y_c - 2Y_a) \\ jY_c & Y_a Y_c - 1 \end{bmatrix}. \tag{5.4}$$

For perfect matching and directivity, there must be no reflection and $A + B - C - D$ in (5.2a) should be zero. This condition is achieved by setting the B term equal to the C term since the A term is already equal to the D term. This gives two possible solutions, but the smaller one should be chosen to reduce junction effects and increase bandwidth [12]. The smaller value is

$$Y_a = \frac{1 - \sqrt{1 - Y_c^2}}{Y_c}. \tag{5.5}$$

Based on (5.3b) and (5.4), the voltages out of ports ② and ③ are calculated as

$$A_2 = \frac{1}{2} S_{21e} + \frac{1}{2} S_{21o} = \frac{Y_a Y_c - 1}{[(Y_a Y_c - 1) - jY_c][(Y_a Y_c - 1) + jY_c]}, \tag{5.6a}$$

$$A_3 = \frac{1}{2} S_{21e} - \frac{1}{2} S_{21o} = \frac{jY_c}{[(Y_a Y_c - 1) - jY_c][(Y_a Y_c - 1) + jY_c]}. \tag{5.6b}$$

For perfect directivity and equal power division, the two waves at ports ② and ③ are of equal amplitude and out of phase by 90°. This is only an example, and $|A_2|^2 + |A_3|^2 = 1$ is concluded for any lossless circuit. Thus, when this coupler is matched and perfectly directive, the voltage amplitude out of port ② is equal to the magnitude of the A term, and the amplitude out of port ③ is equal to the magnitude of the C term in (5.3) and (5.4).

For a four-branch-line coupler, a similar procedure may be followed:

$$\begin{bmatrix} A & B \\ C & D \end{bmatrix}_{e4} = \begin{bmatrix} -Y_a(Y_c^2 - 1) + Y_c & -j[Y_a^2(Y_c^2 - 1) + 2Y_a(-Y_c) + 1] \\ j(Y_c^2 - 1) & -Y_a(Y_c^2 - 1) + Y_c \end{bmatrix}. \tag{5.7}$$

As mentioned before, the matrix for the odd mode may be found by replacing $-Y_a$ and $-Y_c$ by Y_a and Y_c. The amplitude out of port ③ is now equal to the A term in the matrix, and the amplitude out of port ② is equal to the C term. This is true for those with an even number of branch lines, and the former case is true for those with an odd number of branch lines. For the coupler with five branch lines depicted in Fig. 5.2, the matrix is

$$\begin{bmatrix} A & B \\ C & D \end{bmatrix}_{e5} =$$

$$\begin{bmatrix} -Y_a(-Y_c^3 + 2Y_c) - (Y_c^2 - 1) & j[(Y_a^2(-Y_c^3 + 2Y_c) + 2Y_a(Y_c^2 - 1) - Y_c] \\ j(-Y_c^3 + 2Y_c) & -Y_a(-Y_c^3 + 2Y_c) - (Y_c^2 - 1) \end{bmatrix}. \tag{5.8}$$

For the general coupler with n branch lines each a quarter wavelength long and spaced a quarter wavelength away from adjacent branches on a uniform transmission line, the matrix can be found in a similar way. Consideration of the A terms of (5.3b), (5.7), and (5.8), reveals that they are related by Tchebysheff (Chebyshev) polynomials [13], the first few of which have the following values:

$$S_0(x) = 1, \tag{5.9a}$$

$$S_1(x) = x, \tag{5.9b}$$

$$S_2(x) = x^2 - 1, \tag{5.9c}$$

$$S_3(x) = x^3 - 2x, \tag{5.9d}$$

$$S_4(x) = x^4 - 3x^2 + 1, \tag{5.9e}$$

$$S_5(x) = x^5 - 4x^3 + 3x, \tag{5.9f}$$

$$S_6(x) = x^6 - 5x^4 + 6x^2 - 1, \tag{5.9g}$$

$$S_{n+1}(x) = xS_n(x) - S_{n-1}(x). \tag{5.9h}$$

For a general branch-line coupler with n center branch lines having the same admittance Y_c and the two end branch lines having the same admittance Y_a, its even-mode matrix is

$$\begin{bmatrix} A & B \\ C & D \end{bmatrix}_{e(n+2)} = \begin{bmatrix} -Y_a S_n(-Y_c) - S_{n-1}(-Y_c) & -j[{Y_a}^2 S_n(-Y_c) + 2Y_a S_{n-1}(-Y_c) + S_{n-2}(-Y_c)] \\ jS_n(-Y_c) & -Y_a S_n(-Y_c) - S_{n-1}(-Y_c) \end{bmatrix}. \tag{5.10}$$

The values of $S_n(Y_c)$ are available for integral values of n from 2 to 12. To design a matched and perfectly directive coupler with a given coupling ratio, Y_c, the relative admittance value, should be calculated. For this, $n + 2$ branch lines are chosen and the voltage amplitudes out of ports ② and ③ are determined with unity amplitude fed into port ①. For an even number of branch lines, the magnitude of the C term in even-mode $ABCD$ matrix $S_n(-Y_c)$ is set equal to A_2 and Y_c is calculated. If the number of the center branch lines is odd, the value of $S_n(-Y_c)$ is set equal to A_3 and Y_c is calculated. For a matched and perfectly directive coupler, Y_a is calculated as

$$Y_a = \left| \frac{|\sqrt{1 - S_n^2(-Y_c)}| - |S_{n-1}(-Y_c)|}{S_n(-Y_c)} \right|. \tag{5.11}$$

This equation is found from the general matrix above by equating to unity the sum of the squares of the A and C terms in the matrix; again, the smaller value is used for reducing junction effects and increasing bandwidths. Here, $A^2 + C^2 = 1$, or $A_2^2 + A_3^2 = 1$ in (5.6), indicates the unitary property and that the square of the denominator of A_2 or A_3 is unity.

5.4 BRANCH-LINE HYBRIDS FOR IMPEDANCE TRANSFORMING

So far, a general branch-line coupler (hybrid) terminated in equal impedances has been discussed. For some applications it may be useful to employ impedance-transforming hybrids that are terminated in arbitrary impedances but are symmetric. Two types of branch-line hybrids for impedance transforming [14,15] are introduced here using the method discussed above.

Figure 5.5(a) shows a two-section branch-line hybrid where ports ① and ④ are terminated in impedances Z_{01}, while ports ② and ③ are terminated in Z_{02}. Since the termination impedances are different, the characteristic impedances of the branch lines are different, like Z_a, Z_c, and Z_d, and that of the main line is Z_b. Assuming that it is lossless, its even- and odd-mode equivalent circuits are as depicted in Fig. 5.5(b) and (c), respectively. Their $ABCD$ matrices are found as

$$\begin{bmatrix} A & B \\ C & D \end{bmatrix}_{\text{even}} = \begin{bmatrix} 1 & 0 \\ j/Z_a & 1 \end{bmatrix} \begin{bmatrix} 0 & jZ_b \\ j/Z_b & 0 \end{bmatrix} \begin{bmatrix} 1 & 0 \\ j/Z_c & 1 \end{bmatrix} \begin{bmatrix} 0 & jZ_b \\ j/Z_b & 0 \end{bmatrix} \begin{bmatrix} 1 & 0 \\ j/Z_d & 1 \end{bmatrix}, \tag{5.12a}$$

$$\begin{bmatrix} A & B \\ C & D \end{bmatrix}_{\text{even}} = \begin{bmatrix} -1 + \dfrac{Z_b^2}{Z_d Z_c} & -\dfrac{jZ_b^2}{Z_c} \\ j\left(\dfrac{Z_b^2}{Z_a Z_c Z_d} - \dfrac{1}{Z_a} - \dfrac{1}{Z_d}\right) & -1 + \dfrac{(Z_b)^2}{Z_a Z_c} \end{bmatrix}, \tag{5.12b}$$

$$\begin{bmatrix} A & B \\ C & D \end{bmatrix}_{\text{odd}} = \begin{bmatrix} A & -B \\ -C & D \end{bmatrix}_{\text{even}}. \tag{5.12c}$$

When a two-port passive circuit is terminated in different impedances, Z_{01} and Z_{02}, its transmitted and reflected coefficients are found, referring to (2.72) and (2.75), as

$$\Gamma = S_{11} = \frac{AZ_{02} + B - CZ_{01}Z_{02} - DZ_{01}}{AZ_{02} + B + CZ_{01}Z_{02} + DZ_{01}}, \tag{5.13a}$$

$$T_{21} = S_{21} = \frac{2\sqrt{Z_{01}}\sqrt{Z_{02}}}{AZ_{02} + B + CZ_{01}Z_{02} + DZ_{01}}. \tag{5.13b}$$

The corresponding reflected and transmitted scattering parameters of the even and odd modes are

$$S_{11e} = \frac{A_eZ_{02} + B_e - C_eZ_{01}Z_{02} - D_eZ_{01}}{A_eZ_{02} + B_e + C_eZ_{01}Z_{02} + D_eZ_{01}}, \tag{5.14a}$$

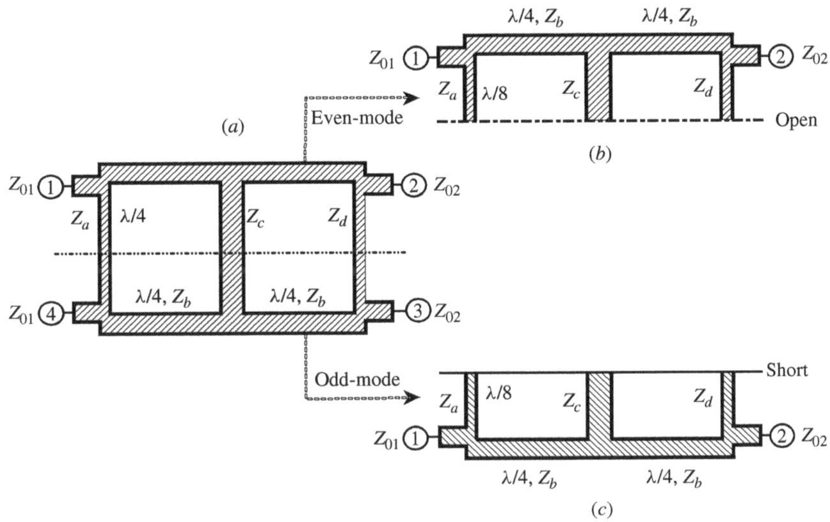

FIGURE 5.5 (*a*) Two-section branch-line hybrid for impedance transforming: (*b*) even-mode equivalent circuit; (*c*) odd-mode equivalent circuit.

$$S_{11o} = \frac{A_e Z_{02} - B_e + C_e Z_{01} Z_{02} - D_e Z_{01}}{A_e Z_{02} - B_e - C_e Z_{01} Z_{02} + D_e Z_{01}}, \tag{5.14b}$$

$$S_{21e} = \frac{2\sqrt{Z_{01}}\sqrt{Z_{02}}}{A_e Z_{02} + B_e + C_e Z_{01} Z_{02} + D_e Z_{01}}, \tag{5.14c}$$

$$S_{21o} = \frac{2\sqrt{Z_{01}}\sqrt{Z_{02}}}{A_e Z_{02} - B_e - C_e Z_{01} Z_{02} + D_e Z_{01}}, \tag{5.14d}$$

where the subscripts e and o indicate even- and odd-mode excitations and $B_o = -B_e$ and $C_o = -C_e$ are used.

Thus, when an amplitude of unity is fed into port ①, the scattering parameters at ports ② and ③ are calculated as

$$A_2 = [(S_{21e} + S_{21o})/2] = \frac{2\sqrt{Z_{01}}\sqrt{Z_{02}}(A_e Z_{02} + D_e Z_{01})}{(A_e Z_{02} + B_e + C_e Z_{01} Z_{02} + D_e Z_{01})(A_e Z_{02} - B_e - C_e Z_{01} Z_{02} + D_e Z_{01})}, \tag{5.15a}$$

$$A_3 = [(S_{21e} - S_{21o})/2] = \frac{-2\sqrt{Z_{01}}\sqrt{Z_{02}}(B_e + C_e Z_{01} Z_{02})}{(A_e Z_{02} + B_e + C_e Z_{01} Z_{02} + D_e Z_{01})(A_e Z_{02} - B_e - C_e Z_{01} Z_{02} + D_e Z_{01})}. \tag{5.15b}$$

Four unknown values, Z_a, Z_b, Z_c, and Z_d, are to be determined, and perfect matching, directivity, and lossless conditions should be used for the determination.

For a matched branch-line hybrid, there should be no reflection with even and odd modes, and the following result is thus obtained:

$$A_e Z_{02} + B_e = C_e Z_{01} Z_{02} + D_e Z_{01}, \tag{5.16a}$$

$$A_e Z_{02} + C_e Z_{01} Z_{02} = B_e + D_e Z_{01}, \tag{5.16b}$$

from which

$$A_e = \frac{D_e Z_{01}}{Z_{02}}, \tag{5.17a}$$

$$B_e = C_e Z_{01} Z_{02} \tag{5.17b}$$

are given.

The output power ratio between output ports ② and ③ is written as

$$k^2 = \left|\frac{A_3}{A_2}\right|^2 = \left|-\frac{B_e + C_e Z_{01} Z_{02}}{A_e Z_{02} + D_e Z_{01}}\right|^2 = \left|-\frac{C_e Z_{01}}{A_e}\right|^2, \tag{5.18}$$

where the relations $A_e = D_e Z_{01}/Z_{02}$ and $B_e = C_e Z_{01} Z_{02}$ in (5.17) are used. For a lossless branch-line hybrid, a unitary property, $|A_2|^2 + |A_3|^2 = 1$, should be satisfied, from which the A_e term is found as

$$A_e = \sqrt{\frac{Z_{01}}{(1 + k^2) Z_{02}}}, \tag{5.19}$$

where the positive sign of the A_e term is taken; otherwise, the input and output impedances are limited to certain values from (5.12*b*). After finding the A_e term, the other terms in (5.12*b*) are determined. The relation between the A_e and D_e terms is given in (5.17*a*), from which the D_e term is determined from (5.19). The relation between the A_e and C_e terms is to be found from the power division ratio in (5.18), so the C_e term is given as

$$C_e = \frac{-jkA_e}{Z_{01}}. \tag{5.20}$$

Finally, using (5.17), (5.19), and (5.20), the *ABCD* parameters of the even-mode circuit are given as

$$\begin{bmatrix} A & B \\ C & D \end{bmatrix}_{\text{even}} = \frac{1}{\sqrt{1 + k^2}} \begin{bmatrix} \sqrt{\dfrac{Z_{01}}{Z_{02}}} & -jk\sqrt{Z_{01} Z_{02}} \\ \dfrac{-jk}{\sqrt{Z_{01} Z_{02}}} & \sqrt{\dfrac{Z_{02}}{Z_{01}}} \end{bmatrix}. \tag{5.21}$$

Now, all the unknown values are derived from (5.12*b*) and (5.21) as

$$Z_a = \frac{jB_e}{1 + D_e}, \tag{5.22a}$$

$$Z_d = \frac{jB_e}{1 + A_e}, \tag{5.22b}$$

$$\frac{Z_b^2}{Z_c} = jB_e, \tag{5.22c}$$

where A_e, B_e and D_e are as given in (5.21). The three equations in (5.22) are the design equations for two-section branch-line hybrids with impedance transformations and arbitrary power divisions. One of many cases in choosing Z_b and Z_c is $Z_b = Z_c$.

Based on the design equations of (5.22), two cases were simulated. When $k = 1$ (equal power division), $Z_{01} = 60\ \Omega$, and $Z_{02} = 50\ \Omega$, the characteristic impedances are $Z_a = 23.54\ \Omega$, $Z_b = 46.\ \Omega$, $Z_c = 54.78\ \Omega$, and $Z_d = 21.8\ \Omega$. When $k = \sqrt{2}$ (a 3-dB power-split ratio), $Z_{01} = 40\ \Omega$, and $Z_{02} = 60\ \Omega$, the impedances are $Z_a = 23.4\ \Omega$, $Z_b = 44.2\ \Omega$, $Z_c = 49\ \Omega$, and $Z_d = 27.2\ \Omega$. Figure 5.6(*a*) shows the simulation results for the first case and Figure 5.6(*b*) those for the second. The two types of simulation results show that the branch-line hybrids are perfectly matched and that the power division ratios and impedance transforming desired have been achieved.

The method discussed above may be applied to the single-section branch-line hybrids commonly used in microwave integrated circuits. Figure 5.7 shows a single-section branch-line hybrid for impedance transforming and its even- and odd-mode equivalent circuits. Design values of the branch- and main-line impedances can be obtained readily using the symmetrical four-port network analyses explained above. In Fig. 5.7, Z_{01} and Z_{02} are the input and output impedances, and Z_a, Z_b, and Z_c are the branch- and main-line impedances. The *ABCD* matrices of the even- and odd-mode circuits are

$$\begin{bmatrix} A & B \\ C & D \end{bmatrix}_{\text{even}} = \begin{bmatrix} 1 & 0 \\ \dfrac{j}{Z_a} & 1 \end{bmatrix} \begin{bmatrix} 0 & jZ_b \\ \dfrac{j}{Z_b} & 0 \end{bmatrix} \begin{bmatrix} 1 & 0 \\ \dfrac{j}{Z_c} & 1 \end{bmatrix}$$

$$= \begin{bmatrix} \dfrac{-Z_b}{Z_c} & jZ_b \\ j\left(\dfrac{1}{Z_b} - \dfrac{Z_b}{Z_c Z_a}\right) & \dfrac{-Z_b}{Z_a} \end{bmatrix}, \tag{5.23a}$$

$$\begin{bmatrix} A & B \\ C & D \end{bmatrix}_{\text{odd}} = \begin{bmatrix} -A & B \\ C & -D \end{bmatrix}_{\text{even}}. \tag{5.23b}$$

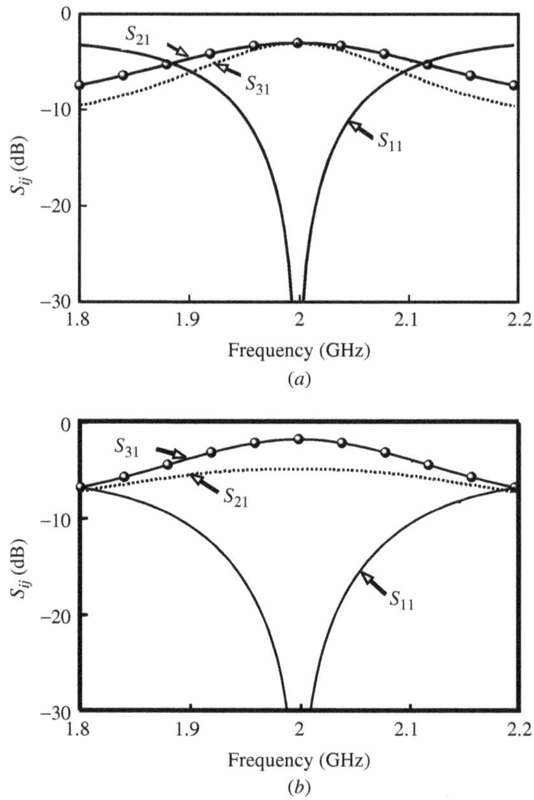

FIGURE 5.6 Simulation of double-section branch-line hybrids: (*a*) impedance transforming of 60 Ω into 50 Ω with an equal power division ($k = 1$); (*b*) impedance transforming of 40 Ω into 60 Ω with a power-split ratio of 3 dB ($k = \sqrt{2}$).

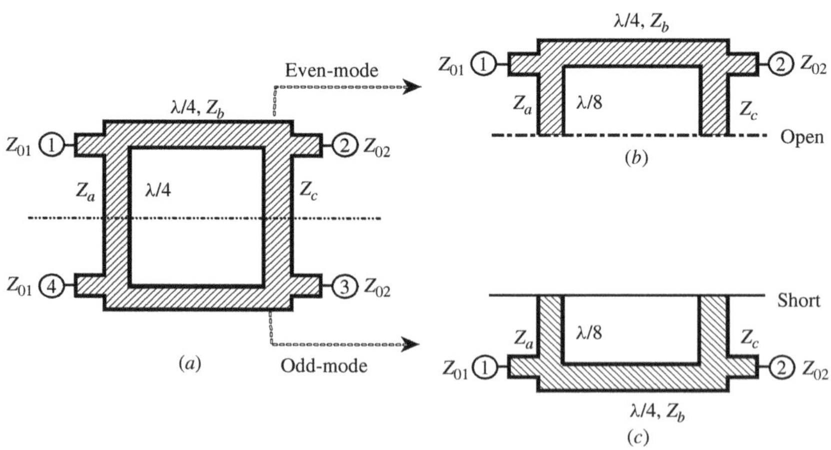

FIGURE 5.7 (*a*) Single-section branch-line hybrid; (*b*) even-mode equivalent circuit; (*c*) odd-mode equivalent circuit.

In a similar way, the amplitudes out of ports ①, ②, and ③ are found as

$$A_1 = [(S_{11e} + S_{11o})/2] = \frac{A_e Z_{02} + B_e - C_e Z_{01} Z_{02} - D_e Z_{01}}{(A_e Z_{02} + B_e + C_e Z_{01} Z_{02} + D_e Z_{01})} + \frac{A_e Z_{02} - B_e + C_e Z_{01} Z_{02} - D_e Z_{01}}{(A_e Z_{02} - B_e - C_e Z_{01} Z_{02} + D_e Z_{01})}, \tag{5.24a}$$

$$A_2 = [(S_{21e} + S_{21o})/2] = \frac{-2\sqrt{Z_{01}}\sqrt{Z_{02}}(B_e + C_e Z_{01} Z_{02})}{(A_e Z_{02} + B_e + C_e Z_{01} Z_{02} + D_e Z_{01})(A_e Z_{02} - B_e - C_e Z_{01} Z_{02} + D_e Z_{01})}, \tag{5.24b}$$

$$A_3 = [(S_{21e} + S_{21o})/2] = \frac{2\sqrt{Z_{01}}\sqrt{Z_{02}}(A_e Z_{02} + D_e Z_{01})}{(A_e Z_{02} + B_e + C_e Z_{01} Z_{02} + D_e Z_{01})(A_e Z_{02} - B_e - C_e Z_{01} Z_{02} + D_e Z_{01})}. \tag{5.24c}$$

For a matched branch-line hybrid, $A_e = D_e Z_{01}/Z_{02}$ and $B_e = C_e Z_{01} Z_{02}$ are obtained like the results in (5.17). The output power ratio between output ports ② and ③ is written as

$$k^2 = \left|\frac{A_3}{A_2}\right|^2 = \left|-\frac{A_e Z_{02} + D_e Z_{01}}{B_e + C_e Z_{01} Z_{02}}\right|^2 = \left|-\frac{A_e}{C_e Z_{01}}\right|^2, \tag{5.25}$$

where the matching conditions $A_e = D_e Z_{01}/Z_{02}$ and $B_e = C_e Z_{01} Z_{02}$ are used. For a passive branch-line hybrid, adopting a unitary property, $|A_2|^2 + |A_3|^2 = 1$, yields

$$|A_e|^2 + |C_e Z_{01}|^2 = \frac{Z_{01}}{Z_{02}}. \tag{5.26}$$

The C_e term is found from (5.25) and (5.26) as

$$C_e = j\frac{1}{\sqrt{Z_{02} Z_{01}(1 + k^2)}}. \tag{5.27}$$

From (5.17), (5.25), and (5.26), the *ABCD* parameters of the even-mode circuit are given as

$$\begin{bmatrix} A & B \\ C & D \end{bmatrix}_{\text{even}} = \frac{1}{\sqrt{1 + k^2}} \begin{bmatrix} -k\sqrt{\dfrac{Z_{01}}{Z_{02}}} & j\sqrt{Z_{01} Z_{02}} \\ \dfrac{j}{\sqrt{Z_{01} Z_{02}}} & -k\sqrt{\dfrac{Z_{02}}{Z_{01}}} \end{bmatrix}. \tag{5.28}$$

Using (5.23*a*) and (5.28), the characteristic impedances of the branch lines are given as

$$Z_a = \frac{Z_{01}}{k}, \tag{5.29a}$$

$$Z_b = \sqrt{\frac{Z_{01}Z_{02}}{1+k^2}}, \tag{5.29b}$$

$$Z_c = \frac{Z_{02}}{k}. \tag{5.29c}$$

Based on the design equations of (5.29), two cases have been simulated. When $k = 1$, $Z_{01} = 60$, and $Z_{02} = 50$, the characteristic impedances are $Z_a = 60\ \Omega$, $Z_b = 38.73\ \Omega$, and $Z_c = 50\ \Omega$. When $k = \sqrt{2}$, $Z_{01} = 40$, and $Z_{02} = 60$, they are $Z_a = 28.28\ \Omega$, $Z_b = 28.28\ \Omega$, and $Z_c = 42.43\ \Omega$. Figure 5.8 shows the simulation results where the first case is plotted in Fig. 5.8(*a*) and the second in Fig. 5.8(*b*). In comparison to the double-section branch-line hybrids in Fig 5.6, the single-section branch-line hybrids in Fig. 5.8 appear to be better.

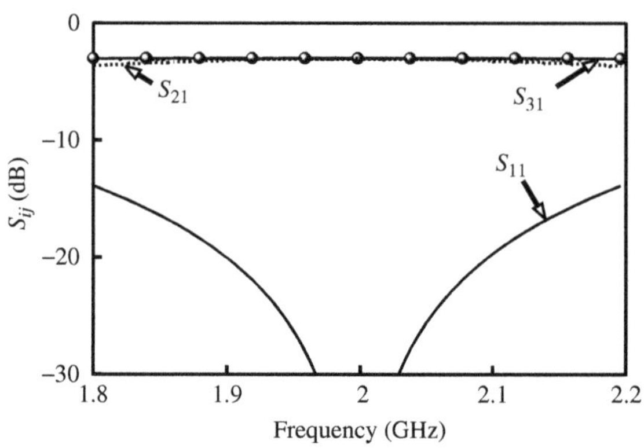

(*a*)

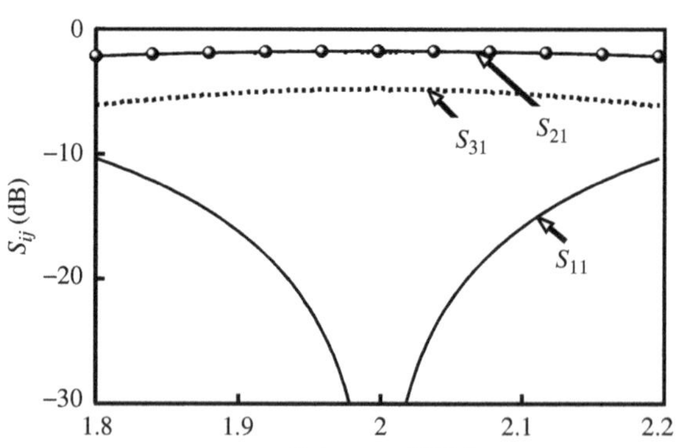

(*b*)

FIGURE 5.8 Simulation of single-section branch-line hybrids: (*a*) impedance transforming of 60 Ω into 50 Ω with an equal power division ($k = 1$); (*b*) impedance transforming of 40 Ω into 60 Ω with a power-split ratio of 3 dB ($k = \sqrt{2}$).

5.5 ASYMMETRIC FOUR-PORT HYBRIDS

So far, symmetric branch-line hybrids have been discussed. Asymmetric branch-line hybrids terminated in arbitrary impedances will be treated in conjunction with an asymmetric ring hybrid. When power is fed into one of four ports, there are three possible cases, depending on their power division characteristics. One is defined as the asymmetric ring hybrid [19–21] and the other two are asymmetric branch-line hybrids [22]. To distinguish the two different asymmetric branch-line hybrids, the first is called a *conventional-direction asymmetric branch-line hybrid* (CABH) and the second an *anti-conventional-direction asymmetric branch-line hybrid* (AABH), depending on the power division directions. Their design equations will be derived using the method that was suggested for the analysis of asymmetric ring hybrids [21]. To show that the design equations derived are reasonable and adaptable, a CABH with an equal power division and termination impedances of 50 Ω, 42 Ω, 55 Ω, and 63 Ω will be simulated, and a uniplanar CABH with a power-split ratio of 3 dB and termination impedances of 30 Ω, 60 Ω, 40 Ω, and 50 Ω will be fabricated and tested. As another example, a uniplanar AABH with an equal power division and termination impedances of 50 Ω, 41.68 Ω, 55.56 Ω, and 62.5 Ω will be fabricated in CPW technology on Al_2O_3 substrate ($\varepsilon_r = 9.9$, $h = 635$ μm). The CABH and AABH can both have arbitrary termination impedances and arbitrary power division ratios, and the results will show good agreement with predictions.

5.5.1 Analyses of Asymmetric Four-Port Hybrids

Figure 5.9 shows three four-port hybrids terminated in arbitrary real impedances R_a, R_b, R_c, and R_d. If one port of the ring hybrid is isolated from the excited power, there are three possible cases, depending on the power division directions. Figure 5.9 shows three representations when power is fed into port ①, where the ratio of d_1 to d_2 is the scattering parameter ratio. In Fig. 5.9(a), the power excited at port ① is split between ports ② and ④, and port ③ is isolated. The two signals at ports ② and ④ are in phase or out of phase by 180°, depending on the lengths of the transmission-line sections in the ring. This application is defined as a ring hybrid terminated in arbitrary impedances. It was discussed sufficiently earlier. Whereas the two output signals of the asymmetric ring hybrid mentioned previously are in phase or 180° out of phase, those in Fig. 5.9(b) and (c) are out of phase by Θ_2 and Θ_3, respectively. These two types of hybrids are called *asymmetric branch-line hybrids*.

There are several differences between conventional branch-line hybrids and asymmetric branch-line hybrids. For conventional branch-line hybrids, R_a, R_b, R_c, and R_d, which describe the termination impedances, are all equal or in some cases $R_a = R_d$ and $R_b = R_c$; the transmission-line characteristic impedances are $Z_1 = Z_3$ and $Z_2 = Z_4$; as there are symmetry planes, even- and odd-mode analyses are possible; nevertheless, derivation of the design equations is somewhat cumbersome. In the case of asymmetric branch-line hybrids, R_a, R_b, R_c, and R_d

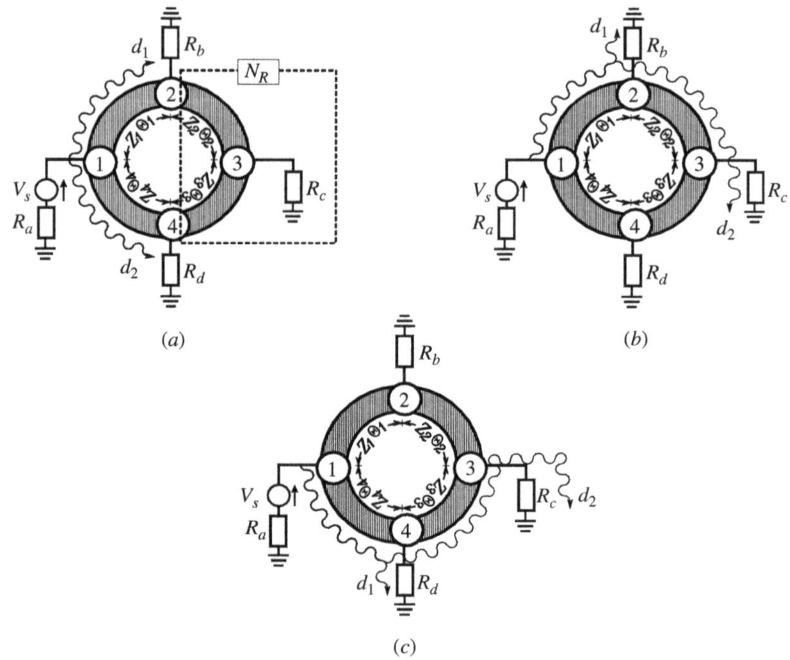

FIGURE 5.9 Three asymmetric four-port hybrids when power is fed into Port ①: (*a*) power division between ports ② and ④; (*b*) power division between ports ② and ③; (*c*) power division between ports ③ and ④. (From Ref. 22 with permission from IEEE.)

all differ from each other; the transmission-line characteristic impedances Z_1, Z_2, Z_3, and Z_4 differ from each other; and as there is no symmetry plane, conventional methods of even- and/or odd-mode excitation analyses cannot be used. Therefore, a new method is necessary to derive the design equations [22].

The power-split direction of the branch-line hybrid in Fig. 5.9(*b*) is the same as that of a conventional branch-line hybrid and thus is a *conventional-direction asymmetric branch-line hybrid.* That of the branch-line hybrid in Fig. 5.9(*c*) is, on the other hand, opposite, as can be seen in Fig. 5.9(*c*), and so is an *anti-conventional-direction asymmetric branch-line hybrid.* It is shown later that the design equations of these two types of hybrids differ. Conventional branch-line hybrids have been investigated separately from ring hybrids. However, Fig. 5.9 demonstrates that the CABH, AABH, and ring hybrid are in principle the same, and that the type of hybrid they are depends on which port is isolated.

5.5.2 Conventional–Direction Asymmetric Branch-Line Hybrids

If port ④ is terminated in a reflection coefficient Γ_4 in Fig. 5.9(*b*), the CABH is expressed by

$$\begin{bmatrix} B_1 \\ B_2 \end{bmatrix} = \begin{bmatrix} P_{11} & P_{12} \\ P_{21} & P_{22} \end{bmatrix} \begin{bmatrix} A_1 \\ A_2 \end{bmatrix}, \tag{5.30}$$

where

$$B_1 = \begin{bmatrix} b_1 \\ b_2 \\ b_3 \end{bmatrix}, \qquad A_1 = \begin{bmatrix} a_1 \\ a_2 \\ a_3 \end{bmatrix}, \qquad B_2 = [b_4], \qquad A_2 = [a_4].$$

When power is fed into port ①, a straightforward calculation in (4.3) results in

$$\begin{bmatrix} b_1 \\ b_2 \\ b_3 \end{bmatrix} = \begin{bmatrix} S_{11} & S_{12} & S_{13} \\ S_{21} & S_{22} & S_{23} \\ S_{31} & S_{32} & S_{33} \end{bmatrix} \begin{bmatrix} a_1 \\ 0 \\ 0 \end{bmatrix} + \frac{S_{41}a_1\Gamma_4}{1 - \Gamma_4 S_{44}} \begin{bmatrix} S_{14} \\ S_{24} \\ S_{34} \end{bmatrix}. \qquad (5.31)$$

If $|S_{41}| = 0$ is assumed, (5.31) indicates that there is no correlation with Γ_4 in terms of the excitation at port ①, so under this assumption, the CABH may be interpreted as shown in Fig. 5.10(*a*). The branch-line and ring hybrids are connected line couplers, and their mechanism may be explained as an ohmic connection rather than as a type of coupled mode [24]. Therefore, an equivalent circuit with excitation at port ① may be constructed as shown in Fig. 5.10. In the case of Fig. 5.10(*a*), when the power is excited at port ①, the voltage across a load R_b is equal to that across the transmission line and a load R_c at port ③. Therefore, if the ratio of $|S_{21}|$ to $|S_{31}|$ is required to be that of d_1 to d_2, as indicated in Fig. 5.9(*b*), the impedance looking into the transmission line connected with the load R_c, In ③, should be $(d_1^2/d_2^2)R_b$, as written in Fig. 5.10(*b*).

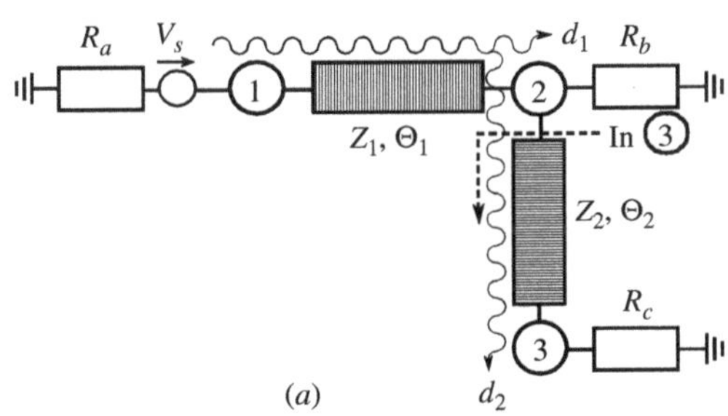

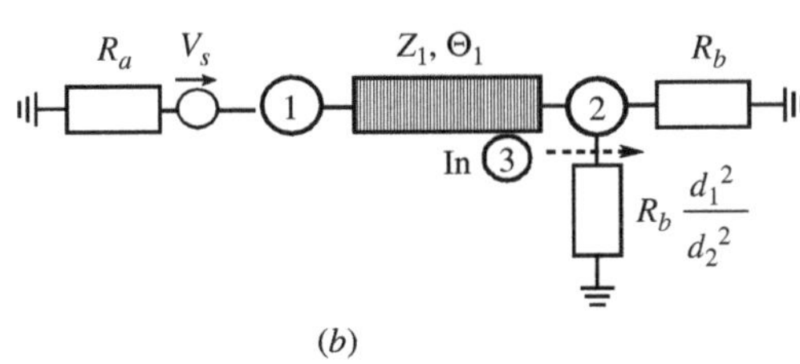

FIGURE 5.10 Excitation at port ① of Fig. 5.9(*b*): (*a*) under the assumption that $|S_{41}| = 0$; (*b*) simplified circuit. (From Ref. 22 with permission from IEEE.)

For a Y-junction device [25–36] such as a Wilkinson three-port power divider, an isolation resistor is necessary for isolation between the two outputs and perfect matching at the outputs. Following the same principle as that shown in Fig. 5.9(*a*), a circuit of N_R consisting of two transmission-line sections and a termination impedance R_c is connected with output ports ② and ④ for their isolation and perfect matching [37]. Similarly, when the power is fed into port ① in Fig. 5.9(*b*), the two transmission-line sections with Z_3 and Z_4, including the termination impedance R_d, are connected with ports ① and ③ so that they are matched and port ④ is isolated. Therefore, the power excited at port ① is delivered to port ②, and the power delivered is divided into load R_b at port ② and load R_c at port ③, as shown in Fig. 5.10(*a*). In practical applications of Fig. 5.9(*b*), a very small amount of the power excited at port ① reflects into port ④, and an extremely small amount of the power delivered at port ③ travels into port ④. To isolate port ④ from the power excited at port ①, the two waves should be out of phase and the ratio of $|S_{41}|$ to $|S_{43}|$ should be d_2 to d_1 [22]. For the asymmetric ring hybrid in Fig. 5.9(*a*), an isolated port ③ is placed between ports ② and ④. Therefore, the two output signals are either in phase or out of phase. In the case of the CABH, if port ④ must be isolated from the power excited at port ①, the phase difference $(\Theta_1 + \Theta_2 + \Theta_3 - \Theta_4)$ should be $\pm$ 180°. To transform a real impedance into another real impedance, the length of the impedance transformer line is 90° or odd multiples of 90°. Since all the transmission-line sections in Fig. 5.9 have the function of impedance transforming, all their lengths should be 90° or odd multiples of 90°, so that Θ_2 in Fig. 5.10(*a*) is naturally 90° or odd multiples of 90°. Thus, the signals at ports ② and ③ are 90° or odd multiples of 90° out of phase. The characteristic impedances of the CABH in Fig. 5.9(*b*) are, assuming that $\Theta_1 = \Theta_2 = \pi/2$ (and $|S_{41}| = 0$), obtained as

$$Z_1 = \sqrt{\frac{d_1^2}{d_1^2 + d_2^2}}\sqrt{R_a R_b}, \tag{5.32a}$$

$$Z_2 = \frac{d_1}{d_2}\sqrt{R_b R_c}. \tag{5.32b}$$

To satisfy the assumed condition of $|S_{41}| = 0$, the part removed in Fig. 5.10(*a*) must be connected. Due to reciprocity, the circuit in Fig. 5.11(*a*) may be considered as being excited at port ④, and the simplified equivalent circuit is that shown in Fig. 5.11(*b*). The characteristic impedances of Z_3 and Z_4 are derived similarly as

$$Z_3 = \sqrt{\frac{d_1^2}{d_1^2 + d_2^2}}\sqrt{R_c R_d}, \tag{5.33a}$$

$$Z_4 = \frac{d_1}{d_2}\sqrt{R_d R_a}. \tag{5.33b}$$

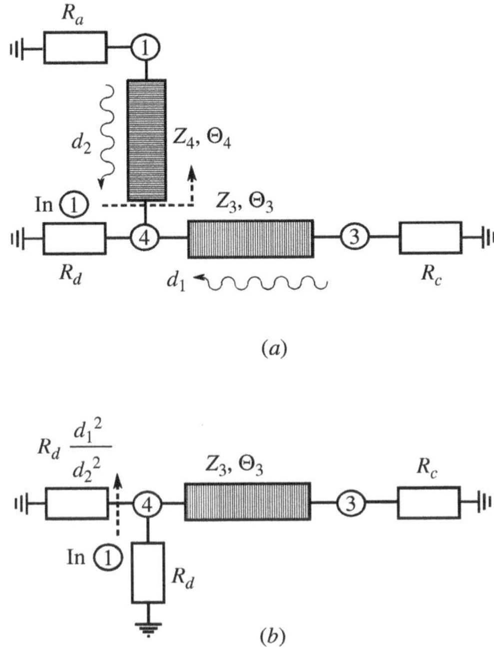

FIGURE 5.11 Remaining arms of Fig. 5.10: (*a*) circuit for isolation; (*b*) simplified circuit.

In (5.32) and (5.33), the ratio of d_1 to d_2 is the scattering parameter ratio between the two output ports in Fig. 5.9(*b*), and R_a, R_b, R_c, and R_d are real termination impedances. From the design equations in (5.32) and (5.33), for $R_a = R_b = R_c = R_d$ and $d_1 = d_2$, the results are those of the well-known 3-dB branch-line hybrid [24]. For $d_1 = d_2$, the results are equal to those reported by Ahn and Wolff [38,39]. If $R_a = R_d = Z_{01}$, $R_b = R_c = Z_{02}$, and the coupling factor is defined as $|S_{31}|/|S_{21}| = d_2/d_1 = k$, the results are equal to those in (5.29), showing that the asymmetric branch-line hybrid can be used as an impedance transformer like that in Fig. 5.7.

On the basis of the design equations derived, an asymmetric hybrid terminated in $R_a = 50\ \Omega$, $R_b = 41.67\ \Omega$, $R_c = 55.56\ \Omega$, and $R_d = 62.5\ \Omega$ with $d_1 = d_2$ was simulated assuming ideal transmission lines, and the results are as plotted in Fig. 5.12. The power division and isolation characteristics are plotted in Fig. 5.12(*a*), matchings in Fig. 5.12(*b*), and phase responses in Fig. 5.12(*c*). When power is fed into port ①, the simulation results are $|S_{11}| = -155.9$ dB, $|S_{21}| = -3.01$ dB, $|S_{31}| = -3.01$ dB, and $|S_{41}| = -151.820$ dB at a center frequency of 3 GHz. The reflected scattering parameters are $|S_{11}| = -152.82$ dB, $|S_{22}| = -146.659$ dB, $|S_{33}| = -144.122$ dB, and $|S_{44}| = -140.840$ dB, showing that all ports are perfectly matched at a design frequency of 3 GHz.

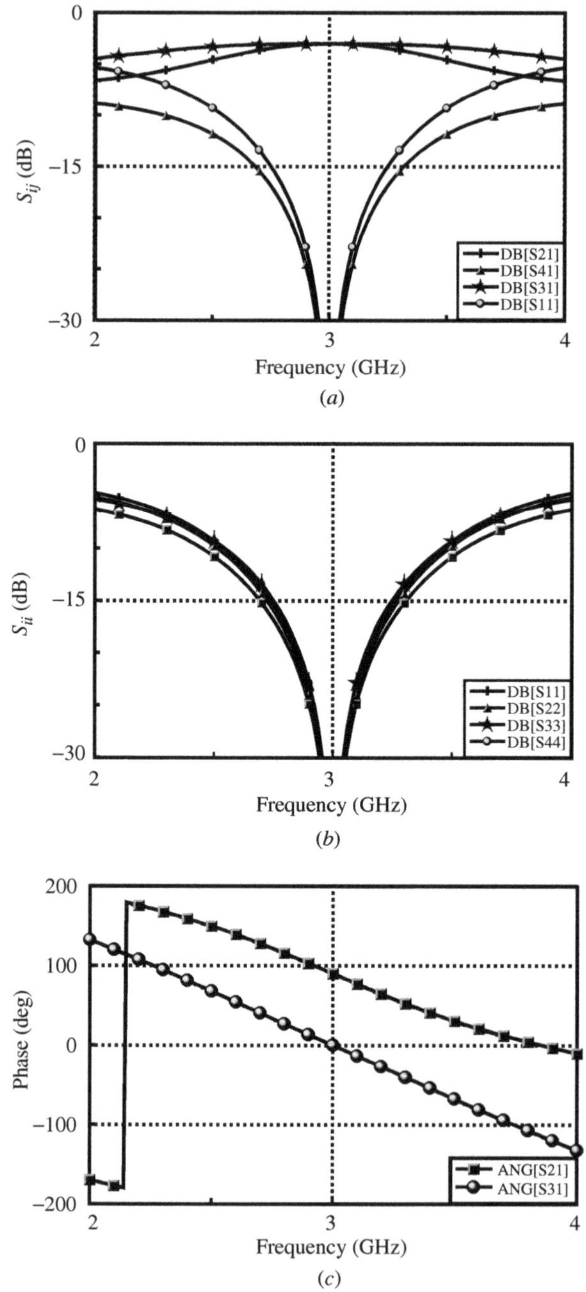

FIGURE 5.12 Simulation of a CA3BH with $R_a = 50\ \Omega$, $R_b = 41.67\ \Omega$, $R_c = 55.56\ \Omega$, $R_d = 62.5\ \Omega$, and $d_1 = d_2$: (*a*) scattering parameters with excitation at port ①; (*b*) reflected scattering parameters; (*c*) phase frequency responses.

TABLE 5.1 Fabrication Data for a CABH with a Split Ratio of 3-dB Power and $\Theta_1 = \Theta_2 = \Theta_3 = \Theta_4 = 90^\circ$ at a Design Frequency of 3 GHz[a]

Termination Impedance	Coplanar Feeding Transformer Line	Coplanar Branch Transmission Line
$R_a = 30\ \Omega$	$Z_{01} = 38.73\ \Omega$	$Z_1 = 34.63\ \Omega$
	$w = 710\ \mu m$	$w = 1049\ \mu m$
	$g = 109\ \mu m$	$g = 99\ \mu m$
	$l = 11{,}222\ \mu m$	$l = 11{,}268\ \mu m$
$R_b = 60\ \Omega$	$Z_{02} = 54.77\ \Omega$	$Z_2 = 69.20\ \Omega$
	$w = 1261\ \mu m$	$w = 289\ \mu m$
	$g = 482\ \mu m$	$g = 339\ \mu m$
	$l = 12{,}185\ \mu m$	$l = 11{,}001\ \mu m$
$R_c = 40\ \Omega$	$Z_{03} = 44.72\ \Omega$	$Z_3 = 36.50\ \Omega$
	$w = 512\ \mu m$	$w = 879\ \mu m$
	$g = 156\ \mu m$	$g = 104\ \mu m$
	$l = 10{,}879\ \mu m$	$l = 11{,}207\ \mu m$
$R_d = 50\ \Omega$	$Z_{04} = 50.00\ \Omega$	$Z_4 = 54.71\ \Omega$
	$w = 407\ \mu m$	$w = 216\ \mu m$
	$g = 174\ \mu m$	$g = 138\ \mu m$
	$l = 10{,}869\ \mu m$	$l = 10{,}681\ \mu m$

[a] w, Center strip width; g, gap width; l, line length

Uniplanar CABH On the basis of the design equations derived, a uniplanar CABH was fabricated on Al_2O_3 substrate ($\varepsilon_r = 9.9$, $h = 635\ \mu m$) with coplanar waveguide technology. It was terminated in 30 Ω, 60 Ω, 40 Ω, and 50 Ω, and the power-split ratio was 3 dB. The fabrication data are given in Table 5.1, where Z_{01}, Z_{02}, Z_{03}, and Z_{04} are the characteristic impedances of impedance transformers to transform termination impedances into 50 Ω. The layout of the CABH is given in Fig. 5.13(*a*), showing that the widths of the transmission-line sections differ from each other. The power division and isolation are plotted in Fig. 5.13(*b*), and the reflected scattering parameters are given in Fig. 5.13(*c*). $|S_{21}|$ is predicted to be −1.764 dB from the calculation of $10 \log[d_1^2/(d_1^2 + d_2^2)]$, but the $|S_{21}|$ measured is −1.9259 dB. The difference between predicted and measured results comes from the losses from transmission lines, dielectric material, connectors, and so on. Since the uniplanar CABH is designed with a power-split ratio of 3 dB, the predicted and measured $|S_{31}|$ values are −4.764 dB and −4.957 dB, respectively. Measured isolations are $|S_{41}| = -34.22$ dB and $|S_{23}| = -36.62$ dB, and reflection coefficients are $|S_{11}| = -31.53$ dB, $|S_{22}| = -22.90$ dB, $|S_{33}| = -29.92$ dB, and $|S_{44}| = -29.26$ dB at a center frequency of 3 GHz.

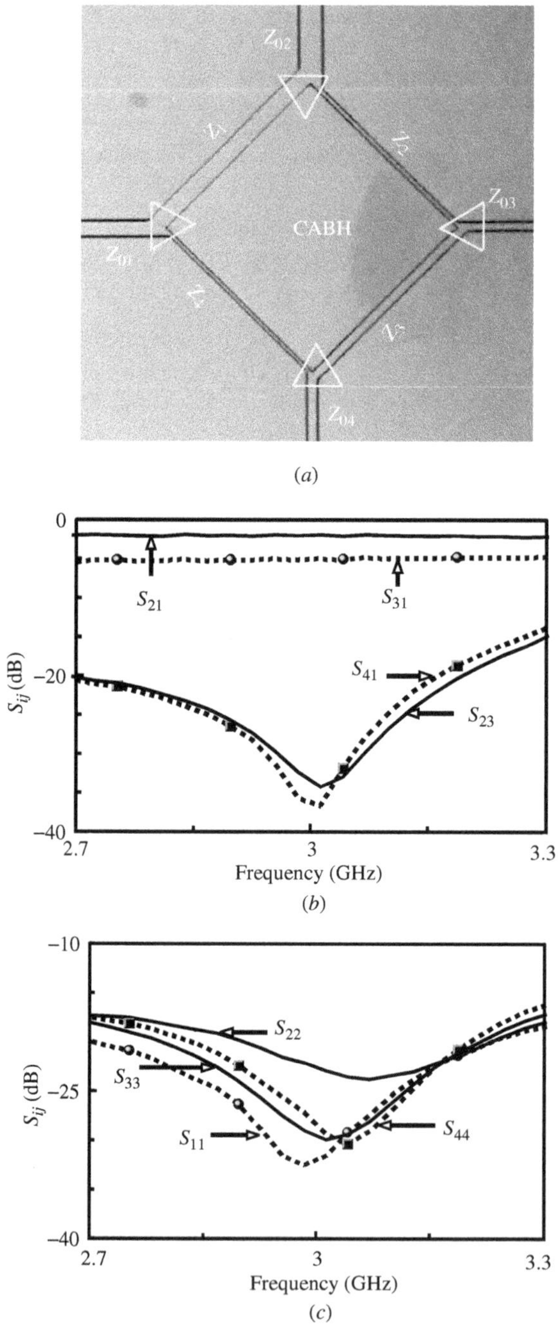

FIGURE 5.13 CABH with a power-split ratio of 3 dB, $R_a = 30\ \Omega$, $R_b = 60\ \Omega$, $R_c = 40\ \Omega$, and $R_d = 50\ \Omega$: (*a*) layout; (*b*) results for the power division and isolations; (*c*) results for return losses at all ports.

5.5.3 Anti-Conventional-Direction Asymmetric Branch-Line Hybrids

For the AABH in Fig. 5.9(c), if port ② is terminated in a reflection coefficient Γ_2, the scattering parameters with excitation at port ① are written as

$$\begin{bmatrix} b_1 \\ b_3 \\ b_4 \end{bmatrix} = \begin{bmatrix} S_{11} & S_{13} & S_{14} \\ S_{31} & S_{33} & S_{34} \\ S_{41} & S_{43} & S_{44} \end{bmatrix} \begin{bmatrix} a_1 \\ 0 \\ 0 \end{bmatrix} + \frac{S_{21}a_1\Gamma_2}{1-\Gamma_2 S_{22}} \begin{bmatrix} S_{12} \\ S_{32} \\ S_{42} \end{bmatrix}. \tag{5.34}$$

In (5.34), if $|S_{21}| = 0$ is assumed, the AABH may again be characterized as shown in Fig. 5.14(a). The characteristic impedances Z_3 and Z_4 in Fig. 5.9(c) are derived similarly as

$$Z_3 = \frac{d_1}{d_2}\sqrt{R_c R_d}, \tag{5.35a}$$

$$Z_4 = \sqrt{\frac{d_1^2}{d_1^2 + d_2^2}}\sqrt{R_d R_a}. \tag{5.35b}$$

The part removed in Fig. 5.14(a) may be considered the same as that in Fig. 5.14(b) and the characteristic impedances Z_1 and Z_2, similarly, are

$$Z_1 = \frac{d_1}{d_2}\sqrt{R_a R_b}, \tag{5.36a}$$

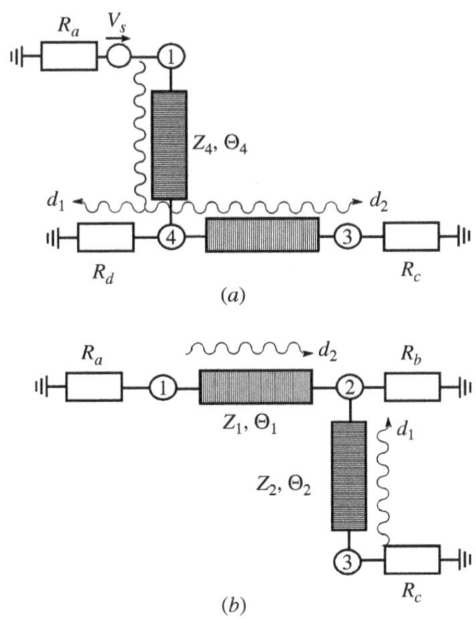

FIGURE 5.14 Excitation at port ① of Fig. 5.9(c): (a) under the assumption that $|S_{21}| = 0$; (b) remaining arms of the AABH.

$$Z_2 = \sqrt{\frac{d_1^2}{d_1^2 + d_2^2}}\sqrt{R_b R_c}. \tag{5.36b}$$

Uniplanar AABH To prove that the design equations derived are applicable, a uniplanar AABH was designed at a center frequency of 3 GHz and fabricated on Al_2O_3 substrate ($\varepsilon_r = 9.9$, $h = 635$ μm) with coplanar technology. Table 5.2 gives the design data. The AABH was terminated in 50 Ω, 41.6 Ω, 55.5 Ω, and 62.5 Ω, and its power-split ratio was 0 dB (equal power division). Since it is terminated in arbitrary impedances, impedance transformers are needed for a general measurement system; Z_{01}, Z_{02}, Z_{03}, and Z_{01} in Table 5.2 are the impedance transformers. The circuit layout is shown in Fig. 5.15, where Z_1, Z_2, Z_3, and Z_4 are the branch-line impedances. Figure 5.16 compares the measured results with predictions for the power divisions [Fig. 5.16(*a*)], isolations [Fig. 5.16(*b*)], and reflection coefficients [Fig. 5.16(c)].

Table 5.3 gives design examples for a CABH and an AABH. Both have termination impedances of 50 Ω, 41.68 Ω, 55.56 Ω, and 62.5 Ω and a power-split ratio of 0 dB. Nevertheless, Table 5.3 shows that their design data differ depending on the power division direction, even if the resulting performances are the same (equal power split and the same termination impedances).

TABLE 5.2 Design Data for an AABH with an Equal Power-Split Ratio and $\Theta_1 = \Theta_2 = \Theta_3 = \Theta_4 = 90°$ at a Design Frequency of 3 GHz[a]

Termination Impedance	Coplanar Feeding Transformer Line	Coplanar Branch Line
Port ①;	$Z_{01} = 50$ Ω	$Z_1 = 45.64$ Ω
50 Ω	$w = 516$ μm	$w = 670$ μm
	$g = 205$ μm	$g = 202$ μm
	$l = 11{,}140$ μm	$l = 11{,}174$ μm
Port ②;	$Z_{02} = 45.64$ Ω	$Z_2 = 34.02$ Ω
41.6 Ω	$w = 686$ μm	$w = 1124$ μm
	$g = 200$ μm	$g = 98$ μm
	$l = 11{,}200$ μm	$l = 11{,}196$ μm
Port ③;	$Z_{03} = 52.7$ Ω	$Z_3 = 58.92$ Ω
55.5 Ω	$w = 422$ μm	$w = 282$ μm
	$g = 205$ μm	$g = 206$ μm
	$l = 11{,}083$ μm	$l = 10{,}986$ μm
Port ④;	$Z_{04} = 55.9$ Ω	$Z_4 = 39.53$ Ω
62.5 Ω	$w = 340$ μm	$w = 596$ μm
	$g = 205$ μm	$g = 101$ μm
	$l = 11{,}030$ μm	$l = 10{,}977$ μm

[a] w, Center strip width; g, gap width; *l*, line length.

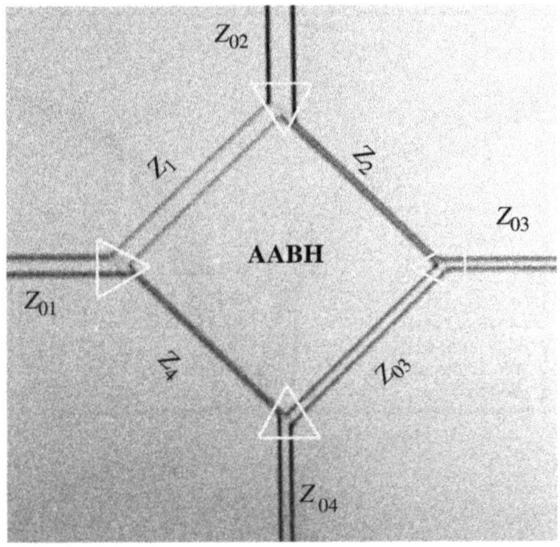

FIGURE 5.15 Layout of a 3-dB AABH.

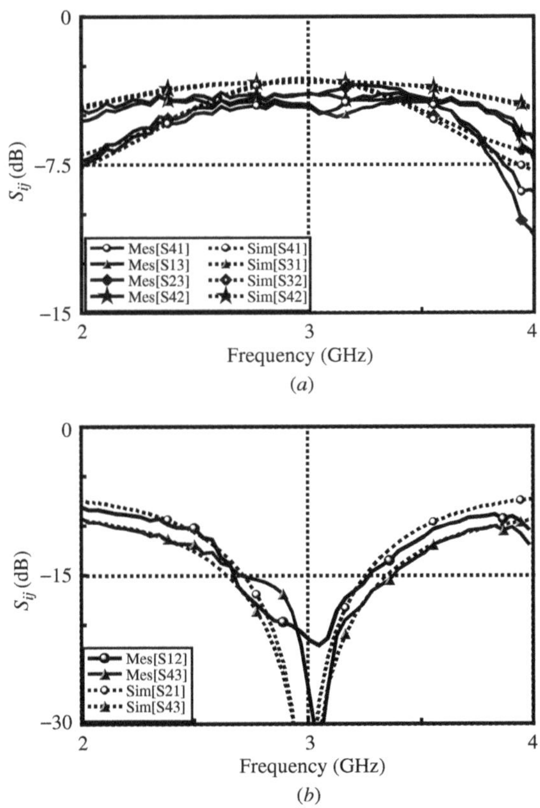

FIGURE 5.16 Measured and predicted results for an AABH with an equal power division and termination impedances of 50 Ω, 41.68 Ω, 55.56 Ω, and 62.5 Ω: (*a*) power division responses; (*b*) isolation characteristics; (*c*) reflection coefficients.

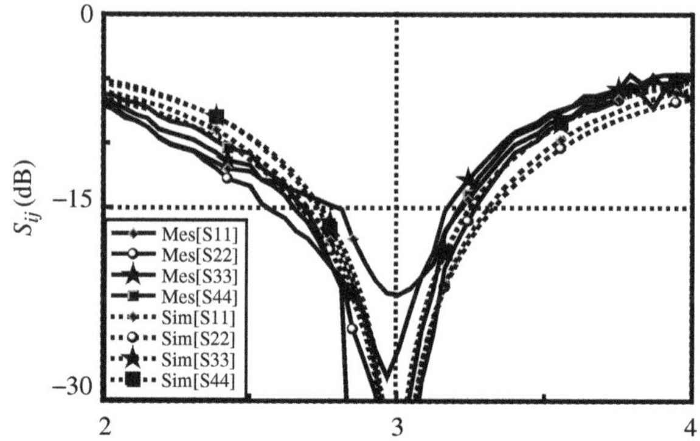

(*c*)

FIGURE 5.16 (*continued*)

TABLE 5.3 Comparison of CABH and AABH Data with the Same Termination Impedances and Power-Split Ratio

Termination: Impedance:	Port ① 50 Ω	Port ② 41.6 Ω	Port ③ 55.5 Ω	Port ④ 62.5 Ω
	Characteristic Impedances of Branch Lines (Ω)			
	Z_1	Z_2	Z_3	Z_4
CABH	32.275	48.113	41.67	55.90
AABH	45.644	34.021	58.93	39.53
	Characteristic Impedances of Transformer Lines (Ω)			
	Z_{01}	Z_{02}	Z_{03}	Z_{04}
CABH	50	45.64	45.64	55.9
AABH	50	45.64	45.64	55.9

EXERCISES

5.1 Derive (5.2).

5.2 Derive (5.19).

5.3 For the three-branch-line coupler in Fig. 5.4, determine specific values of Y_a and Y_c for the 50-Ω termination impedances.

5.4 Design the branch-line hybrid in Fig. 5.5 when $Z_{01} = 30\ \Omega$, $Z_{02} = 50\ \Omega$, and $k = 2$.

5.5 Design the branch-line hybrid in Fig. 5.7 when $Z_{01} = 30\ \Omega$, $Z_{02} = 50\ \Omega$, and $k = 2$.

5.6 Plot and compare the simulation results of Exercises 5.4 and 5.5.

5.7 For asymmetric branch-line hybrids:

(a) Design a CABH and an AABH in Fig. 5.9(*b*) and (*c*) when $R_a = R_d = 30\ \Omega$, $R_b = R_c = 50\ \Omega$, and $d_1/d_2 = 2$.

(b) Compare the results of Exercises 5.4, 5.6 and 5.7(a).

REFERENCES

1. W. W. Mumford, Directional Couplers, *Proc. IRE*, Vol. 35, February 1947, pp. 159–165.
2. R. Levy, Analysis of Practical Branch-Guide Directional Couplers, *IEEE Trans. Microwave Theory Tech.* Vol. 17, May 1969, pp. 289–290.
3. W. A. Tyrrel, Hybrid Circuits for Microwaves, *Proc. IRE*, Vol. 35, November 1947, pp. 1294–1306.
4. T. Morita and L. S. Sheingold, A Coaxial Magic-T, *IRE Trans. Microwave Theory Tech.*, Vol. 1, November, 1953, pp. 17–23.
5. V. I. Albanese and W. P. Peyser, An Analysis of a Broad-Band Coaxial Hybrid Ring, *IRE Trans. Microwave Theory Tech.*, Vol. 6, October 1958, pp. 869–373.
6. W. V. Tyminski and A. E. Hylas, A Wide-Band Hybrid Ring for UHF, *Proc. IRE*, Vol. 41, January 1953, pp. 81–87.
7. S. March, Wideband Stripline Hybrid Ring, *IEEE Trans. Microwave Theory Tech.*, Vol. 16, June 1968, pp. 361–362.
8. C. Y. Pon, Hybrid-Ring Directional Coupler for Arbitrary Power Divisions, *IRE Trans. Microwave Theory Tech.*, Vol. 9, November 1961, pp. 529–535.
9. T. Hirota, Y. Tarusawa, and H. Ogawa, Uniplanar MMIC Hybrids: A Proposed New MMIC Structure, *IEEE Trans. Microwave Theory Tech.*, Vol. 35, June 1987, pp. 576–581.
10. C.-H. Ho, L. Fan, and K. Chang, Broad-Band Uniplanar Hybrid-Ring and Branch-Line Couplers, *IEEE Trans. Microwave Theory Tech.*, Vol. 41, December 1993, pp. 2116–2124.
11. C.-H. Ho, L. Fan, and K. Chang, New Uniplanar Coplanar Waveguide Hybrid-Ring Couplers and Magic-T's, *IEEE Trans. Microwave Theory Tech.*, Vol. 42, December 1994, pp. 2440–2448.
12. J. Reed, The Multiple Branch Waveguide Coupler, *IRE Trans. Microwave Theory Tech.*, Vol. 6, October 1958, pp. 398–403.
13. G. Matthaei, L. Young, and E. M. T. Jones, *Microwave Filters, Impedance-Matching Networks, and Coupled Structures*, Artech House, Dedham, MA, 1980, p. 816.
14. R. K. Gupta, S. E. Anderson, and W. Getsinger, Impedance-Transforming 3-dB Hybrids, *IEEE Trans. Microwave Theory Tech.*, Vol. 35, December 1987, pp. 1303–1307.
15. S. Kumar, C. Tannous, and T. Danshin, A Multisection Broadband Impedance Transforming Branch-Line Hybrid, *IEEE Trans. Microwave Theory Tech.*, Vol. 43, November 1995, pp. 2517–2523.
16. H. J. Riblet, Mathematical Theory of Directional Couplers, *Proc. IRE*, Vol. 35, December 1947, pp. 1307–1313.

17. R. Levy and L. F. Lind, Synthesis of Symmetrical Branch-Guide Directional Couplers, *IEEE Trans. Microwave Theory Tech.*, Vol. 4, February 1968, pp. 80–89.
18. L. F. Lind, Synthesis of Asymmetrical Branch-Guided Directional Coupler-Impedance Transformers, *IEEE Trans. Microwave Theory Tech.*, Vol. 17, January 1969, pp. 45–48.
19. H.-R. Ahn, I.-S. Chang, and S.-W. Yun, Miniaturized 3-dB Ring Hybrid Terminated by Arbitrary Impedances, *IEEE Trans. Microwave Theory Tech.*, Vol. 42, December 1994, pp. 2216–2221.
20. H.-R. Ahn, I. Wolff, and I.-S. Chang, Arbitrary Termination Impedances, Arbitrary Power Division and Small-Sized Ring Hybrids, *IEEE MTT-S Dig.*, June 1997, pp. 285–288.
21. H.-R. Ahn, I. Wolff, and I.-S. Chang, Arbitrary Termination Impedances, Arbitrary Power Division and Small-Sized Ring Hybrids, *IEEE Trans. Microwave Theory Tech.*, Vol. 44, December 1997, pp. 2241–2247.
22. H.-R. Ahn and I. Wolff, Asymmetric Four-Port and Branch-Line Hybrids, *IEEE Trans. Microwave Theory Tech.*, Vol. 48, September 2000, pp. 1585–1588.
23. J. Reed and G. J. Wheeler, A Method of Analysis of Symmetrical Four-Port Networks, *IRE Trans. Microwave Theory Tech.*, Vol. 4, October 1956, pp. 346–352.
24. J. A. G. Malherbe, *Microwave Transmission Line Couplers*, Artech House, Dedham, MA, 1988, p. 23.
25. H.-R. Ahn and I. Wolff, Three-Port 3-dB Power Divider Terminated in Different Impedances and Its Application to MMIC's, *IEEE Trans. Microwave Theory Tech.*, Vol. 47, June 1999, pp. 786–794.
26. H.-R. Ahn and I. Wolff, Three-Port 3-dB Power Divider Terminated in Arbitrary Impedances, *IEEE MTT-S Dig.*, Baltimore, June 1998, pp. 781–784.
27. H.-R. Ahn and I. Wolff, General Design Equations of Three-Port Power Dividers, *IEEE MTT-S Dig.*, Boston, June 2000, pp. 1137–1140.
28. H.-R. Ahn and I. Wolff, General Design Equations of Three-Port Power Dividers, Small-Sized Impedance Transformers, and Their Applications to Small-Sized Three-Port 3-dB Power Dividers, *IEEE Trans. Microwave Theory Tech.*, Vol. 49, July 2001, pp. 1277–1288.
29. E. J. Wilkinson, An *n*-Way Hybrid Power Divider, *IRE Trans. Microwave Theory Tech.*, Vol. 8, January 1960, pp. 116–118.
30. L. I. Parad and R. L. Moynihan, Split-Tee Power Divider, *IRE Trans. Microwave Theory Tech.*, Vol. 8, January 1965, pp. 91–95.
31. S. B. Cohn, A Class of Broadband Three-Port TEM-Mode Hybrids, *IRE Trans. Microwave Theory Tech.*, Vol. 16, February 1968, pp. 110–116.
32. R. B. Ekinge, A New Method of Synthesizing Matched Broad-Band TEM-Mode Three-Ports, *IEEE Trans. Microwave Theory Tech.*, Vol. 19, January 1971, pp. 81–88.
33. S. Rosloniec, Three-Port Hybrid Power Dividers Terminated in Complex Frequency-Dependent Impedances, *IEEE Trans. Microwave Theory Tech.*, Vol. 44, August 1996, pp. 1490–1493.
34. H. Hayashi, H. Okazaki, A. Kanda, T. Hirota, and M. Muraguch, Millimeter-Wave-Band Amplifier and Mixer MMIC's Using a Broad-Band 45° Power Divider/Combiner, *IEEE Trans. Microwave Theory Tech.*, Vol. 46, June 1998, pp. 811–818.

35. B. Kopp, Asymmetric Lumped Element Power Splitters, *IEEE MTT-S Dig.*, 1989, pp. 333–336.
36. D. Köther, B. Hopf, T. Sporkmann, and I. Wolff, MMIC Wilkinson Couplers for Frequencies Up to 110 GHz, *IEEE MTT-S Dig.*, 1995, pp. 663–665.
37. H.-R. Ahn, Comment on Converting Baluns into Broad-Band Impedance-Transforming 180° Hybrids, *IEEE Trans. Microwave Theory Tech.*, Vol. 52, January 2004, pp. 228–230.
38. H.-R. Ahn and I. Wolff, 3-dB Branch-Line Hybrids with Arbitrary Termination Impedance Values, *IEICE Trans. Electrons*, Vol. E82-C, No. 7, July 1999, pp. 1324–1326.
39. H.-R. Ahn and I. Wolff, 3-dB Branch-Line Hybrid Terminated in Arbitrary Impedances, *IEE Electron. Lett.*, Vol. 34, No. 11, May 1998, pp. 1109–1110.

CHAPTER SIX

Conventional Three-Port Power Dividers

6.1 INTRODUCTION

A large number of different types of power dividers, with and without isolation between the output ports, are used for various applications. Power dividers that provide isolation between their output ports have either quadrature outputs or in-phase or out-of-phase outputs. A special class of couplers, the hybrids, which include ring hybrids [1–8] and three-port power dividers [9–13], provide in-phase or out-of-phase responses. Those that provide quadrature outputs are branch-line hybrids [14–17]. Four-port power dividers, ring hybrids, and branch-line hybrids have been discussed extensively in previous chapters.

The use of three-port hybrids is especially important for array antenna systems that utilize a power-splitting network. The powers from a large number of semiconductor devices, each giving only a small amount of power at microwave frequencies, are conveniently added by such a tree of three-port hybrids. The history of the three-port power divider began in 1960 when Wilkinson [18] described a device that separated one signal into n equiphase–equiamplitude signals. Theoretically perfect isolation between all output ports was achieved at one frequency. In 1965, Parad and Moynihan [19] presented a hybrid with arbitrary amplitude difference of the output signals. A perfect three-port hybrid property was again achieved at one frequency. In 1968, Cohn [20] presented a class of equal-power dividers with isolation and matching at any number of frequencies. In 1971, Ekinge [21] described three-port hybrids, which consist of n sections in cascade, where each section is composed of two coupled lossless transmission lines of electrical length Θ and an intermediate resistor. His analysis seems to be similar to that of Cohn. However, Cohn treated the equal-power-split three-port hybrid, whereas Ekinge described the arbitrary-power-split three-port hybrid.

Since that time, studies on three-port hybrids have continued [22–25]. However, they now focus on symmetrical structures that have available mirror symmetries. Recently, Ahn and Wolff [10] developed a new design method for asymmetric three-port hybrids and a perfect isolation condition which was not analyzed completely Refs. in 19,21, and 22. The asymmetric three-port power dividers are treated in Chapters 7 and 8; in this chapter we cover only conventional three-port power dividers.

6.2 THREE-PORT 3-dB POWER DIVIDERS

The problem of dividing an input signal into a number of equiphase–equiamplitude output signals is a familiar one to radio-frequency engineers, particularly those working in the field of array antennas. It is often desired that the equiphase–equiamplitude condition be obtained in a manner that is fundamentally independent of frequency, and, in addition, that a fairly high degree of isolation exists between the various output terminals over a specified frequency band.

Wilkinson [18] described a device that separated one signal into n equiphase–equiamplitude signals ($n \geq 2$). With $n = 2$, his circuit may be reduced to a three-port 3-dB power divider terminated in equal impedances. Such a device is depicted in Fig. 6.1(a). Power entering the shunt port emerges with equal amplitude and phase at the other two ports. Each of the three ports has a nearly unitary voltage standing-wave ratio (VSWR), and isolation of the output ports is high. In addition to its application as a power divider, reciprocity allows this type of hybrid junction to function as a power combiner of two signals. Simulation results are plotted in Fig. 6.1(b) and (c). For this simulation, the termination impedances are all unity and the characteristic impedance of the transmission lines is thus $Z_1 = \sqrt{2}$. Also, the electrical length of the transmission line is 90° at a frequency of 3 GHz. The simulated isolation and power division are plotted in Fig. 6.1(b), and the reflection coefficients (VSWRs) are shown in Fig. 6.1(c). In the band range where the VSWR value is less than 1.4 : 1, the performance is quite good, but at 2 : 1 VSWR bands, isolation is only 14.7 dB and the input port VSWR is 1.42. Its performance, although excellent, may be improved by adding a quarter-wave transformer in front of a T-junction. Another way to improve the performance of a three-port power divider was suggested by Cohn [20]. His design contains a multiplicity of cascaded pairs of transmission-line sections and interconnecting resistors. Compared to the earlier designs [18,19], an enormous improvement in VSWR and isolation is obtained over a given bandwidth, even when two pairs of lines and resistors are used. The bandwidth capability improves without limit as the number of line sections and resistors is increased.

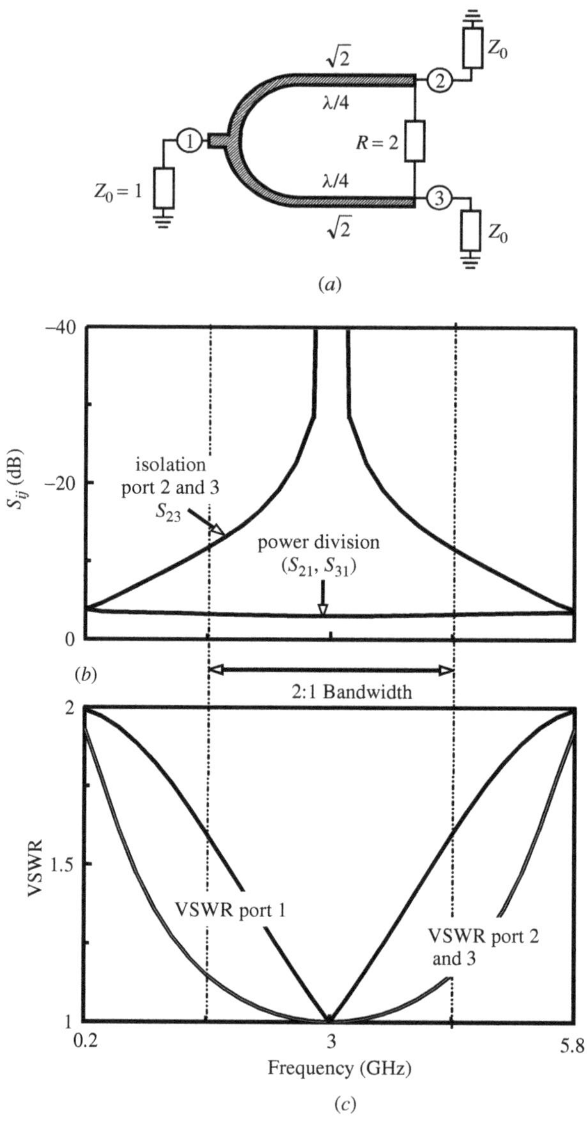

FIGURE 6.1 Basic three-port hybrid and its frequency responses: (*a*) basic three-port power divider; (*b*) isolation and power division responses; (*c*) reflection coefficients.

6.3 THREE-PORT POWER DIVIDERS WITH ARBITRARY POWER DIVISIONS

In the design of a microwave distributed network, a power divider providing two different output signals is often required. A ring hybrid may be used for this purpose. However, the magnitude and phase of its output signals vary by about 1.5 dB and 4° over a 1.5 : 1 band. The three-port power divider with arbitrary power divisions suggested by Parad and Moynihan [19] is, when constructed in a microstrip or coplanar-waveguide structure, a simple, compact, and broadband device. It provides two isolated signals with unequal amplitudes, and each output port is matched. It is similar to an N-way power divider, which provides N equiphase, equiamplitude, and isolated ports. In fact, the power divider with arbitrary power divisions may be developed from an N-way power divider as follows: Connect n of the output ports together to form one port, connect the remaining $N - n$ output ports together to form the other port, and connect quarter-wavelength transformers to the two resulting output ports to adjust their impedance level. A power divider with two equiphase output signals and a power ratio of n to $N - n$ is thus derived.

The detailed configuration is depicted in Fig. 6.2, where the length of the transmission-line sections is Θ and an isolation resistor is connected between ports ② and ③. When power is fed into port ① in Fig. 6.2, circuits equivalent to those in Fig. 6.3 can be derived. Figure 6.3(*a*) is drawn in terms of ports ①, ②, and ③, and Fig. 6.3(*b*) includes only port ①. The divider is to be designed so that all ports are matched and the power division ratio is K^2, as shown in Fig. 6.3(*a*). Assuming that the voltages at ports ② and ③ are equal when measured at equal distances from port ① and that the equivalent impedances looking into the transmission-line sections connected with ports ② and ③ are Z_{i2} and Z_{i3}, respectively, a relation for port ① matched is given as

$$\frac{Z_{i2}Z_{i3}}{Z_{i2} + Z_{i3}} = Z_0. \tag{6.1}$$

For the output power ratio at ports ② and ③, assuming that discontinuities caused by corners and junction effect are negligible, the relation between Z_{i2} and Z_{i3}, is given as

$$Y_{i2} : Y_{i3} = I_{i2} : I_{i3} = 1 : K^2, \tag{6.2}$$

where $Y_{i2} = 1/Z_{i2}$, $Y_{i3} = 1/Z_{i3}$, and I_{i2} and I_{i3} are the currents entering Z_{i2} and Z_{i3}, as shown in Fig. 6.3(*b*). The input impedances Z_{i2} and Z_{i3} are computed from (6.1) and (6.2) as

$$Z_{i2} = (1 + K^2)Z_0, \tag{6.3a}$$

$$Z_{i3} = \frac{1 + K^2}{K^2} Z_0. \tag{6.3b}$$

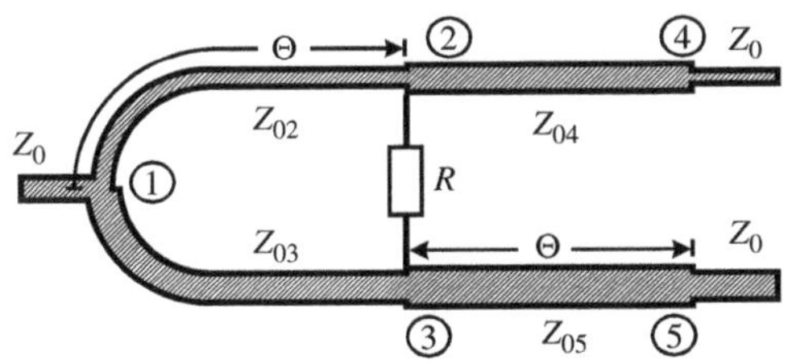

FIGURE 6.2 Three-port power divider with arbitrary power divisions.

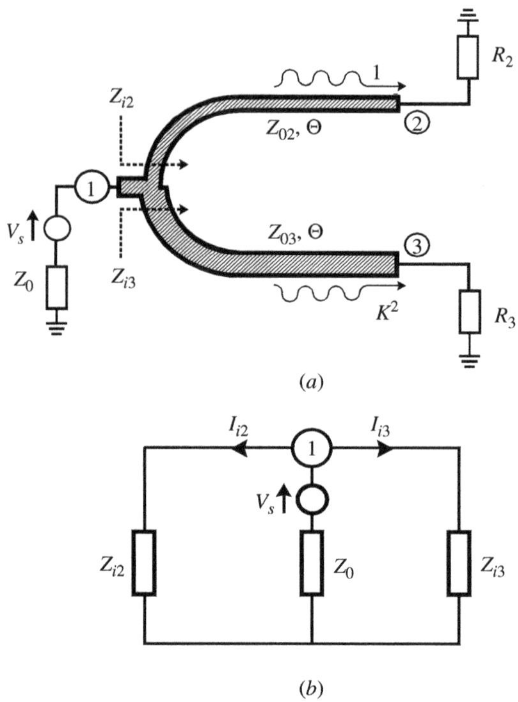

FIGURE 6.3 Equivalent circuits: (*a*) including ports ①, ②, and ③; (*b*) in terms of port ①.

Again, for an output power ratio of K^2, with $I_{i2}/I_{i3} = 1/K^2$, the ratio of R_2 to R_3 in Fig. 6.3(*a*) is, obtained as

$$R_2 : R_3 = K^2 : 1. \tag{6.4}$$

There are many possible ways to in choose one set of R_2 and R_3, and one case of $R_2 = KZ_0$ and $R_3 = Z_0/K$ is determined. The reason for the case is discussed in Section 8.2.2. The isolation resistance is equal to $R_2 + R_3$, and the transmission-line sections in Fig. 6.2 are, when their electrical length is 90°, just impedance transformers. Those between ports ① and ② and between ① and ③ transform the input impedances Z_{i2} and Z_{i3} into R_2 and R_3, respectively, and those with Z_{03} and Z_{04} transform R_2 and R_3 into the output impedances Z_0. Thus, from (6.1)–(6.4), the design equations are derived as

$$Z_{02} = Z_0\sqrt{K(1 + K^2)}, \tag{6.5a}$$

$$Z_{03} = Z_0\sqrt{\frac{1 + K^2}{K^3}}, \tag{6.5b}$$

$$Z_{04} = Z_0\sqrt{K}, \tag{6.5c}$$

$$Z_{05} = \frac{Z_0}{\sqrt{K}}, \tag{6.5d}$$

$$R = \frac{Z_0(1 + K^2)}{K}. \tag{6.5e}$$

In this design, the voltages at ports ② and ③ are equal. Hence, theoretically, a resistor may be placed across these two ports without causing any power dissipation. However, if the power is fed from port ② or ③, energy will be dissipated in the resistor. The resistor is indispensable for perfect matching at ports ② and ③ and perfect isolation between the two ports. The three-port power divider in Fig. 6.2 requires four consecutive quarter-wavelength transformers to obtain optimum performance, because the output ports are not terminated in the system impedance. If concerned only with size reduction, poorer isolation can be traded off for smaller size. In this case, the voltages at ports ② and ③ will not be equal, but all the termination impedances are equal to the system impedance level [24].

Consider the three-port power divider shown in Fig. 6.4, where k represents the percentage of power in the first arm. The first step for the design is to determine the values of the impedance levels for the two transmission lines. A relation between Z_{02} and Z_{03} can be determined from

$$k = \frac{Z_{03}}{Z_{02} + Z_{03}}. \tag{6.6}$$

For a three-port power divider perfectly matched at port ①, the relation in (6.1) holds and expressions relating the equivalent impedances Z_{i2} and Z_{i3} are written as

$$(Z_{02})^2 = Z_0 Z_{i2}, \tag{6.7a}$$

$$(Z_{03})^2 = Z_0 Z_{i3}. \tag{6.7b}$$

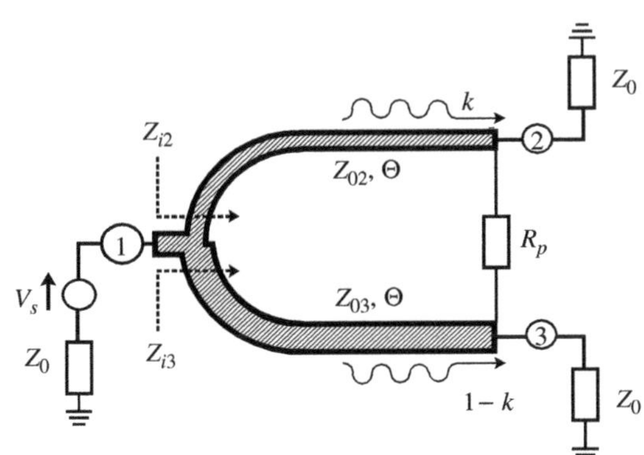

FIGURE 6.4 Three-port power divider with arbitrary power divisions.

The final design equations are given from (6.1), (6.6), and (6.7) as

$$Z_{03} = Z_0 \sqrt{\left[1 + \left(\frac{k}{1-k}\right)^2\right]}, \tag{6.8a}$$

$$Z_{02} = \frac{1-k}{k} Z_{03}, \tag{6.8b}$$

$$R_P = 2Z_0. \tag{6.8c}$$

In the design equations of (6.6)–(6.8), if the power-split ratio is 2, $k = \frac{1}{3}$. Isolation resistors should be approximated using (6.8c).

6.4 SYMMETRIC ANALYSES OF ASYMMETRIC THREE-PORT POWER DIVIDERS

A method of designing a broadband three-port power divider has been presented by Ekinge [21]. The three-port power divider consists of n sections in cascade, and each section is composed of two coupled lossless transmission lines of electrical length Θ and an intermediate resistor. It is analyzed by means of a convenient symmetrical four-port analysis with even- and/or odd-mode excitations [26], where a new definition of the odd mode is introduced. This definition simplifies the treatment of asymmetrical three- and four-ports considerably, with one half of the network identical to the other apart from an impedance scaling factor. The even-mode networks are identical to cascaded impedance transformers, and the odd-mode networks contain all isolation resistors.

Figure 6.5 is a schematic diagram of the circuit. It consists of coupled lossless transmission-line sections in cascade. All of the sections are of same length, reactive coupling is distributed along the sections, and resistive coupling is concentrated in one of the endpoints of the sections. The three-port power divider is conveniently split into equivalent even- and odd-mode two-port circuits, and its design is closely related to the design of the two circuits. The four-port circuit shown in Fig. 6.5 is used for even- and odd-mode analyses.

By definition, voltage sources V_t and V_b are equal in even-mode analysis (i.e., there should be zero current in the intermediate resistances). This is obtained if the voltage distributions on the top and bottom sides are identical in the x-direction. A magnetic wall can then be placed between the lines, and the two even-mode circuits are shown in Fig. 6.6(a) and (b). For the even-mode characteristic impedances $\{Z_{ek}^t\}_1^n$ and $\{Z_{ek}^b\}_1^n$ to satisfy the even-mode condition, the hypotheses are given as

$$Z_{ek}^b = k_e Z_{ek}^t, \qquad k = 1, 2, 3, \ldots, n \tag{6.9a}$$

$$\frac{R_d}{R_a} = \frac{R_c}{R_b} = k_e, \tag{6.9b}$$

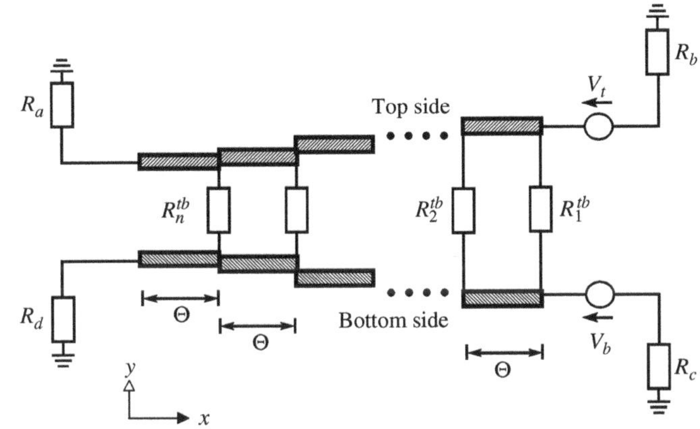

FIGURE 6.5 Four-port power divider consisting of n sets of coupled transmission lines in cascade.

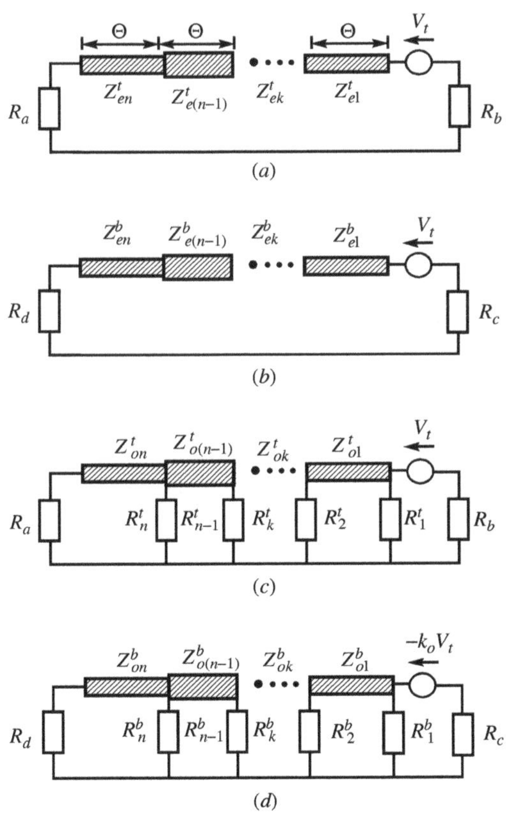

FIGURE 6.6 Equivalent two-port networks: (a, b) even-mode networks; (c, d) odd-mode networks.

where k_e is a positive real constant and the subscripts t and b indicate the top and bottom sides.

In the odd-mode analysis, the voltage sources V_t and V_b are by definition related by $V_b = -k_o V_t$, where k_o is a positive real constant. This definition is not standard, but a symmetrical four-port can be analyzed by the method defined above. It is assumed that the top and bottom sides have equal but opposite current distributions in the x-direction. An electric wall can then be placed between the two sides and the two odd-mode equivalent circuits are drawn in Fig. 6.6 (c) and (d). The impedance level distinguishes the two odd-mode networks from each other; that is,

$$Z^b_{ok} = k_o Z^t_{ok}, \qquad k = 1, 2, 3, \ldots, n \tag{6.10a}$$

$$R^b_k = k_o R^t_k, \tag{6.10b}$$

$$\frac{R_d}{R_a} = \frac{R_c}{R_b} = k_o, \tag{6.10c}$$

where $\{Z^t_{ok}\}^n_1$ and $\{Z^b_{ok}\}^n_1$ are the odd-mode characteristic impedances and the intermediate resistances are $\{R^{tb}\}^n_1 = \{R^t + R^b\}^n_1$. Equations (6.9$b$) and (6.10$c$) demand that $k_e = k_o = k$.

The four-port power divider terminated in asymmetrical impedances in Fig. 6.5 can be analyzed such that the top-side network is identical to the bottom-side network except for an impedance scaling factor k, as shown in Fig. 6.7. Thus, it is not symmetric, but even- and odd-mode excitation analyses can be used to solve the problems.

The actual hybrid is shown in Fig. 6.8(a). In the even mode, $V_t = V_b$, the hybrid in Fig. 6.8(a) is identical with that in Fig. 6.5 if (1) $R_a R_d/(R_a + R_d) = R_{ad}$, (2) (6.9$b$) is fulfilled, and (3) the elements of the hybrid satisfy (6.9a). In the odd mode, $V_b = -kV_t$, Fig. 6.8(a) is identical to Fig. 6.5 if (6.10a) and (6.10b) are satisfied and $R_a = R_d = 0$. Therefore, the even- and odd-mode networks of the three-port hybrid terminated in R_{ad}, R_b, and R_c are drawn in Fig. 6.8(b) and (c), where only the top-side networks are shown because the top and bottom sides are distinguished only by the impedance level. The reflected and transmitted scattering parameters can be obtained from these two equivalent networks, and

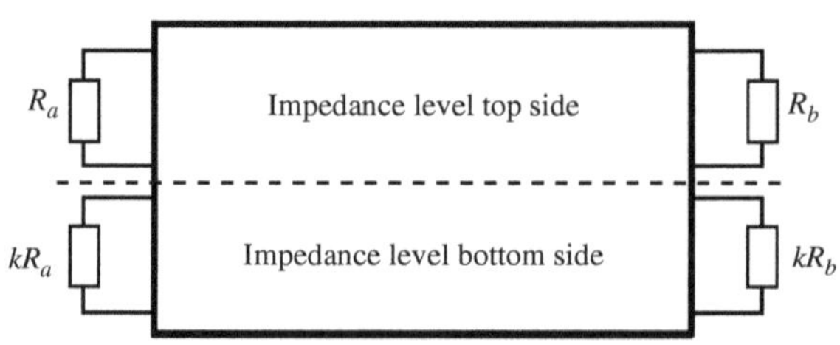

FIGURE 6.7 Four-port power divider terminated in various impedances.

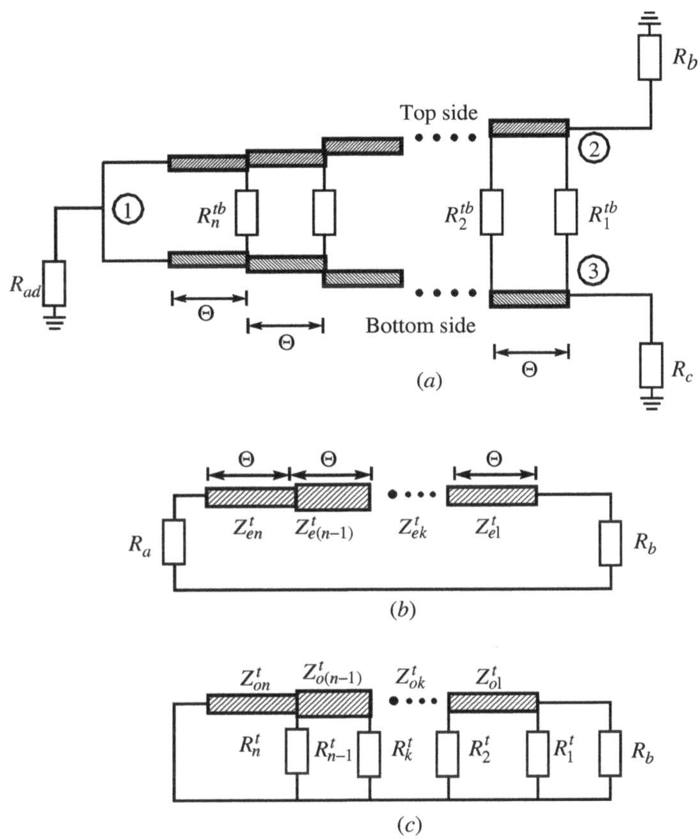

FIGURE 6.8 Three-port power divider terminated in asymmetric impedances: (*a*) actual three-port power divider; (*b*) even-mode equivalent circuit; (*c*) odd-mode equivalent circuit.

the final scattering parameters at each port are calculated by superposition as explained in Section 2.5. Additional details are given in the next section.

6.5 THREE-PORT 3-dB POWER DIVIDERS TERMINATED IN COMPLEX FREQUENCY-DEPENDENT IMPEDANCES

As microwave devices are used more at low and medium power levels, three-port power dividers are being used for a greater variety of applications. As a rule, these dividers are terminated in equal frequency–independent resistances, usually 50 Ω. Typical examples of such designs are described in the literature [18–20]. In contrast, algorithms have been presented for designing similar dividers whose ports are terminated asymmetrically [21]. In some applications, although it may be advantageous to apply complex impedances instead of frequency-independent constant resistances. For example, they are especially desirable for broadband equiphase array antennas. The design of broadband microwave dividers terminated in complex impedances is presented here, and an optimization algorithm for the design of a modified Wilkinson power divider is described.

The proposed power dividers are composed of transmission-line sections and lumped-element resistors. As the characteristic impedances of these transmission-line sections are limited on both sides by the assumed values $Z_{0\,\min}$ and $Z_{0\,\max}$, they may be realized easily [22]. Schematic diagrams of two- and four-section power dividers are shown in Figs. 6.9(*a*) and 6.10(*a*), respectively. These circuits have mirror-reflection symmetry (with respect to plane $S-S'$), so they can be analyzed by means of even- and odd-mode excitation analyses. The even- and odd-mode equivalent circuits are shown in Figs. 6.9(*b*) and (*c*) and 6.10(*b*) and (*c*). The scattering parameters of the power dividers may be expressed in terms of even- and odd-mode scattering parameters as

$$S_{11}(f) = S_{11e}(f), \tag{6.11a}$$

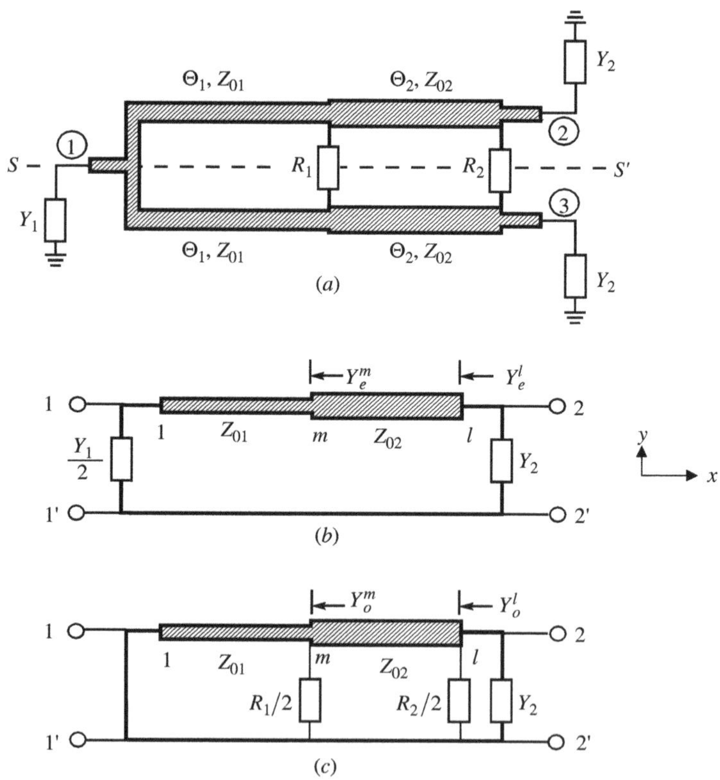

FIGURE 6.9 Two-section power divider: (*a*) overall divider structure; (*b*) even-mode equivalent circuit; (*c*) odd-mode equivalent circuit.

$$S_{12}(f) = S_{21}(f) = S_{13}(f) = S_{31}(f) = \frac{S_{12e}(f)}{\sqrt{2}}, \tag{6.11b}$$

$$S_{22}(f) = S_{33}(f) = \frac{S_{22e}(f) + S_{22o}(f)}{2}, \tag{6.11c}$$

$$S_{23}(f) = S_{32}(f) = \frac{S_{22e}(f) - S_{22o}(f)}{2}. \tag{6.11d}$$

It is evident that the even-mode equivalent circuits in Figs. 6.9(*b*) and 6.10(*b*) are impedance transformers transforming a complex admittance $Y_1(f)/2$ into $Y_2(f)$. From (6.11*a*), the reflection coefficient at port ①, $S_{11}(f)$, is determined unequivocally by only the even-mode scattering parameters. That is, $S_{11}(f)$ is determined by the complex admittances $Y_1(f)/2$, $Y_2(f)$, and the electrical parameters of the matching circuit connecting them. In other words, the $S_{11}(f)$ is independent of the isolation resistances, so it may be optimized first without any influence on the isolation resistors.

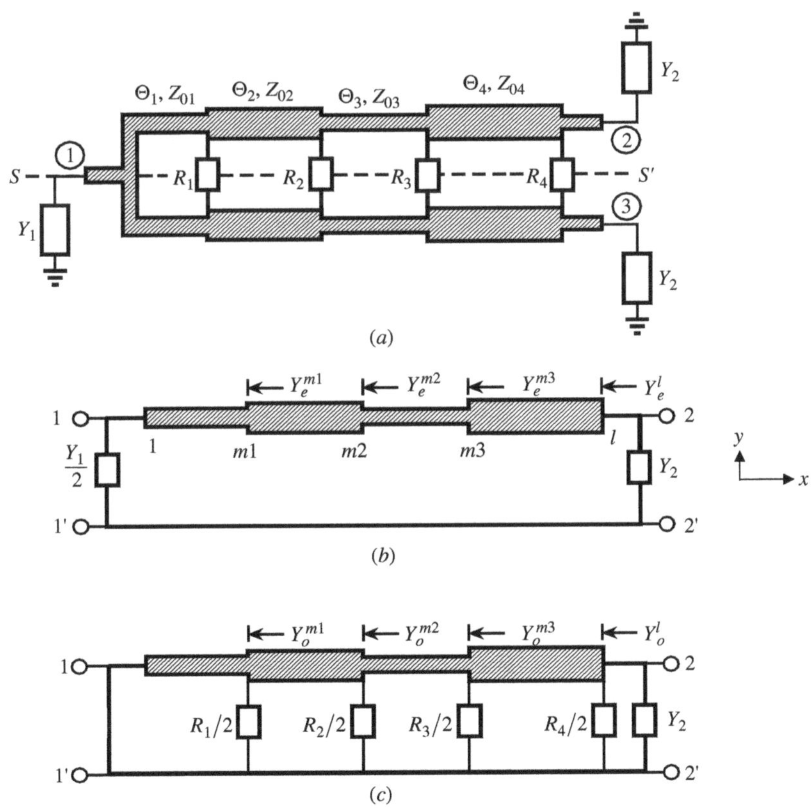

FIGURE 6.10 Four-section power divider: (*a*) overall divider structure; (*b*) even-mode equivalent circuit; (*c*) odd-mode equivalent circuit.

The next problem in the design process is to calculate the isolating resistors such that $S_{23}(f)$, $S_{22}(f)$, and $S_{33}(f)$ will be optimal. This optimization problem may be solved successfully numerically by using the minimization technique presented in (6.12):

$$|S_{23}(f)| = \frac{|[S_{22e}(f) - S_{22o}(f)]|}{2}, \tag{6.12}$$

where

$$S_{22e}(f) = -\frac{Y_e^p(f) - Y_2^*(f)}{Y_e^p(f) + Y_2(f)},$$

$$S_{22o}(f) = -\frac{Y_o^p(f) - Y_2^*(f)}{Y_o^p(f) + Y_2(f)},$$

and $Y_e^p(f)$ and $Y_o^p(f)$ are the even- and odd-mode admittances looking toward port ① at $x = p$, where p can be either l or m in Fig. 6.9. The admittance $Y_2^*(f)$ is the complex conjugate of the termination admittance of port ②. From (6.12), the scattering parameter $|S_{23}(f)|$ reaches its minimum when admittances $Y_e^p(f) = G_e^p(f) + jB_e^p(f)$ and $Y_o^p(f) = G_o^p(f) + jB_o^p(f)$ are equal to each other. Thus, to minimize the maximum value of the function $|S_{23}(f)|$, these admittances must be kept close over the frequency band required. That problem, formulated in mathematical terms, may be written as

$$\min_{G} \max_{f_1 \le f \le f_2} \frac{[G_e^p(f) - G_o^p(f)]^2 + [B_e^p(f) - B_o^p(f)]^2}{[G_e^p(f)]^2 + [B_e^p(f)]^2}, \tag{6.13}$$

where G is a vector of the lumped-element conductances or resistances being sought. The problem can be solved effectively using the two-stage optimization procedure that follows. First, a value of the conductance $G_1 = 1/R_1$ is chosen such that the admittances $Y_e^m(f) = G_e^m(f) + jB_e^m(f)$ and $Y_o^m(f) = G_o^m(f) + jB_o^m(f)$ will be as close as possible in the sense of the following criterion, (6.14), where $Y_e^m(f)$ and $Y_o^m(f)$ are the even- and odd-mode admittances looking into a line $x = m$ in Fig. 6.9:

$$\min_{G_1} \max_{f_1 \le f \le f_2} \frac{[G_e^m(f) - G_o^m(f)]^2 + [B_e^m(f) - B_o^m(f)]^2}{[G_e^m(f)]^2 + [B_e^m(f)]^2}. \tag{6.14}$$

Next, when the first approximation $G_1^{(1)}$ of the conductance G_1 is known, it is possible to evaluate the first approximation of the conductance $G_2 = 1/R_2$ in the same manner. Like problem (6.14), the one-dimensional optimization problem is

$$\min_{G_2} \max_{f_1 \le f \le f_2} \frac{[G_e^l(f) - G_o{}^l(f)]^2 + [B_e^l(f) - B_o^l(f)]^2}{[G_e^l(f)]^2 + [B_e^l(f)]^2}, \tag{6.15}$$

where $G_e^l(f)$, $B_e^l(f)$, $G_o^l(f)$, and $B_o^l(f)$ are the real and imaginary part of admittances $Y_e^l(f)$ and $Y_o^l(f)$, respectively. This first approximation of the vector G serves as a starting point for the second stage of the optimization procedure. Further minimization of the function $|S_{23}(f, G)|$ within a given frequency band can be continued by optimization methods.

The number of divider sections is evaluated on the basis of the return loss characteristic $S_{11}(f)$. If the return loss calculated for the two-section divider does not satisfy the requirements given, more sections are needed. For the four-stage divider in Fig. 6.10, the isolation resistors are determined in the same way. The characteristic impedances of the transmission-line sections are limited by extreme impedances $Z_{0\,\min}$ and $Z_{0\,\max}$, which are assumed freely at the beginning of the design process. The algorithm begins under the assumption that the return loss characteristic $S_{11}(f)$ is independent of the isolated resistors. Therefore, the initial values of the isolated resistors are not optimum.

6.6 THREE-PORT 45° POWER DIVIDER/COMBINER

Millimeter-wave communication systems are attractive because they are able to offer wider bandwidth and larger capacity. The wide bandwidth makes high-bit-rate transmissions possible, and the large capacity is needed for multimedia applications. In these systems, equipment should be more compact, lighter, less expensive, and so on. To satisfy these requirements, the application of MMIC technology to millimeter-wave circuits is highly desirable. Furthermore, to reduce the size of the integrated circuits, additional isolators, filters, or switches should be avoided. To reduce the transmitter intermodulation without inserting an isolator between the RF circuit and the antenna, two choices can be made: use of a 90° hybrid and or use of a 45° divider/combiner [23].

Figure 6.11 shows a 45° power divider consisting of a Wilkinson power divider and a 45° delay line. If the power is excited at port ①, the power is divided between ports ④ and ③, and the waves at ports ④ and ③ are in phase. For the two waves to have a 45° phase difference, a 45° delay line is connected at port ④.

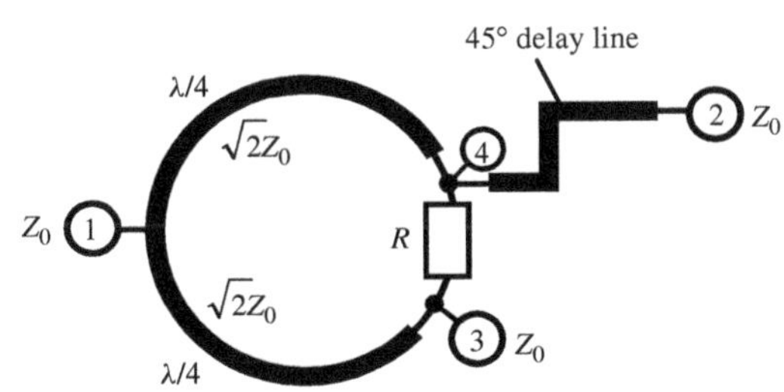

FIGURE 6.11 A 45° power divider/combiner.

EXERCISES

6.1 Design the three-port power divider in Fig. 6.2 for a power-split ratio of 3 dB and termination impedances of $R_a = 50\ \Omega$, $R_b = 60\ \Omega$, and $R_c = 40\ \Omega$.

6.2 Design the three-port power divider in Fig. 6.8 for a power-split ratio of 3 dB and termination impedances of $R_{ad} = 50\ \Omega$, $R_b = 60\ \Omega$, and $R_c = 40\ \Omega$.

6.3 Find S_{11e}, S_{11o}, S_{12e}, and S_{12o} when $Z_{01} = 50\ \Omega$, $\Theta_1 = 30^\circ$, $Z_{02} = 70.71\ \Omega$, $\Theta_2 = 28.56^\circ$, $Z_1 = 35\ \Omega$, and $Z_2 = 50\ \Omega$ in Fig. 6.9(*b*), where $Z_1 = 1/Y_1$ and $Z_2 = 1/Y_2$.

6.4 Calculate all the scattering parameters of the three-port power divider in Fig. 6.9(*a*) under the conditions given in Exercise 6.3.

6.5 Design the 45° power divider in Fig. 6.11 when the termination impedances at ports ①, ② and ③ are 30 Ω, 50 Ω, and 50 Ω, respectively.

REFERENCES

1. H.-R. Ahn, I.-S. Chang, and S.-W. Yun, A Study on the Miniaturization of 3-dB Ring Hybrid and Power Divider Using Lumped-Element Circuit, *J. KITE*, Vol. 28-A, No. 1, January 1991, pp. 15–22.
2. H.-R. Ahn, I.-S. Chang, and S.-W. Yun, A Study on the Miniaturization of 3-dB Ring Hybrid Having Arbitrary Termination Impedances Using Lumped-Element Circuit, *MTT Korean Chapter KITE-S Dig.*, Vol. 14, No. 2, October 1991, pp. 104–109.
3. H.-R. Ahn, I.-S. Chang, and S.-W. Yun, A Study on the Miniaturization of 3-dB Ring Hybrid Having Arbitrary Termination Impedances Using Lumped Equivalent Circuit, *J. KITE*, Vol. 29-A, No. 3, March 1992, pp. 25–32.
4. H.-R. Ahn, I.-S. Chang, and S.-W. Yun, A Study on the Miniaturization of 3-dB Ring Hybrid Having Arbitrary Termination Impedances Using Lumped Equivalent Circuit, *J. Telecommun. Rev.*, No. 5, May 1992, pp. 112–125.
5. H.-R. Ahn, I.-S. Chang, and S.-W. Yun, Lumped Element 3-dB 180° Hybrid with Asymmetrically Terminated Impedances, *J. KITE*, Vol. 31-A, No. 6, June 1994, pp. 18–25.
6. H.-R. Ahn, I.-S. Chang, and S.-W. Yun, Miniaturized 3-dB Ring Hybrid Terminated by Arbitrary Impedances, *IEEE Trans. Microwave Theory Tech.*, Vol. 42, December 1994, pp. 2216–2221.
7. H.-R. Ahn, I. Wolff, and I.-S. Chang, Arbitrary Termination Impedances, Arbitrary Power Division and Small-Sized Ring Hybrids, *IEEE MTT-S Dig.*, June 1997, pp. 285–288.
8. H.-R. Ahn, I. Wolff, and I.-S. Chang, Arbitrary Termination Impedances, Arbitrary Power Division and Small-Sized Ring Hybrids, *IEEE Trans. Microwave Theory Tech.*, Vol. 44, December 1997, pp. 2241–2247.
9. H.-R. Ahn and I. Wolff, Asymmetric Four-Port and Branch-Line Hybrids, *IEEE Trans. Microwave Theory Tech.*, Vol. 48, September 2000, pp. 1585–1588.

10. H.-R. Ahn and I. Wolff, Three-Port 3-dB Power Divider Terminated by Different Impedances and Its Application to MMIC's, *IEEE Trans. Microwave Theory Tech.*, Vol. 47, June 1999, pp. 786–794.
11. H.-R. Ahn and I. Wolff, Three-Port 3-dB Power Divider Terminated by Arbitrary Impedances, *IEEE MTT-S Dig.*, Baltimore, June 1998, pp. 781–784.
12. H.-R. Ahn and I. Wolff, General Design Equations of Three-Port Power Dividers, *IEEE MTT-S Dig.*, Boston, June 2000, pp. 1137–1140.
13. H.-R. Ahn and I. Wolff, General Design Equations of Three-Port Power Dividers, Small-Sized Impedance Transformers, and Their Applications to Small-Sized Three-Port 3-dB Power Divider, *IEEE Trans. Microwave Theory Tech.*, Vol. 49, July 2001, pp. 1277–1288.
14. H.-R. Ahn and I. Wolff, 3-dB Branch-Line Hybrids with Arbitrary Termination Impedance Values, *IEICE Trans. Electrons*, Vol. E82-C, No. 7, July 1999, pp. 1324–1326.
15. H.-R. Ahn and I. Wolff, Arbitrary Power Division Branch-Line Hybrid Terminated by Arbitrary Impedances, *IEE Electron. Lett.*, Vol. 35, No. 7, April 1999, pp. 572–273.
16. H.-R. Ahn and I. Wolff, 3-dB Branch-Line Hybrid Terminated by Arbitrary Impedances, *IEE Electron. Lett.*, Vol. 34, No. 11, May 1998, pp. 1109–1110.
17. H.-R. Ahn and I. Wolff, Asymmetric Four-Port Hybrids, Asymmetric 3-dB Branch-Line Hybrids, *Asia–Pacific Microwave Conf. Proc.*, Yokohama, Japan, December 1998, pp. 677–680.
18. E. J. Wilkinson, An *n*-Way Hybrid Power Divider, *IRE Trans. Microwave Theory Tech.*, Vol. 8, January 1960, pp. 116–118.
19. L. I. Parad and R. L. Moynihan, Split-Tee Power Divider, *IRE Trans. Microwave Theory Tech.*, Vol. 8, January 1965, pp. 91–95.
20. S. B. Cohn, A Class of Broadband Three-Port TEM-Mode Hybrids, *IRE Trans. Microwave Theory Tech.*, Vol. 16, February 1968, pp. 110–116.
21. R. B. Ekinge, A New Method of Synthesizing Matched Broad-Band TEM-Mode Three-Ports, *IEEE Trans. Microwave Theory Tech.*, Vol. 19, January 1971, pp. 81–88.
22. S. Rosloniec, Three-Port Hybrid Power Dividers Terminated in Complex Frequency-Dependent Impedances, *IEEE Trans. Microwave Theory Tech.*, Vol. 44, August 1996, pp. 1490–1493.
23. H. Hayashi, H. Okazaki, A. Kanda, T. Hirota, and M. Muraguch, Millimeter-Wave-Band Amplifier and Mixer MMIC's Using a Broad-Band 45° Power Divider/Combiner, *IEEE Trans. Microwave Theory Tech.*, Vol. 46, June 1998, pp. 811–818.
24. B. Kopp, Asymmetric Lumped Element Power Splitters, *IEEE MTT-S Dig.*, 1989, pp. 333–336.
25. D. Köther, B. Hopf, T. Sporkmann, and I. Wolff, MMIC Wilkinson Couplers for Frequencies Up to 110 GHz, *IEEE MTT-S Dig.*, 1995, pp. 663–665.
26. J. Reed and G. J. Wheeler, A Method of Analysis of Symmetrical Four-Port Networks, *IRE Trans. Microwave Theory Tech.*, Vol. 4, October 1956, pp. 346–352.

CHAPTER SEVEN

Three-Port 3-dB Power Dividers Terminated in Different Impedances

7.1 INTRODUCTION

When three- or four-port power dividers are used with active elements and/or other passive elements, additional matching networks are necessary to obtain the output performances desired. If these power dividers are terminated in different impedances, no matching circuit is required, and the total size of integrated microwave circuits can be reduced. Four-port power dividers terminated in arbitrary impedances were first treated by Ahn, who investigated ring hybrids [1–3] and branch-line hybrids [4–8]. For three-port power dividers terminated in different impedances, there are four references [9–12]. However, they have special load impedances where even- and odd-mode excitation analyses are possible, and all termination impedances should be equal in the case of 3-dB power division [11–12]. A three-port power divider terminated in complex frequency-dependent impedances has been treated by Rosloniec [12]. However, it can be designed effectively using optimization techniques with available mirror-reflected symmetry, so no design equations are presented. To overcome problems with conventional three-port power dividers, exact isolation conditions which were not analyzed in detail in Refs. 13 and 14 are derived, and three-port power dividers terminated in different impedances are analyzed in more detail [10].

In this chapter we treat three-port power dividers terminated in various impedances suggested by Ahn and Wolff [10], where perfect isolation conditions are derived. These conditions can be used for any three-port power divider with arbitrary power divisions and arbitrary termination impedances, and on the basis of

Asymmetric Passive Components in Microwave Integrated Circuits, by Hee-Ran Ahn

the isolation conditions, design equations for such a three-port power divider can be derived. If a three-port 3-dB power divider (T3PD) is terminated in different impedances, it does not possess an available symmetry plane for easy analysis of even- and/or odd-mode calculations [15]. For the asymmetric situation, two equivalent circuits are constructed. Based on them, the basis-independent normalized scattering parameters are derived and simulated under all possible load conditions. To verify that the scattering parameters derived are correct, the results calculated are compared with simulation results from a commercial program. The design equations is also used to predict frequency responses of a T3PD terminated in 40 Ω, 50 Ω, and 60 Ω under lossless and no-junction-effect assumptions.

For low loss, a new type of lumped-element three-port 3-dB power divider (LET3PD) and its design equations are presented, and a LET3PD terminated in 60 Ω, 70 Ω, and 80 Ω is simulated. Finally, to make sure that the design equations are reasonable, a coplanar T3PD terminated in 30 Ω, 53 Ω and 47 Ω is fabricated on Al_2O_3 substrate ($\varepsilon_r = 9.9$ $h = 635$ μm) and measured.

7.2 PERFECT ISOLATION CONDITION

Figure 7.1 shows a three-port power divider terminated in arbitrary impedances R_a, R_b, and R_c. A schematic for the entire power divider is depicted in Fig 7.1(*a*), and the circuit for isolation analyses between ports ② and ③ is shown in Fig. 7.1(*b*). As network [*N*] in Fig. 7.1(*b*) has no symmetry plane with $Z_b \neq Z_c$, the relation between scattering and impedance or admittance matrices is more complicated than in the symmetric case [16].

The relations between scattering *S*, admittance *Y*, and impedance *Z* matrices are given in Table 7.1. The current- and voltage-basis scattering parameters are derived with reference to (2.19) and (2.21), where *y* and *z* are reference load admittance and impedance matrices. In any case, the scattering parameter S_{21} is directly proportional to Y_{21} or Z_{21}. Thus, to derive perfect isolation conditions, knowledge of Y_{21} or Z_{21} is necessary. For the structure shown in Fig. 7.1(*b*), circuit [*N*] consists of two parts. One is a isolation resistor and the other consists of two transmission-line sections connected by a termination impedance R_a. The two parts are connected in parallel. Therefore, Y_{21} is easier to derive.

The *ABCD* parameters of the circuit in Fig. 7.1(*b*) without the resistance *R* are expressed using Figs. 2.13(*c*) and 2.15(*a*), as

$$\begin{bmatrix} A & B \\ C & D \end{bmatrix} = \begin{bmatrix} \cos\Theta_b & jZ_b\sin\Theta_b \\ \dfrac{j\sin\Theta_b}{Z_b} & \cos\Theta_b \end{bmatrix} \begin{bmatrix} 1 & 0 \\ \dfrac{1}{R_a} & 1 \end{bmatrix} \begin{bmatrix} \cos\Theta_c & jZ_c\sin\Theta_c \\ \dfrac{j\sin\Theta_c}{Z_c} & \cos\Theta_c \end{bmatrix}. \tag{7.1}$$

Applying $\Theta_b = \Theta_c = 90^\circ$ to (7.1) at the design frequency results in

$$\begin{bmatrix} A & B \\ C & D \end{bmatrix} = -\begin{bmatrix} \dfrac{Z_b}{Z_c} & \dfrac{Z_b Z_c}{R_a} \\ 0 & \dfrac{Z_c}{Z_b} \end{bmatrix}. \tag{7.2}$$

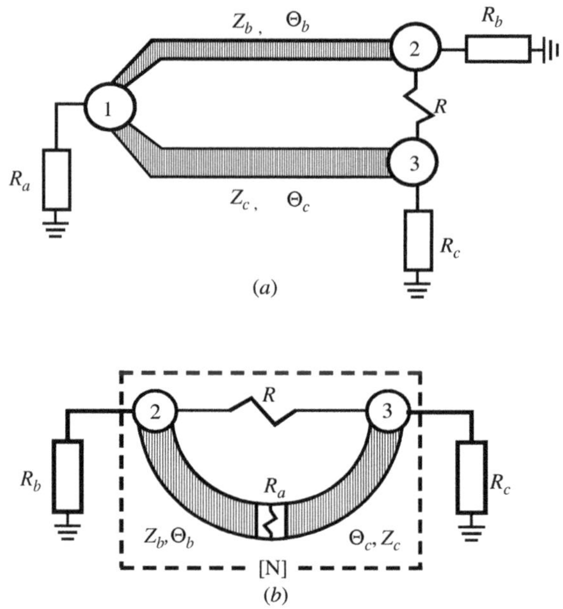

FIGURE 7.1 Three-port power divider terminated in arbitrary impedances: (*a*) power divider; (*b*) schematic diagram for analyzing the isolation between two ports, ② and ③.

From the conversion matrices in Table 2.1, the admittance matrix Y_{under} of the circuit where two transmission lines are connected with the load R_a is

$$\boldsymbol{Y}_{\text{under}} = \begin{bmatrix} \dfrac{R_a}{Z_b^2} & \dfrac{R_a}{Z_b Z_c} \\ \dfrac{R_a}{Z_b Z_c} & \dfrac{R_a}{Z_c^2} \end{bmatrix}. \tag{7.3}$$

In a similar way, the admittance matrix Y_R of the isolation resistor is

$$\boldsymbol{Y_R} = \begin{bmatrix} \dfrac{1}{R} & -\dfrac{1}{R} \\ -\dfrac{1}{R} & \dfrac{1}{R} \end{bmatrix}. \tag{7.4}$$

The resulting admittance matrix Y contributed by ports ② and ③ in Fig. 7.1(*b*) is

$$\boldsymbol{Y} = \begin{bmatrix} \dfrac{R_a}{Z_b^2} + \dfrac{1}{R} & \dfrac{R_a}{Z_b Z_c} - \dfrac{1}{R} \\ \dfrac{R_a}{Z_b Z_c} - \dfrac{1}{R} & \dfrac{R_a}{Z_b^2} + \dfrac{1}{R} \end{bmatrix}. \tag{7.5}$$

TABLE 7.1 Relation Between Impedance Matrix Z, Admittance Matrix Y, and Scattering Matrix S with Arbitrary Termination Impedances

Voltage-Basis Scattering Parameters	Current-Basis Scattering Parameters
$\mathbf{S}^V = -[\mathbf{Y}+\mathbf{y}]^{-1}[\mathbf{Y}-\mathbf{y}_*]$	$\mathbf{S}^I = [\mathbf{Z}+\mathbf{z}]^{-1}[\mathbf{Z}-\mathbf{z}_*]$
$S^V_{11} = -\dfrac{1}{\Delta_y}[(Y_{11}-y_{01*})(Y_{22}+y_{02})-Y_{12}Y_{21}]$	$S^I_{11} = \dfrac{1}{\Delta_z}[(Z_{11}-z_{01*})(Z_{22}+z_{02})-Z_{12}Z_{21}]$
$S^V_{12} = -\dfrac{1}{\Delta_y}Y_{12}(y_{02}+y_{02*})$	$S^I_{12} = \dfrac{1}{\Delta_z}Z_{12}(z_{02}+z_{02*})$
$S^V_{21} = -\dfrac{1}{\Delta_y}Y_{21}(y_{01}+y_{01*})$	$S^I_{21} = \dfrac{1}{\Delta_z}Z_{21}(z_{01}+z_{01*})$
$S^V_{22} = -\dfrac{1}{\Delta_y}[(Y_{11}+y_{01})(Y_{22}-y_{02*})-Y_{12}Y_{21}]$	$S^I_{22} = \dfrac{1}{\Delta_z}[(Z_{11}+z_{01})(Z_{22}-z_{02*})-Z_{12}Z_{21}]$
$S_{11} = S^V_{11}$	$S_{11} = S^I_{11}$
$S_{12} = S^V_{12}\sqrt{\dfrac{\mathrm{Re}(y_{01})}{\mathrm{Re}(y_{02})}}$	$S_{12} = S^I_{12}\sqrt{\dfrac{\mathrm{Re}(z_{01})}{\mathrm{Re}(z_{02})}}$
$S_{21} = S^V_{21}\sqrt{\dfrac{\mathrm{Re}(y_{02})}{\mathrm{Re}(y_{01})}}$	$S_{21} = S^I_{21}\sqrt{\dfrac{\mathrm{Re}(z_{02})}{\mathrm{Re}(z_{01})}}$
$S_{22} = S^V_{22}$	$S_{22} = S^I_{22}$
$\Delta_y = (Y_{11}+y_{01})(Y_{22}+y_{02})-Y_{12}Y_{21}$	$\Delta_z = (Z_{11}+z_{01})(Z_{22}+z_{02})-Z_{12}Z_{21}$

Therefore, the perfect isolation condition is derived as

$$Y_{12} = Y_{21} = \frac{R_a}{Z_b Z_c} - \frac{1}{R} = 0, \tag{7.6}$$

where the subscript number indicates the port number of the two-port network contributed be ports ② and ③ in Fig. 7.19(b).

7.3 ANALYSES

Wilkinson [17] described a device that separates one signal into n equiphase–equiamplitude signals ($n \geq 2$). With $n = 2$, his circuit may be reduced to a T3PD terminated in equal impedances. Figure 7.2 shows a three-port power divider terminated in different impedances R_a, R_b, and R_c. To design a T3PD with different termination impedances, the elements to be determined are the isolation resistor R and the characteristic impedances Z_b and Z_c of the transmission-line sections. From the perfect isolation conditions in (7.6), one relation, $R = Z_b Z_c / R_a$, is derived.

A voltage source V is fed into port ② as shown in Fig. 7.2. The voltages from ports ①, ②, and ③ to ground are V_a, V_b, and V_c, respectively. The currents at port ① are I_a, I'_b and I'_c. As shown in Fig. 7.2, I_a goes out port ① and toward ground, I'_b goes out port ② and enters port ① and I'_c goes out port ① and enters

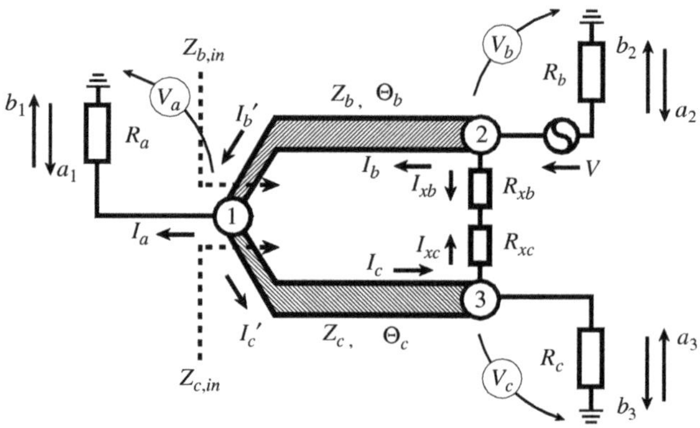

FIGURE 7.2 Three-port power divider terminated in arbitrary impedances.

port ③. Those at port ② are I_b and I_{xb}. I_b goes out port ② and toward port ①, and I_{xb} goes out port ② and toward port ③. Those at port ③ are I_c and I_{xc}. I_c goes out port ① and enters port ③. I_{xc} goes out port ③ and toward port ②. To find other conditions, applying the transmission-line equations (*ABCD* parameters) in Fig. 7.2 yields the following results. For nodes ② and ①,

$$V_b = V_a \cos\Theta_b + jI_b' Z_b \sin\Theta_b, \tag{7.7}$$

$$I_b = I_b' \cos\Theta_b + j\frac{V_a}{Z_b}\sin\Theta_b. \tag{7.8}$$

For nodes ③ and ①,

$$V_a = V_c \cos\Theta_c + jI_c Z_c \sin\Theta_c, \tag{7.9}$$

$$I_c' = I_c \cos\Theta_c + j\frac{V_c}{Z_c}\sin\Theta_c. \tag{7.10}$$

From the node equations:

$$\text{Node ①:} \qquad I_b' = \frac{V_a}{R_a} + I_c', \tag{7.11}$$

$$\text{Node ③:} \qquad I_c = \frac{V_c}{R_c} + I_{xc} \tag{7.12}$$

are derived. Additionally, the following equations hold:

$$\text{Node ② and ground:} \qquad (I_b + I_{xb})R_b = V - V_b; \tag{7.13}$$

$$\text{Nodes ② and ③:} \qquad I_{xb}R_{xb} - I_{xc}R_{xc} = V_b - V_c. \tag{7.14}$$

Applying $\Theta_b = \Theta_c = \pi/2$ to (7.7)–(7.14) and performing additional calculations, three equations for the voltages V_a, V_b, and V_c are derived as

$$\frac{V_b}{jZ_b} = \frac{V_a}{R_a} + j\frac{V_c}{Z_c}, \tag{7.15a}$$

$$\frac{V_b - V_c}{R_{xc} + R_{xb}} = \frac{V_c}{R_c} + j\frac{V_a}{Z_c}, \tag{7.15b}$$

$$\frac{V_b - V_c}{R_{xb} + R_{xc}} + j\frac{V_a}{Z_b} = \frac{V - V_b}{R_b}. \tag{7.15c}$$

The power-split ratio of the three-port power divider is proportional to the admittance ratio between its two output arms [11,13,14,18]. Therefore, the characteristic impedances of Z_b and Z_c in Fig. 7.2 are eventually equal for a 3-dB power divider. Substitution of a perfect-isolation conditions of $V_c = 0$, $R = Z_b Z_c / R_a$, and the equal power division of $Z_b = Z_c$ for (7.15) give

$$\frac{V_b}{V} = \frac{1/R_b}{1/(R_{xc} + R_{xb}) + 1/R + R_a/Z_b^2} = \frac{1}{1 + 2R_b/(R_{xc} + R_{xb})}. \tag{7.16}$$

Alternatively, if a voltage source V is excited at port ③, the following equations may be written:

Nodes ③ and ①:
$$V_c = V_a \cos\Theta_c + jI_c' Z_c \sin\Theta_c, \tag{7.17}$$
$$I_c = I_c' \cos\Theta_c + j\frac{V_a}{Z_c} \sin\Theta_c. \tag{7.18}$$

Nodes ② and ①:
$$V_a = V_b \cos\Theta_b + jI_b Z_b \sin\Theta_b, \tag{7.19}$$
$$I_b' = I_b \cos\Theta_b + j\frac{V_b}{Z_b} \sin\Theta_b. \tag{7.20}$$

From the node equations:

Node ①:
$$I_c' = \frac{V_a}{R_a} + I_b', \tag{7.21}$$

Node ②:
$$I_b = \frac{V_b}{R_b} + I_{xb} \tag{7.22}$$

are derived. Additionally, the following equations hold:

Node ③ and ground:
$$(I_c + I_{xc})R_c = V - V_c, \tag{7.23}$$

Nodes ② and ③:
$$I_{xb}R_{xb} - I_{xc}R_{xc} = V_b - V_c. \tag{7.24}$$

Applying $\Theta_b = \Theta_c = \pi/2$ to (7.17)–(7.24) and performing additional calculations, three equations for voltages V_a, V_b, and V_c are derived:

$$\frac{V_c}{jZ_c} = \frac{V_a}{R_a} + j\frac{V_b}{Z_b}, \tag{7.25a}$$

$$j\frac{V_a}{Z_b} = \frac{V_b - V_c}{R_{xc} + R_{xb}} + \frac{V_b}{R_b}, \tag{7.25b}$$

$$\frac{V_c - V_b}{R_{xb} + R_{xc}} + j\frac{V_a}{Z_c} = \frac{V - V_c}{R_c}. \tag{7.25c}$$

In a similar way, the ratio of V_c to V is derived as

$$\frac{V_c}{V} = \frac{1/R_c}{1/(R_{xc} + R_{xb}) + 1/R + R_a/Z_c^2} = \frac{1}{1 + 2R_c/(R_{xc} + R_{xb})}. \tag{7.26}$$

In the conventional case of equal power division, (7.16) must be equal to (7.26), but here, due to $R_b \neq R_c$, they are not the same. To minimize the difference caused by different R_b and R_c, an optimum load R_{av} is chosen to be either the geometric or arithmetic mean of R_b and R_c. Depending on this choice, from (7.6), (7.16), and (7.26), the characteristic impedances Z_b and Z_c and an isolation resistor R are derived differently.

Case 1: If $R_{\text{av}} = \sqrt{R_b R_c}$, Z_b, Z_c, and the isolation resistor R are calculated as

$$R = R_{xc} + R_{xb} = 2\sqrt{R_b R_c}, \tag{7.27a}$$

$$Z_b = Z_c = \sqrt{2\sqrt{R_b R_c} R_a}. \tag{7.27b}$$

Case 2: If $R_{\text{av}} = (R_b + R_c)/2$, Z_b, Z_c, and the isolation resistor R are derived as

$$R = R_{xc} + R_{xb} = R_b + R_c, \tag{7.28a}$$

$$Z_b = Z_c = \sqrt{(R_b + R_c) R_a}. \tag{7.28b}$$

$R = R_{xc} + R_{xb} = 2\sqrt{R_b R_c}$ in (7.27*a*) may also be derived from the relation $V_b/V + V_c/V = 1$ in (7.16) and (7.26).

Equations (7.27) and (7.28) were derived depending on R_{av} when the power was excited at port ② or ③. In another approach, the design equations are derived from the viewpoint of the input impedance R_a when the power is fed into port ①. The input impedance R_a is divided into $Z_{b,\text{in}}$ and $Z_{c,\text{in}}$ as shown in Fig. 7.2. In a conventional T3PD, $Z_{b,\text{in}}$ is equal to $Z_{c,\text{in}}$. However, this is not true for a T3PD terminated in different impedances R_b and R_c.

With $\Theta_b = \Theta_c = \pi/2$ in Fig. 7.2, Z_b and Z_c are calculated as

$$Z_b^2 = Z_{b,\text{in}} R_b, \tag{7.29a}$$

$$Z_c^2 = Z_{c,\text{in}} R_c. \tag{7.29b}$$

Also, the input impedance R_a is

$$\frac{1}{R_a} = \frac{1}{Z_{b,\text{in}}} + \frac{1}{Z_{c,\text{in}}}. \tag{7.30}$$

From (7.29) and (7.30), the characteristic impedances Z_b and Z_c are easily calculated as $Z_b = Z_c = \sqrt{R_a(R_b + R_c)}$, which is the same result as for Z_b and Z_c in the case of $R_{\text{av}} = (R_b + R_c)/2$ in (7.28).

7.4 SCATTERING PARAMETERS OF THREE-PORT POWER DIVIDERS

Admittance, impedance, hybrid, transmission, and scattering matrices are circuit matrices that can describe a large class of passive microwave components. Some matrices are defined with respect to zero or infinite load at the ports. On the other hand, the scattering matrix is defined in terms of some finite stable loads at the ports. In other words, as they originated from the power concept of transmission-line waves, the scattering matrix is more convenient than voltages or currents to describe microwave devices.

Consider the three-port power divider terminated in different impedances in Fig. 7.2, which is not symmetrical. Due to the asymmetry, calculation of the scattering parameters requires complex manipulation. For easy derivation of one set of scattering parameters, the first consideration will be a two-port circuit with port ① terminated in a fixed load R_a, from which S_{22} and S_{33} can be derived. Then two two-port equivalent circuits are constructed to derive S_{21} and S_{31}. The last scattering coefficient of S_{11} is computed based on the S_{22} and S_{33} derived.

Figure 7.3(a) shows a two-port network contributed by ports ② and ③, where port ① is terminated in R_a. Assuming that $\Theta_b = \Theta_c = \Theta$ and $Z_b = Z_c$, network $[N]$ in Fig. 7.3(a) is symmetric and its even- and odd-mode equivalent circuits are depicted in Fig. 7.3(b) and (c), respectively. When the transmission-line sections are assumed to be lossless, the even- and odd-mode impedances $Z_{\text{O/C}}$ and $Z_{\text{S/C}}$ of $[N]$ in Fig. 7.3(b) and (c) are

$$Z_{\text{O/C}} = Z_b \frac{2R_a + jZ_b \tan\Theta}{Z_b + j2R_a \tan\Theta}, \tag{7.31a}$$

$$Z_{\text{S/C}} = \frac{jRZ_b \tan\Theta}{j2Z_b \tan\Theta + R}. \tag{7.31b}$$

Now network $[N]$ is characterized by the impedance matrix, which is derived with reference to (2.109) as $Z_{11} = Z_{22} = (Z_{\text{O/C}} + Z_{\text{S/C}})/2$ and $Z_{12} = Z_{21} =$

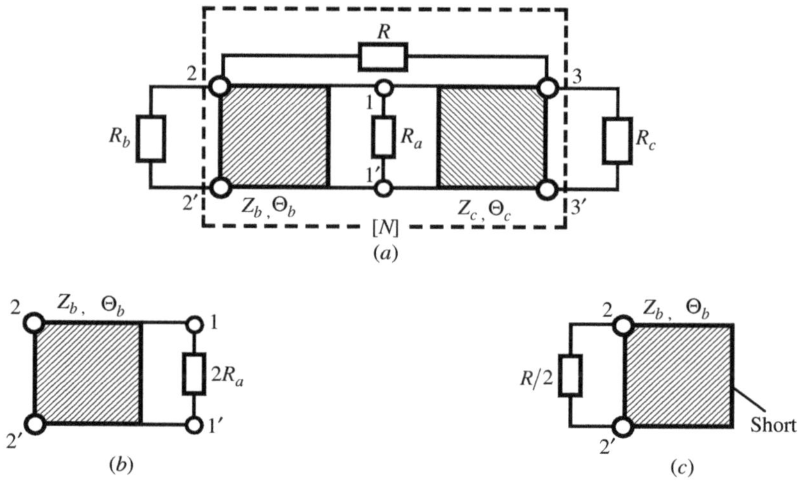

FIGURE 7.3 Scattering parameters between ports ② and ③. (*a*) Two-port network contributed by ports ② and ③; (*b*) even-mode equivalent circuit; (*c*) odd-mode equivalent circuit. (From Ref. 10 with permission from IEEE.)

$(Z_{O/C} - Z_{S/C})/2$. Applying $\Theta = \pi/2$ to (7.31) and performing additional calculations, the impedance matrix $\mathbf{Z}$ of $[N]$ is given as

$$\mathbf{Z} = \begin{bmatrix} \dfrac{R}{4} + \dfrac{Z_b^2}{4R_a} & \dfrac{Z_b^2}{4R_a} - \dfrac{R}{4} \\ \dfrac{Z_b^2}{4R_a} - \dfrac{R}{4} & \dfrac{R}{4} + \dfrac{Z_b^2}{4R_a} \end{bmatrix}. \tag{7.32}$$

Applying the perfect isolation conditions of $R = Z_b^2/R_a$ and equal-power-division conditions of $Z_b = Z_c$ to (7.32), the impedance matrix $\mathbf{Z}$ is again simplified as

$$\mathbf{Z} = \begin{bmatrix} \dfrac{R}{2} & 0 \\ 0 & \dfrac{R}{2} \end{bmatrix}, \tag{7.33}$$

which indicates that the impedance seen from the load impedance R_b or R_c is equally $R/2$. So the scattering parameters S_{22}, S_{33}, S_{23}, and S_{32} are derived easily as

$$S_{22} = \frac{R/2 - R_b}{R/2 + R_b} = \frac{R - 2R_b}{R + 2R_b}, \tag{7.34a}$$

$$S_{33} = \frac{R/2 - R_c}{R/2 + R_c} = \frac{R - 2R_c}{R + 2R_c}, \tag{7.34b}$$

$$S_{23} = 0 = S_{32}, \tag{7.34c}$$

where the perfect isolation condition in (7.6) is applied.

Table 7.2 shows simulated isolation results where s stands for R_a/R_b and t is defined as R_c/R_b. The characteristic impedances of the transmission-line sections and the isolation resistances are calculated on the basis of (7.27), and under ideal conditions, all the data in Table 7.2 are produced by a commercial circuit simulator. Since all the termination impedances differ from 50 Ω, additional impedance transformers may be needed and $R_b = 50\ \Omega$ is fixed. The data in Table 7.2 show that the isolation between ports ② and ③ can be considered perfect for arbitrary loads. Based on (7.34), reflection coefficients at ports ② and ③ were simulated, and the simulation data are also given in Table 7.3, where t again indicates R_c/R_b and R_a is fixed as 50 Ω. Careful study of Table 7.3 shows that ports ② and ③ can be well matched, with similar values of R_b and R_c.

Based on Table 7.2, two two-port equivalent circuits may be constructed. Perfect isolation between ports ② and ③ can be accomplished using arbitrary loads, which indicates that port ③ is, with $V_c = 0$, considered as an electrical short when power is fed into port ②. From the perfect isolation conditions of $R = Z_b^2/R_a$, we can see that the transmission line with Z_b in Fig. 7.2 transforms

TABLE 7.2 Simulated Isolation Results (dB) with Possible Load Conditions[a]

s \ t	0.1	0.5	1	1.5	2
0.1	−158.72	−156.45	−160.67	−158.73	−162.50
0.5	−158.78	−162.65	−180.91	−169.86	−181.47
1	−160.61	−166.26	−167.32	−153.47	−161.81
1.5	−155.87	−162.93	−159.55	−161.50	−165.30
2	−157.41	−158.80	−161.68	−161.77	−163.16
3	−154.48	−157.20	−157.24	−155.44	−158.51
4	−157.09	−157.03	−157.61	−152.34	−157.43

[a] $t = R_c/R_b$; $s = R_a/R_b$.

TABLE 7.3 Reflection Coefficients Calculated for S_{22} and S_{33}

	$\lvert S_{22}\rvert \le -15$ dB	$\lvert S_{33}\rvert \le -15$ dB
$R_{av} = \sqrt{R_b R_c}$	$0.487 \le t \le 2.05$	$0.487 \le t \le 2.052$
$R_{av} = (R_b + R_c)/2$	$0.396 \le t \le 1.865$	$0.536 \le t \le 2.524$
	$\lvert S_{22}\rvert \le -20$ dB	$\lvert S_{33}\rvert \le -20$ dB
$R_{av} = \sqrt{R_b R_c}$	$0.6694 \le t \le 1.494$	$0.6694 \le t \le 1.494$
$R_{av} = (R_b + R_c)/2$	$0.6363 \le t \le 1.444$	$0.6923 \le t \le 1.571$
	$\lvert S_{22}\rvert \le -25$ dB	$\lvert S_{33}\rvert \le -25$ dB
$R_{av} = \sqrt{R_b R_c}$	$0.7983 \le t \le 1.253$	$0.7983 \le t \le 1.253$
$R_{av} = (R_b + R_c)/2$	$0.7870 \le t \le 1.238$	$0.8075 \le t \le 1.270$

a load impedance R_a into the resistance R. Therefore, an equivalent circuit is, in terms of ports ① and ②, depicted as in Fig. 7.4(*a*). However, this circuit is true in terms of port ② but not true in terms of port ①. We can use reciprocity and thereby derive S_{12}. By the same principle, the equivalent circuit between ports ③ and ① is constructed as shown in Fig. 7.4(*b*).

From the equivalent circuit of Fig. 7.4(*a*), the *ABCD* parameters between ports ① and ② are

$$\begin{bmatrix} A & B \\ C & D \end{bmatrix} = \begin{bmatrix} \cos\Theta_b + j\dfrac{Z_b}{R}\sin\Theta_b & jZ_b\sin\Theta_b \\ j\dfrac{1}{Z_b}\sin\Theta_b & 0 \end{bmatrix}. \tag{7.35}$$

From (7.35), the impedance matrix **Z** is, with reference to Table 7.1, derived as

$$\mathbf{Z} = \frac{-jZ_b}{\sin\Theta_b}\begin{bmatrix} \cos\Theta_b + j\dfrac{Z_b}{R}\sin\Theta_b & 1 \\ 1 & 0 \end{bmatrix}. \tag{7.36}$$

Referring to Table 7.1, Δ_z in the case of Fig. 7.4(*a*) is

$$\Delta_z = R_a(2R_b + R). \tag{7.37}$$

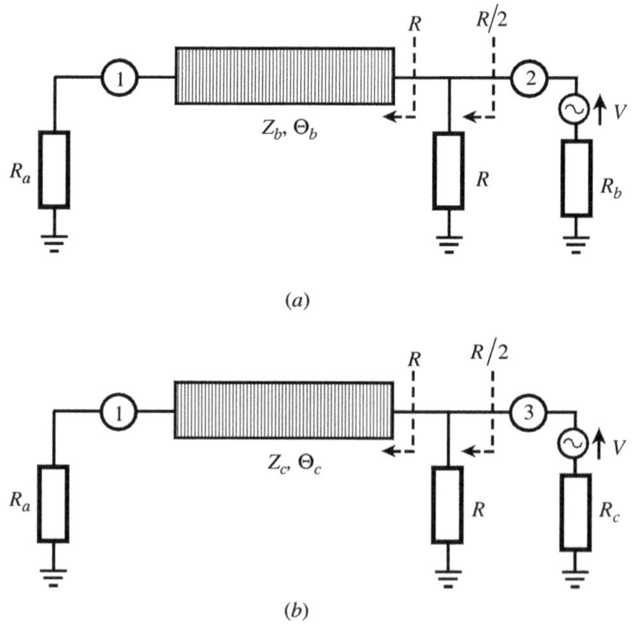

FIGURE 7.4 Equivalent two-port circuits for scattering parameters. (*a*) circuit for S_{12} between ports ① and ②; (*b*) circuit for S_{13} between ports ① and ③. (From Ref. 10 with permission from IEEE.)

The reference impedance matrix z is determined by a load condition, so that in Fig. 7.4(a) it is given by

$$z = \begin{bmatrix} R_a & 0 \\ 0 & R_b \end{bmatrix}. \tag{7.38}$$

Referring to Table 7.1, the current-basis scattering parameter S^I_{12} is determined as $S^I_{21} = -1/\Delta_z Z_{21}(z_{11} + z_{11*})$ and gives

$$S^I_{12} = \frac{-j2\sqrt{RR_a}}{2R_b + R}. \tag{7.39}$$

Finally, the basis-independent scattering parameter S_{21} is calculated as

$$S_{21} = \frac{-j2\sqrt{RR_b}}{R + 2R_b} = S_{12}. \tag{7.40}$$

In Fig. 7.4(b), the reference impedance matrix z is

$$z = \begin{bmatrix} R_a & 0 \\ 0 & R_c \end{bmatrix}. \tag{7.41}$$

Also, in the case of Fig. 7.4(b), Δ_z is

$$\Delta_z = R_a(2R_c + R). \tag{7.42}$$

A basis-independent scattering parameter S_{31} can be derived in the same way as

$$S_{13} = \frac{-j2\sqrt{RR_c}}{R + 2R_c} = S_{31}. \tag{7.43}$$

From the two circuits in Fig. 7.4, S_{31} and S_{21} were derived. Only one element S_{11} now remains unknown for a set of complete scattering parameters. $\boldsymbol{a}$ is an incident normalized wave vector and $\boldsymbol{b}$ is a reflected wave vector in Fig. 7.2. If the incident wave a_1 is excited at port ①, half of the power is delivered to port ②, and part of it will be reflected from port ② into port ①. In the same way, half of the power excited at port ① will be delivered to port ③, and part of it will be reflected into port ①. The sum of the two reflected power waves is b_1. Additional reflections will occur, but they may be neglected from the second reflection on.

Setting the reflection coefficients S_{22L} or S_{33L} looking into the load R_b or R_c, respectively, the result for b_1 is

$$\begin{aligned} b_1 &= \left(-j\frac{1}{\sqrt{2}}\right) S_{22L} \left(-j\frac{1}{\sqrt{2}}\right) a_1 + \left(-j\frac{1}{\sqrt{2}}\right) S_{33L} \left(-j\frac{1}{\sqrt{2}}\right) a_1 \\ &= \frac{1}{2}(S_{22} + S_{33})a_1, \end{aligned} \tag{7.44}$$

where $S_{22} = -S_{22L} = b_2/a_2$ and $S_{33} = -S_{33L} = b_3/a_3$ for real load impedances R_b and R_c. So S_{11} is

$$S_{11} = \frac{1}{2}\left(\frac{R - 2R_b}{R + 2R_b} + \frac{R - 2R_c}{R + 2R_c}\right). \tag{7.45}$$

From the derived scattering elements in (7.40), (7.43), and (7.45), one set of scattering parameters are: For $R_{av} = \sqrt{R_b R_c}$,

$$[S] = \begin{bmatrix} 0 & \dfrac{-j\sqrt{2\sqrt{t}}}{\sqrt{t}+1} & \dfrac{-j\sqrt{2\sqrt{t}}}{\sqrt{t}+1} \\ \dfrac{-j\sqrt{2\sqrt{t}}}{\sqrt{t}+1} & \dfrac{\sqrt{t}-1}{\sqrt{t}+1} & 0 \\ \dfrac{-j\sqrt{2\sqrt{t}}}{\sqrt{t}+1} & 0 & \dfrac{1-\sqrt{t}}{\sqrt{t}+1} \end{bmatrix}, \tag{7.46}$$

and for $R_{av} = (R_b + R_c)/2$,

$$[S] = \begin{bmatrix} \dfrac{(t-1)^2}{(3+t)(3t+1)} & \dfrac{-j2\sqrt{(1+t)}}{3+t} & \dfrac{-j2\sqrt{(t^2+t)}}{1+3t} \\ \dfrac{-j2\sqrt{(1+t)}}{3+t} & \dfrac{t-1}{3+t} & 0 \\ \dfrac{-j2\sqrt{(t^2+t)}}{1+3t} & 0 & \dfrac{1-t}{1+3t} \end{bmatrix}, \tag{7.47}$$

where $t = R_c/R_b$.

When $t = 1$, perfect matching appears at all ports in (7.46) and (7.47), and all scattering parameters are the same as those of the conventional T3PD [17]. The sets of scattering parameters in (7.46) and (7.47) depend solely on the ratio of R_c to R_b, or t. Since design equations (7.46) and (7.47) already include perfect isolation between ports ② and ③ and perfect matching at port ①, perfect isolation and matching always occur regardless of R_b or R_c. All other scattering parameters depend strongly on t and were simulated depending on t. Table 7.3 and Fig. 7.5 give computed responses of the reflection coefficients S_{22} and S_{33} depending on $t = R_c/R_b$, and power division characteristics depending on $t = R_c/R_b$ are listed in Table 7.4. In all cases, using the geometric mean $R_{av} = \sqrt{R_b R_c}$ causes ports ② and ③ to share identical properties, while using the arithmetic mean $R_{av} = (R_b + R_c)/2$ gives each port unique values. The relations between S_{21} and S_{22} and between S_{31} and S_{33} are described in Fig. 7.6(*a*) and (*b*), respectively. These two graphs are created based on (7.46). In Fig. 7.6(*a*), the range of S_{22} and S_{33} is from -15 to -60 dB and that of $t = R_c/R_b$ is from 0.1 to 1.9. Since the S_{22} and S_{33} are negative infinitive in decibles when $t = 1$, the calculation at $t = 1$ is excluded. Figure 7.6(*a*) shows that S_{21} is saturated to -3.010 dB with S_{22} close to -60 dB, and Fig. 7.6(*b*) shows similar results.

To verify that the scattering parameters derived are correct, those in (7.46) and (7.47) were compared with parameters simulated by the circuit simulator ADS. The load condition is the same as that used to generate Table 7.2. The load impedance R_a does not influence the results at the band center. So the load impedances of R_a and R_b were fixed as 40 Ω and 50 Ω, and t was varied. In

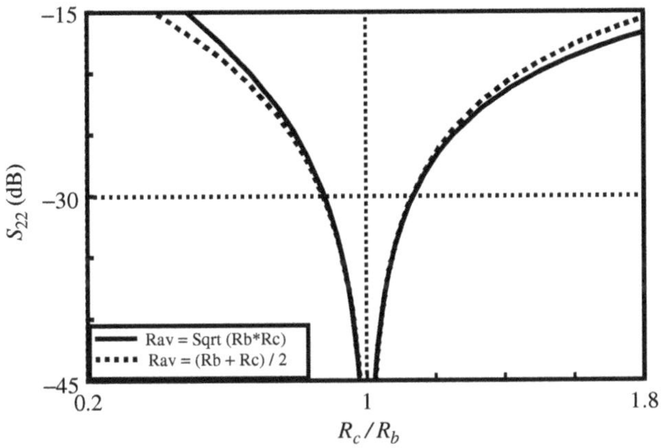

(*a*)

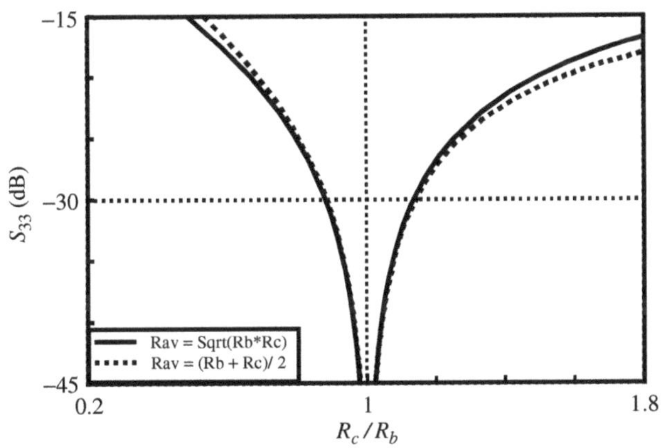

(*b*)

FIGURE 7.5 Scattering parameters (*a*) S_{22} and (*b*) S_{33} versus t (R_c/R_b).

TABLE 7.4 Power Division Characteristics of S_{21} and S_{31}

	$\|S_{21}\| \leq -3.02$ dB	$\|S_{31}\| \leq -3.02$ dB
$R_{av} = \sqrt{R_b R_c}$	$0.8277 \leq t \leq 1.208$	$0.8277 \leq t \leq 1.208$
$R_{av} = (R_b + R_c)/2$	$0.8195 \leq t \leq 1.1983$	$0.8345 \leq t \leq 1.2201$
	$\|S_{21}\| \leq -3.05$ dB	$\|S_{31}\| \leq -3.05$ dB
$R_{av} = \sqrt{R_b R_c}$	$0.682 \leq t \leq 1.466$	$0.682 \leq t \leq 1.466$
$R_{av} = (R_b + R_c)/2$	$0.652 \leq t \leq 1.422$	$0.703 \leq t \leq 1.535$
	$\|S_{21}\| \leq -3.1$ dB	$\|S_{31}\| \leq -3.1$ dB
$R_{av} = \sqrt{R_b R_c}$	$0.562 \leq t \leq 1.7787$	$0.562 \leq t \leq 1.7787$
$R_{av} = (R_b + R_c)/2$	$0.500 \leq t \leq 1.667$	$0.5998 \leq t \leq 2.001$

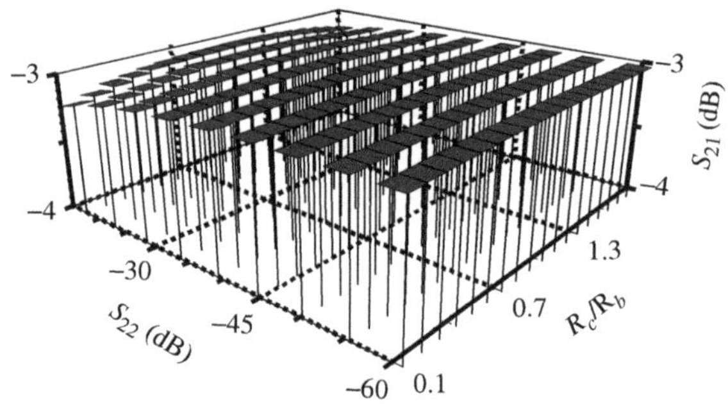

(*a*)

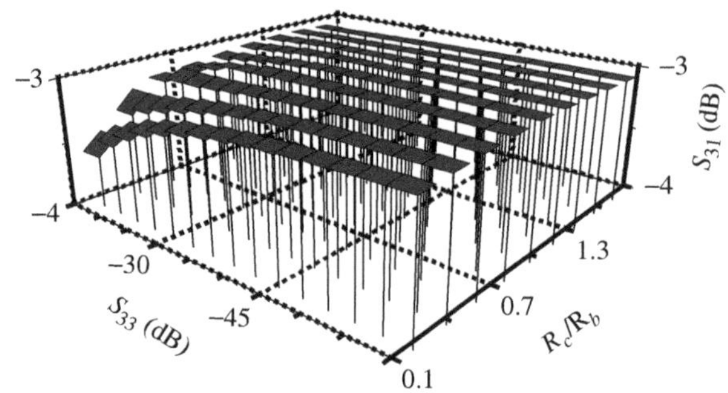

(*b*)

FIGURE 7.6 Relations between power division and matching responses at ports ② and ③: (*a*) relation between S_{22} and S_{12} with changes in t (R_c/R_b); (*b*) relation between S_{33} and S_{13} with changes in t (R_c/R_b).

TABLE 7.5 Comparison of Calculated Results and ADS Simulation (dB)

	$\|S_{11}\|$	$\|S_{21}\|$	$\|S_{31}\|$	$\|S_{22}\|$	$\|S_{33}\|$
		$(R_a, R_b, R_c) = (40\ \Omega, 50\ \Omega, 1.5\ \Omega)$			
ADS	−151.47	−5.992	−5.992	−3.040	−3.040
Case 1	$-\infty$	−5.9915	−5.9915	−3.0395	−3.0395
		$(R_a, R_b, R_c) = (40\ \Omega, 50\ \Omega, 35\ \Omega)$			
ADS	−42.106	−3.039	−3.051	−21.822	−20.285
Case 2	−42.106	−3.0389	−3.0511	−21.822	−20.285
		$(R_a, R_b, R_c) = (40\ \Omega, 50\ \Omega, 75\ \Omega)$			
ADS	−143.45	−3.055	−3.055	−19.912	−19.912
Case 1	$-\infty$	−3.054	−3.054	−19.911	−19.911
		$(R_a, R_b, R_c) = (40\ \Omega, 50\ \Omega, 100\ \Omega)$			
ADS	−30.881	−3.188	−3.100	−13.979	−16.902
Case 2	−30.881	−3.188	−3.099	−13.979	−16.902

Table 7.5, the first rows compare the ADS simulation results with those calculated from (7.46), the second rows compare ADS results with those calculated from (7.47), and so on. Table 7.5 shows that the results calculated from (7.46) or (7.47) are identical with those from a commercial ADS program.

Three-port 3-dB power dividers had been required to have the property $R_a = R_b = R_c$ [11], or at least $R_b = R_c$ from (7.16) and (7.26). However, even though they are terminated in different impedances, they can be used for equal power division. The purpose of this chapter is the design of 3-dB power dividers terminated in different impedances. Since the load impedance R_b or R_c is not the optimal load R_{av} except when $R_b = R_c$, ports ② and ③ are not perfectly matched. However, this is not a problem in practical applications. If the performances of the T3PD terminated in different impedances satisfy the requirement within a bandwidth, the T3PD can be used without a problem. In such a design, limiting factors are S_{22} and S_{33}, as given in Table 7.3. If both of them are less than -20 dB within the bandwidth, it can be acceptable. From (7.46) and Tables 7.3 to 7.5, even though the two load impedances R_b and R_c differ from each other, S_{22} and S_{33} are the same in magnitude and out of phase by 180°. Therefore, S_{11} is always zero and S_{12} is equal to S_{13}. This means that a T3PD terminated in different impedances is not symmetrical but a potential symmetric line exists somewhere by the design method in (7.27) when power is excited at port ①. However, one cannot say that the design suggested in (7.27) is better than that suggested in (7.28), because (7.28) may be useful in real applications where different levels of power are combined.

Figure 7.7 shows the simulated performances of a T3PD terminated in $R_a = 40\ \Omega$, $R_b = 50\ \Omega$, and $R_c = 60\ \Omega$ on the basis of (7.27). The T3PD was simulated under ideal conditions. Calculated design data are: the isolation resistance R is 109.54 Ω, the optimum load impedance of R_{av} is 54.77 Ω, and the characteristic impedances of Z_b and Z_c are both 66.195 Ω at a center frequency of 3 GHz. In this case, t is 1.2 and $|S_{22}| = |S_{33}| = -26.830$ dB, $|S_{21}| = |S_{31}| = -3.019$ dB,

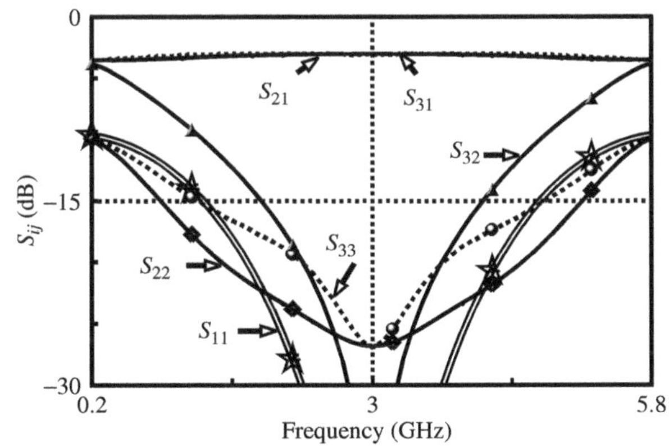

FIGURE 7.7 Simulated scattering parameters of a T3PD terminated in 40 Ω, 50 Ω, and 60 Ω.

$|S_{23}| = -154.347$ dB, and $|S_{11}| = -148.415$ dB at the band center. The isolation is perfect, and perfect matching appears at port ① at the band center. However, $|S_{22}|$ has a wider bandwidth than $|S_{11}|$, as shown in Fig. 7.7. This design shows that the limiting factor to determine the bandwidth is $|S_{23}|$, not $|S_{22}|$ or $|S_{33}|$.

7.5 LUMPED-ELEMENT THREE-PORT 3-dB POWER DIVIDERS

Figure 7.8 shows lumped-element models for a transmission line with electrical length Θ, where Z_0 or Y_0 is the characteristic impedance or admittance of the transmission line. Figure 7.8(*a*) shows the lumped-element model for $0^\circ < \Theta < 180^\circ$, and Fig. 7.8(*b*) shows that for $-180 < \Theta < 0^\circ$. From the results in Fig. 7.8, the application of a capacitance between two ports is advantageous for a low-insertion-loss LET3PD terminated in different impedances. Therefore, a LET3PD for low loss is suggested in Fig. 7.9.

There have been conventional lumped-element three-port power dividers [14,19,20]. However, those in Refs. 14 and 19 used inductances between the input and output ports, and the termination impedances were equal. In Ref. 20, capacitances were used between two ports, but only symmetric structures could be used. The equivalent circuit suggested in Fig. 7.9, on the other hand, uses capacitances between the input and output ports; its design equations will be presented for a T3PD terminated in different impedances. All elements in Fig. 7.9 are derived on the basis of (7.27), and in the ideal situation, they are given as

$$C_a = C_b = \frac{1}{\sqrt{2R_a\sqrt{R_bR_c}\ \omega^2}}, \tag{7.48a}$$

$$L_b = L_c = \sqrt{\frac{2R_a\sqrt{R_bR_c}}{\omega^2}}, \tag{7.48b}$$

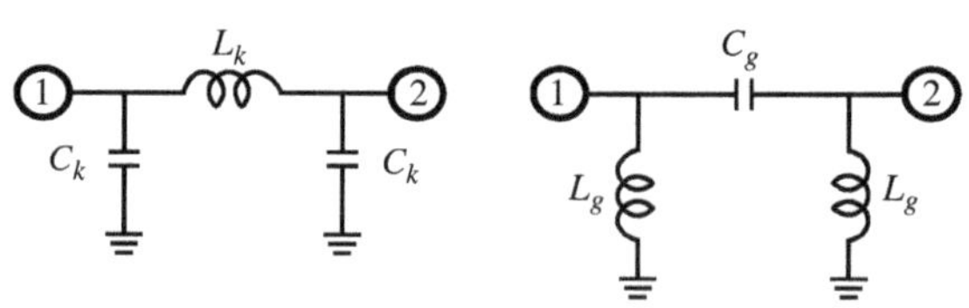

$$L_k = Z_0\frac{\sin\Theta}{2\pi f} \qquad C_g = Y_0\frac{\csc\Theta}{2\pi f}$$

$$C_k = Y_0\frac{\tan(\Theta/2)}{2\pi f} \qquad L_g = Z_0\frac{\cot(\Theta/2)}{2\pi f}$$

(*a*) (*b*)

FIGURE 7.8 Lumped-element transmission-line model for electrical length Θ: (*a*) for $0^\circ < \Theta \leq 180^\circ$; (*b*) for $-180^\circ \leq \Theta < 0^\circ$.

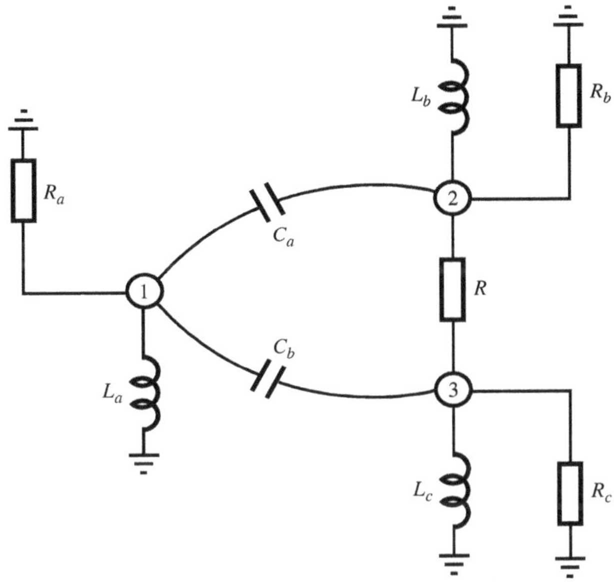

FIGURE 7.9 LET3PD terminated in different impedances.

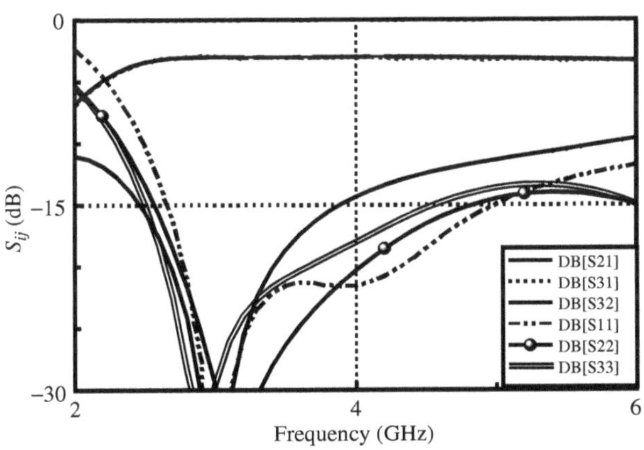

FIGURE 7.10 Simulation results for a LET3PD terminated in 60 Ω, 70 Ω, and 80 Ω.

$$L_a = \sqrt{\frac{R_a\sqrt{R_bR_c}}{2\omega^2}}, \tag{7.48c}$$

$$R = 2\sqrt{R_bR_c}, \tag{7.48d}$$

where $\omega = 2\pi f_c$ and f_c is a center frequency. Using these equations, LET3PD terminated in 60 Ω, 70 Ω, and 80 Ω was designed and analyzed at a center frequency of 3 GHz. The lumped elements are $L_a = 2.514$ nH, $L_b = L_c = 5.207$ nH, $C_a = C_b = 0.5598$ pF, and the resistance $R = 149.67$ Ω. Figure 7.10

shows its simulation results. It was designed at 3 GHz, but the power division characteristics and the matching properties at ports ② and ③ show good performances in the higher-frequency region.

An LET3PD can provide any microwave equipment with the advantages of small losses, small size, and no spurious response. Lumped-element modeling approximates the transmission-line section well at the design frequency, but it has a narrower bandwidth. To increase the bandwidth, two sections of 45° each may be cascaded.

7.6 COPLANAR THREE-PORT 3-dB POWER DIVIDERS

On the basis of the derived design equations (7.28), a coplanar T3PD terminated in 30 Ω, 53 Ω, and 47 Ω was fabricated on an Al_2O_3 substrate ($\varepsilon_r = 9.9$, $h = 635$ μm). Its layout and packages are shown in Fig. 7.11(*a*) and (*b*), respectively. Since the T3PD is not terminated in 50 Ω, additional impedance

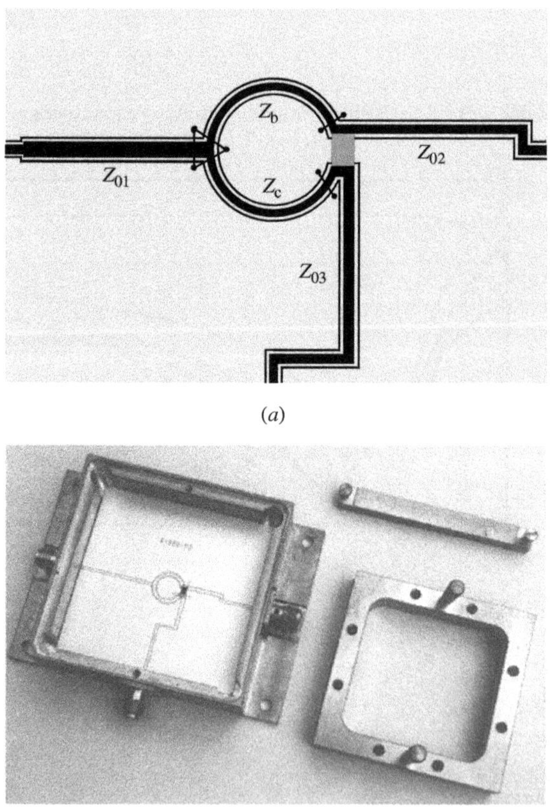

(*a*)

(*b*)

FIGURE 7.11 Coplanar T3PD terminated in 30 Ω, 53 Ω, and 47 Ω: (*a*) detailed layout for measuring; (*b*) packages of the three-port power divider. (From Ref. 10 with permission from IEEE.)

TABLE 7.6 Experimental Data on a Coplanar T3PD Terminated in Impedances of 30 Ω, 53 Ω, and 47 Ω at a Center Frequency of 3 GHz[a]

Termination Impedance	Coplanar Feeding Transformer Line	Coplanar Transmission Line
Port ①;	Z_{01}; 38.73 Ω	Z_b; 54.77 Ω
$R_a = 30$ Ω	$w = 756$ μm	$w = 422$ μm
	$g = 120$ μm	$g = 234$ μm
	$l = 11,219$ μm	$l = 11,018$ μm
Port ②;	Z_{02}; 51.48 Ω	Z_c; 54.77 Ω
$R_b = 53$ Ω	$w = 518$ μm	$w = 422$ μm
	$g = 226$ μm	$g = 234$ μm
	$l = 11,023$ μm	$l = 11,018$ μm
Port ③;	Z_{03}; 48.48 Ω	
$R_c = 47$ Ω	$w = 602$ μm	
	$g = 226$ μm	
	$l = 11,186$ μm	

[a] w, Center strip width; g, gap width; l, line length.

transformers are needed for a general measurement system. Z_{01}, Z_{02}, and Z_{03} in Fig. 7.11(a) are the impedance transformers, and a chip resistor R is used for the resistance. Table 7.6 gives information on the termination impedances, transmission lines, transformer lines, and characteristic impedances of Z_b and Z_c. Figure 7.12 compares simulated results with measurements; frequency responses of power division and isolation are plotted in Fig. 7.12(a) and those of reflection coefficients in Fig. 7.12(b). Figure 7.12 shows that isolation is better than 15 dB in a 2.6 : 1 bandwidth, equal power division is performed in the range 0.2 to 5.8 GHz, and reflection coefficients are less than −15 dB in a 2.5 : 1 bandwidth.

In this chapter a new design method was presented for three-port 3-dB power dividers terminated in different impedances, and two types of scattering parameters were derived using two equivalent circuits. Based on the design equations derived, a new type of lumped-element three-port 3-dB power divider (LET3PD), which can be used for low-insertion loss, was also suggested. By the design method suggested, the total size of any integrated microwave circuit can be reduced and any type of three-port power divider can be analyzed easily.

EXERCISES

7.1 Verify that S_{12} and S_{21} in Table 7.1 are the same as those in Table 2.2.

7.2 Design a three-port power divider based on Fig. 7.1(a) with equal power division and termination impedances of $R_a = 50$ Ω, $R_b = 40$ Ω, and $R_c = 60$ Ω.

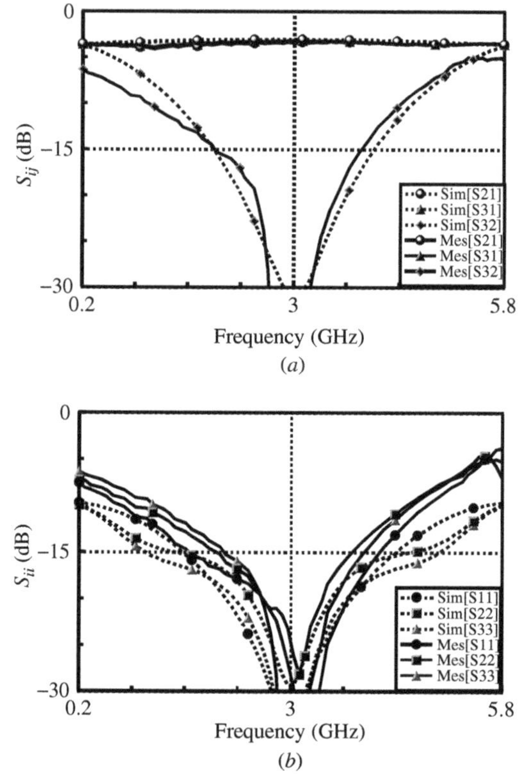

FIGURE 7.12 Measured and simulated results of a coplanar T3PD terminated in 30 Ω, 53 Ω, and 47 Ω: (*a*) power division and isolation characteristics; (*b*) reflection coefficients.

7.3 Derive the scattering parameters of a three-port power divider terminated in R_a, R_b, and R_c using (2.29*b*) or (2.30*b*).

7.4 Design a three-port power divider based on Fig. 7.9 with equal power division and termination impedances of $R_a = 50\ \Omega$, $R_b = 40\ \Omega$, and $R_c = 60\ \Omega$ at 3 GHz.

REFERENCES

1. H.-R. Ahn, I.-S. Chang, and S.-W. Yun, Miniaturized 3-dB Ring Hybrid Terminated by Arbitrary Impedances, *IEEE Trans. Microwave Theory Tech.*, Vol. 42, December 1994, pp. 2216–2221.
2. H.-R. Ahn, Ingo Wolff, and Ik-Soo Chang, Arbitrary Termination Impedances, Arbitrary Power Division and Small-Sized Ring Hybrids, *IEEE MTT-S Dig.*, June 1997, pp. 285–288.
3. H.-R. Ahn, I. Wolff, and I.-S. Chang, Arbitrary Termination Impedances, Arbitrary Power Division and Small-Sized Ring Hybrids, *IEEE Trans. Microwave Theory Tech.*, Vol. 44, December 1997, pp. 2241–2247.

4. H.-R. Ahn and I. Wolff, Asymmetric Four-Port and Branch-Line Hybrids, *IEEE Trans. Microwave Theory Tech.*, Vol. 48, September 2000, pp. 1585–1588.
5. H.-R. Ahn and I. Wolff, 3-dB Branch-Line Hybrids with Arbitrary Termination Impedance Values, *IEICE Trans. Electrons*, Vol. E82-C, No. 7, July 1999, pp. 1324–1326.
6. H.-R. Ahn and I. Wolff, Arbitrary Power Division Branch-Line Hybrid Terminated by Arbitrary Impedances, *IEE Electron. Lett.*, Vol. 35, No. 7, April 1999, pp. 572–273.
7. H.-R. Ahn and I. Wolff, 3-dB Branch-Line Hybrid Terminated by Arbitrary Impedances, *IEE Electron. Lett.*, Vol. 34, No. 11, May 1998, pp. 1109–1110.
8. H.-R. Ahn and I. Wolff, Asymmetric Four-Port Hybrids, Asymmetric 3-dB Branch-Line Hybrids, *Asia–Pacific Microwave Conf. Proc.*, Yokohama, Japan, December 1998, pp. 677–680.
9. H.-R. Ahn and I. Wolff, Three-Port 3-dB Power Divider Terminated by Arbitrary Impedances, *IEEE MTT-S Dig.*, Baltimore, June 1998, pp. 781–784.
10. H.-R. Ahn and I. Wolff, Three-Port 3-dB Power Divider Terminated by Different Impedances and Its Application to MMIC's, *IEEE Trans. Microwave Theory Tech.*, Vol. 47, June 1999, pp. 786–794.
11. R. B. Ekinge, A New Method of Synthesizing Matched Broad-Band TEM-Mode Three-Ports, *IEEE Trans. Microwave Theory Tech.*, Vol. 19, January 1971, pp. 81–88.
12. S. Rosloniec, Three-Port Hybrid Power Dividers Terminated in Complex Frequency-Dependent Impedances, *IEEE Trans. Microwave Theory Tech.*, Vol. 44, August 1996, pp. 1490–1493.
13. L. I. Parad and R. L. Moynihan, Split-Tee Power Divider, *IRE Trans. Microwave Theory Tech.*, Vol. 8, January 1965, pp. 91–95.
14. B. Kopp, Asymmetric Lumped Element Power Splitters, *IEEE MTT-S Dig.*, 1989, pp. 333–336.
15. S. B. Cohn, A Class of Broadband Three-Port TEM-Mode Hybrids, *IRE Trans. Microwave Theory Tech.*, Vol. 16, February 1968, pp. 110–116.
16. J. Helszajn, *Passive and Active Microwave Circuits,* Wiley, New York, 1978, pp. 101–102.
17. E. J. Wilkinson, An *n*-Way Hybrid Power Divider, *IRE Trans. Microwave Theory Tech.*, Vol. 8, January 1960, pp. 116–118.
18. C. Y. Pon, Hybrid-Ring Directional Coupler for Arbitrary Power Divisions, *IRE Trans. Microwave Theory Tech.*, Vol. 9, November 1961, pp. 529–535.
19. D. Köther, B. Hopf, T. Sporkmann, and I. Wolff, MMIC Wilkinson Couplers for Frequencies Up to 110 GHz, *IEEE MTT-S Dig.*, 1995, pp. 663–665.
20. T. Ohira, Y. Suzuki, H. Ogawa, and H. Kamitsuna, Megalithic Microwave Signal Processing for Phased-Array Beamforming and Steering, *IEEE Trans. Microwave Theory Tech.*, Vol. 45, December 1997, pp. 2324–2332.

CHAPTER EIGHT

General Design Equations for N-Way Arbitrary Power Dividers

8.1 INTRODUCTION

Three-port power dividers with equal power division were treated in earlier chapters. In this chapter, general design equations are derived for N-way power dividers with both arbitrary termination impedances and arbitrary power divisions. To derive the design equations, the case of three-port power dividers will first be investigated [1] and expanded to general N-way power dividers [2]. For the derivation of design equations, an arbitrary design impedance Z_{Ad}, which is arbitrary and positive, is introduced. Therefore, many sets of design equations will be available and called *general design equations*. It will be shown later that the performance of the three-port power divider using Z_{Ad} is, in terms of bandwidth, generally better than that of a conventional three-port power divider, as described by Wilkinson [3].

Based on the general design equations derived, a uniplanar three-port power divider with a power-split ratio of 3 dB and termination impedances of 50 Ω, 60 Ω, and 70 Ω will be fabricated, and the isolation resistance will be determined so that a commercially available resistor can be used by choosing the appropriate value of Z_{Ad}. In this way, Z_{Ad} may be determined for special purposes. As an example, a Z_{Ad} -based method is used to increase bandwidth.

In a manner similar to that for a T3PD (three-port 3-dB power divider), the general design equations will also be derived for N-way arbitrary power dividers, which consist of N transmission lines and N resistors and may be terminated in arbitrary impedances. If a transmission line is designed with too high

Asymmetric Passive Components in Microwave Integrated Circuits, by Hee-Ran Ahn

a characteristic impedance, microstrip fabrication of the component is difficult. The use of Z_{Ad} can solve this problem. To verify the design method, a three-way power divider with a power-split ratio of $1:2:4$ will be fabricated using microstrip technology, and it will be shown that the results measured are in good agreement with those predicted.

8.2 GENERAL DESIGN EQUATIONS FOR THREE-PORT POWER DIVIDERS

A three-port power divider for arbitrary power divisions and arbitrary termination impedances is depicted in Fig. 8.1(a). It consists of two pairs of transmission lines and a bridging isolation circuit (IC) between port i② and i③ and is terminated in arbitrary real impedances R_a, R_b, and R_c. The length of the pair of transmission lines with characteristic impedances Z_{02} and Z_{03} is Θ_1 and that with Z_{04} and Z_{05} is Θ_2, to satisfy the original definition that the two output signals are in phase. When the power is excited at port ①, a circuit equivalent to that in Fig. 8.1(b) may be drawn. If the power excited at port ① is split between ports i② and i③ depending on the power division ratio $1:K^2$ as indicated in Fig. 8.1(b), the input impedance $Z_{2,\text{in}}$ looking into the transmission line with Z_{02} is $R_a(1+K^2)$ and $Z_{3,\text{in}}$ is $R_a(1+K^2)/K^2$, as mentioned in(6.3). The relation is valid for both three-port power dividers [4–7] and ring hybrids. Therefore,

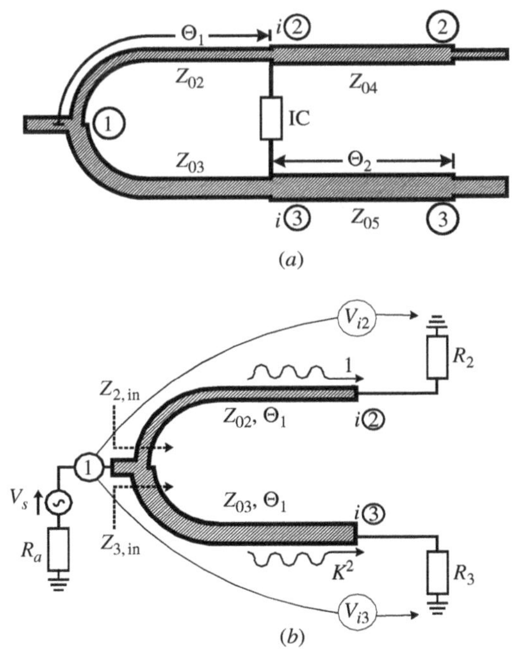

FIGURE 8.1 Three-port power divider with arbitrary power divisions and arbitrary termination impedances: (a) entire circuit of a three-port power divider; (b) equivalent circuit when power is fed into port ①. (From Ref. 1 with permission from IEEE.)

both asymmetric ring hybrids and asymmetric three-port power dividers have the following characteristics: The power excited at port ① is divided into two output ports, i② and i③, where the two signals are in phase; the voltage V_{i2} (from port ① via port i② to ground) is equal to that of V_{i3} (from port ① via port i③ to ground) in Fig. 8.1(*b*); since the ratio of $Z_{2,\text{in}}$ to $Z_{3,\text{in}}$ is K^2, the current entering the transmission line with Z_{03} is K^2 times that entering that with Z_{02}; and an isolation circuit is necessary for matching at two output ports and for isolation between the two ports.

However, for asymmetric ring hybrids, the characteristic impedance Z_{02} is not related to Z_{03}; since the isolation circuit consists of two other lossless transmission-line sections and is not related to the circuit in Fig. 8.1(*b*), the isolation characteristics may be defined as forward isolation; R_2 and R_3 are independent of the power-split ratio; and the input reflection coefficient S_{11} is not a function of reflection coefficients S_{22} and S_{33} at ports ② and ③. For three-port power dividers, the ratio of R_2 to R_3 should be K^2 to satisfy the condition that the voltage at port i② is equal to that at port i③; since the isolation circuit is related to the two transmission lines with Z_{02} and Z_{03} and the termination impedance R_a, the isolation characteristics may be defined as a backward isolation; and S_{11} is a function of S_{22} and S_{33}, as explained in (7.44).

The ratio of R_2 to R_3 in Fig. 8.1(*b*) yields

$$R_2 = K^2 Z_{Ad}, \tag{8.1a}$$

$$R_3 = Z_{Ad}, \tag{8.1b}$$

where the Z_{Ad} is an arbitrary and positive real resistance.

Depending on the value of Z_{Ad}, many sets of design equations are available. Therefore, Z_{Ad} may be defined as an arbitrary design impedance. Parad and Moynihan [8] suggested $R_2 = KR_a$ and $R_3 = R_a/K$ as a trial value among many possible cases in (8.1). Transmission lines with characteristic impedances Z_{02}, Z_{03}, Z_{04}, and Z_{05} are impedance transformers and they are obtained with $\Theta_1 = \Theta_2 = 90^\circ$ as

$$Z_{02} = \sqrt{Z_{Ad} R_a K^2 (1 + K^2)}, \tag{8.2a}$$

$$Z_{03} = \sqrt{Z_{Ad} R_a \frac{1 + K^2}{K^2}}, \tag{8.2b}$$

$$Z_{04} = K\sqrt{Z_{Ad} R_b}, \tag{8.2c}$$

$$Z_{05} = \sqrt{Z_{Ad} R_c}. \tag{8.2d}$$

The isolation circuit (IC) in Fig. 8.1 consists of resistance only if $\Theta_1 = 90^\circ$, but it will be shown later that the isolation circuit is not directly related to R_2 and R_3 and generally consists of resistance combined with capacitance or inductance. When $\Theta_1 = 90^\circ$, the isolation circuit consists of pure resistance and the resistance R_{es} gives

$$\text{IC} \rightarrow R_{es} = Z_{Ad}(1 + K^2). \tag{8.3}$$

Design equations (8.2) and (8.3) contain Z_{Ad}, which is determined independently. Therefore, they may be called general design equations of three-port power dividers. The design equations of the conventional three-port power divider [8] are only one of many possible sets in the case of $R_a = R_b = R_c = Z_0$. The three-port power dividers are easily realized with microstrip lines, striplines, coplanar waveguides, and so on, but commercially available resistors limit their realization in microwave integrated circuit (MIC) and hybrid microwave integrated circuit (HIC) technology. However, proper choice of the design impedance Z_{Ad} allows three-port power dividers to be realized easily.

When a conventional three-port power divider [8] is terminated in equal impedances of 50 Ω, its power-split ratio is 3 dB, and Z_{Ad} is 35.39 Ω, the corresponding values are $Z_{02} = 102.85$ Ω, $Z_{03} = 51.54$ Ω, $Z_{04} = 59.43$ Ω, $Z_{05} = 42.07$ Ω, and $R_{es} = 106.03$ Ω. In another case of the same power-split

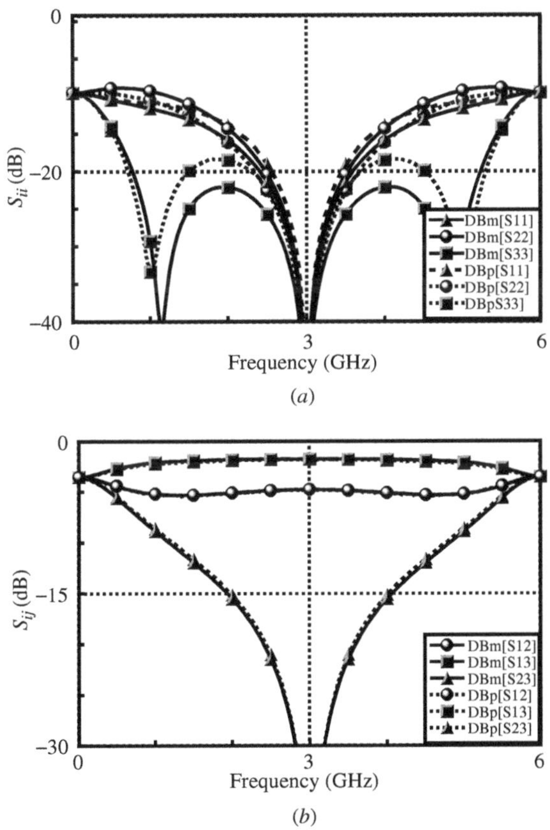

FIGURE 8.2 Results of conventional and suggested three-port power dividers compared (all termination impedances 50 Ω, power-split ratio 3 dB, $K^2 = 1.995$, design center frequency 3 GHz): (*a*) frequency responses of reflection coefficients; (*b*) frequency responses of power division and isolation characteristics.

TABLE 8.1 Results of Conventional and Suggested Three-Port Power Dividers Compared

	Using a General Design Equation	Conventional Power Divider
Z_{Ad} (design impedance)	39.5 Ω	35.39 Ω
Power division	$\|S_{21}\| = -4.76$ dB	$\|S_{21}\| = -4.76$ dB
	$\|S_{31}\| = -1.76$ dB	$\|S_{31}\| = -1.76$ dB
Matching	$\|S_{11}\| = -118.2$ dB	$\|S_{11}\| = -91.0$ dB
	$\|S_{22}\| = -96.16$ dB	$\|S_{22}\| = -78.16$ dB
	$\|S_{33}\| = -95.2$ dB	$\|S_{33}\| = -84.5$ dB
Isolation	$\|S_{23}\| = -101.9$ dB	$\|S_{23}\| = -90.0$ dB
Bandwidth less than −20 dB in reflection coefficients and isolations	1.5 : 1 at port ①	1.3 : 1 at port ①
	1.4 : 1 at port ②	1.6 : 1 at port ②
	6.5 : 1 at port ③	1.6 : 1 at port ③
	1.5 : 1 between ports ② and ③	1.45 : 1 between ports ② and ③

[a] All termination impedances 50 Ω, power-split ratio 3 dB, $K^2 = 1.995$, design center frequency 3 GHz.

ratio, the same termination impedance, but $Z_{Ad} = 39.5\ \Omega$, $Z_{02} = 108.6\ \Omega$, $Z_{03} = 54.45\ \Omega$, $Z_{04} = 62.775\ \Omega$, $Z_{05} = 44.44\ \Omega$, and $R_{es} = 118.3\ \Omega$. The two dividers were designed at a design center frequency of 3 GHz and, under ideal conditions, simulated by a circuit simulator. The reflection coefficients are plotted in Fig. 8.2(*a*), and the power division and isolation frequency responses are shown in Fig. 8.2(*b*). The solid lines represent the case of $Z_{Ad} = 39.5\ \Omega$ and the dashed lines represent those from [8], where $Z_{Ad} = 35.39\ \Omega$. Simulation results of the two power dividers are given in Table 8.1. Perfect matching and isolation appear with the values $|S_{11}| = -118.2$ dB, $|S_{22}| = -96.16$ dB, $|S_{33}| = -95.2$ dB, and $|S_{23}| = -101.8$ dB in the three-port power divider with $Z_{Ad} = 39.5\ \Omega$. From the data in Table 8.1, the frequency dependencies of the three-port power divider with $Z_{Ad} = 39.5\ \Omega$ are, in terms of bandwidth, better than those of a conventional three-port power divider.

8.2.1 Coplanar Three-Port Power Divider Terminated in 50 Ω, 60 Ω, and 70 Ω

On the basis of general design equations (8.2) and (8.3), a coplanar three-port power divider terminated in 50 Ω, 60 Ω, and 70 Ω was fabricated on Al_2O_3 substrate ($\varepsilon_r = 9.9$, $h = 635$ μm). Its power-split ratio was 3 dB and the arbitrary design impedance Z_{Ad} was set to 33.33 Ω to adjust the isolation resistance to a commercially available 100-Ω resistor. The corresponding values were $K^2 = 2$, $Z_{02} = 100\ \Omega$, $Z_{03} = 50\ \Omega$, $Z_{04} = 63.25\ \Omega$, $Z_{05} = 48.3\ \Omega$, and $R_{es} = 100\ \Omega$ at

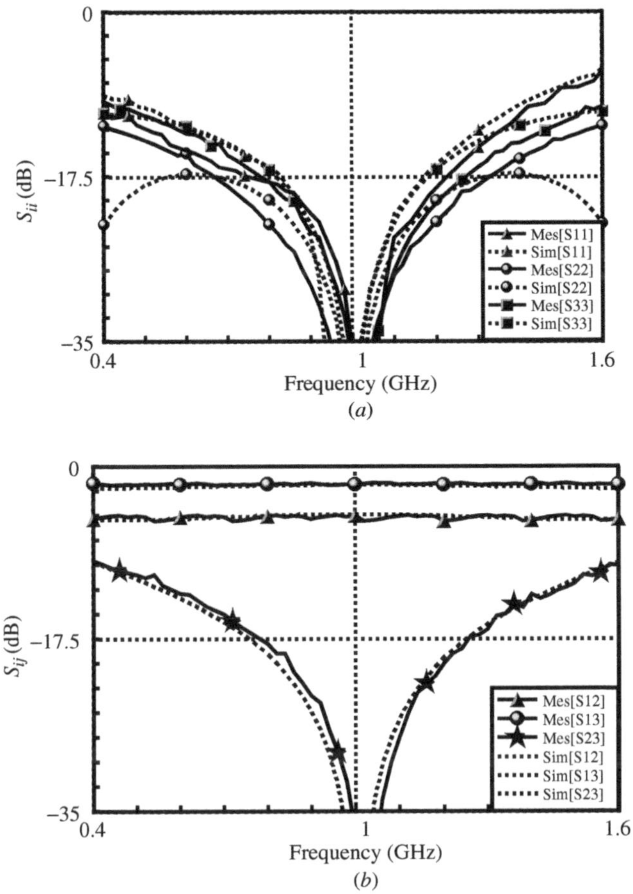

FIGURE 8.3 Results measured for a uniplanar three-port power divider with a power-split ratio of 3 dB and termination impedances of 50 Ω, 60 Ω, and 70 Ω: (*a*) frequency responses of reflection coefficients; (*b*) frequency responses of power division and isolation characteristics.

a center frequency of 1 GHz. The measured matching frequency responses are described in Fig. 8.3(*a*) and power division and isolation frequency responses in Fig. 8.3(*b*), where the solid lines are the measured results and the dotted lines the simulated results. In the case of $K^2 = 2$, the power-split ratio is 3 dB, and the measured result of S_{31} and S_{21} are −1.84 dB and −4.8 dB at 1 GHz. The measured matching and isolation are less than −35 dB within a relative 28% bandwidth.

8.2.2 Determining Z_{Ad}

The arbitrary design impedance Z_{Ad} was determined such that a commercially available 100-Ω resistor could be used for the coplanar three-port power divider mentioned above. Z_{Ad} can also be adjusted for other purposes. In the conventional three-port power divider in Table 8.1, if the bandwidth is defined as a frequency range where the reflection coefficients are less than 20 dB, that of S_{11} is 1.3 : 1, which is the worst. The worst property of S_{11} limits the three-port power divider to a 1.3 : 1 bandwidth. Z_{Ad} can be used to increase the bandwidth. The transmission line with Z_{02} transforms the impedance $Z_{2,\text{in}}$ into R_2, and the transmission line with Z_{03} transforms the impedance $Z_{3,\text{in}}$ into R_3 in Fig. 8.1(*b*). Both impedance transformation ratios are equal and are given as

$$IR_{1i3} = IR_{1i2} = R_a \frac{1 + K^2}{K^2 Z_{Ad}}, \tag{8.4}$$

where IR_{1i3} and IR_{1i2} are the impedance transformation ratios between the two ports. These ratios are not dependent on the termination impedances R_b and R_c and are determined solely by R_a and K^2. Since the two impedance transformation ratios are always equal, the frequency responses of the power division characteristics show wideband performances, as can be seen in Figs. 8.2 and 8.3, and IR_{1i3} and IR_{1i2} determine the bandwidth of the reflection coefficient S_{11} at port ①. The termination impedance R_a and the power-split ratio K^2 are given and cannot be determined arbitrarily in (8.4). Therefore, by a proper choice of the arbitrary design impedance Z_{Ad}, not only the bandwidth of S_{11} but also that of the three-port power divider can be increased. Figure 8.4 shows the relation between K^2, Z_{Ad}, and IR_{1i3}, when $R_a = 50\ \Omega$, $Z_{2,\text{in}} \geq R_2$, and $Z_{3,\text{in}} \geq R_3$. "P.R." on the graph indicates the power-split ratio K^2 in decibels. The results show that the impedance transformation ratio *IR* is inversely proportional to the design impedance Z_{Ad} as shown in (8.8), and that it has the highest values with "P.R. 0 dB," which determines the smallest bandwidth of S_{11}.

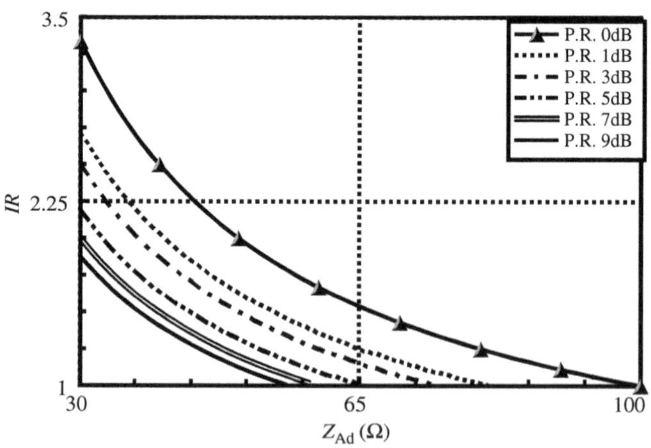

FIGURE 8.4 Impedance transformation ratio IR_{1i2} as a function of power division ratio K^2 and arbitrary design impedance Z_{Ad}.

Since R_2 and R_3 influence S_{22} and S_{33} as shown in (7.34), the determination of Z_{Ad} is limited by the design of the next stage. The two transmission lines with characteristic impedances Z_{04} and Z_{05} have no relation to each other, and the two waves at ports ② and ③ are assumed to be in phase. Nevertheless, the smaller the difference between the two impedance (admittance) transformation ratios IR_{i33} and IR_{i22}, the better the performance that can be expected. To reduce the difference, first find a value of Z_{Ad} that satisfies the minimum difference among the four impedances R_2, R_b, R_3, and R_c. That is,

$$\min \| R_b - R_2| - |R_c - R_3\|. \tag{8.5}$$

However, the value of Z_{Ad} satisfying (8.5) does not guarantee that the difference between IR_{i33} and IR_{i22} is minimized. The value is just an initial value to decide $IR_{i22} = R_2/R_b$ or R_b/R_2 and $IR_{i33} = R_3/R_c$ or R_c/R_3. After determining both IR_{i33} and IR_{i22} greater than unity, the difference is defined as

$$\text{Diff} = \min |IR_{i33} - IR_{i22}|. \tag{8.6}$$

After calculating the value of Z_{Ad} satisfying (8.6), compare S_{11} with S_{22} and S_{33}. If S_{11} is better than the worst of the two, use the value of Z_{Ad}. Otherwise, increase Z_{Ad} until S_{11} is about the same as the worst value. For a conventional three-port power divider with a power-split ratio of 3 dB and equal termination impedances of 50 Ω, the value of Z_{Ad} satisfying (8.10) is 35.39 Ω, which is exactly that suggested by [8]. However, the bandwidth of S_{11} is not wider than that of S_{22} or S_{33}, as shown in Table 8.1. Therefore, Z_{Ad} should be increased to increase the bandwidth of S_{11}.

8.3 GENERAL DESIGN EQUATIONS FOR N-WAY POWER DIVIDERS

The concepts of the general design equations for three-port power dividers discussed so far can be extended to the general N-way power divider. It consists of N quarter-wavelength transmission lines and N resistors. The most popular N-way in-phase power divider was proposed by Wilkinson in 1960 [3]. He described a circularly symmetric power divider, which split a signal into N equiphase–equiamplitude signals with an even or odd number of N. Since that time, there have been many studies on N-way power dividers [9–12], but they deal with equal power dividers and equal termination impedances. If the N-way power dividers are terminated in arbitrary impedances, a reduction in the total size of integrated circuit can be expected as asymmetric ring hybrids, asymmetric branch-line hybrids, and asymmetric three-port power dividers [13–19] do. In this section, the design equations for N-way arbitrary power dividers terminated in arbitrary impedances are derived, and one way to solve fabrication problems is discussed together with the proper choice of the arbitrary design impedance Z_{Ad}. To verify the design equations, a four-way power divider with a power-split ratio of 1 : 2 : 4 : 8 will be simulated, and a three-way power divider with a power-split ratio of 1 : 2 : 4 will be fabricated with microstrip technology.

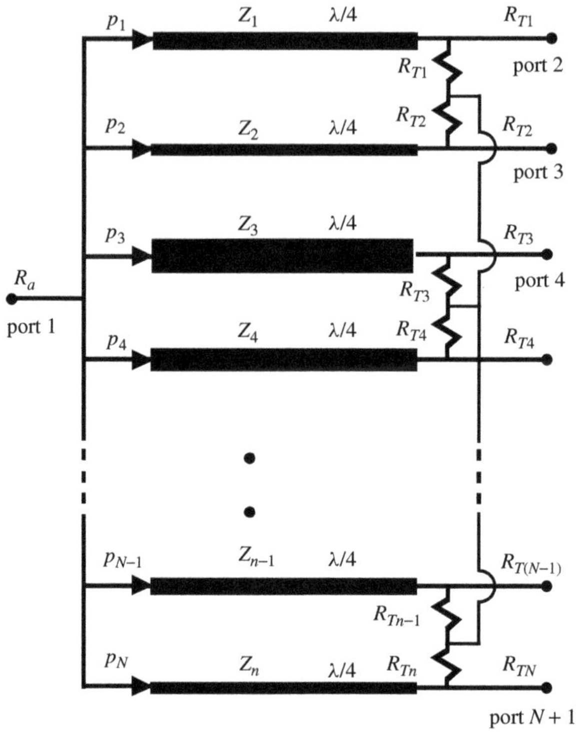

FIGURE 8.5 N-way arbitrary power dividers.

8.3.1 Analyses of N-Way Power Dividers

An N-way arbitrary power divider with an arbitrary power division ratio is depicted in Fig. 8.5, where all the transmission lines are λ/4 long. R_a, $R_1, \ldots, R_N$ are termination impedances, $Z_1, Z_2, \ldots, Z_n$ are characteristic impedances of the transmission lines, and $P_1 : P_2 : P_3 : \cdots : P_n$ is a power-split ratio. The port number is defined as shown in Fig. 8.5. The design formulas are

$$R_{TN} = \frac{P_{\text{tot}}}{P_N} Z_{Ad}, \tag{8.7a}$$

$$Z_N = \sqrt{R_a R_{TN} \frac{P_{\text{tot}}}{P_N}}, \tag{8.7b}$$

$$ZT_N = \sqrt{R_{TN} R_N}, \tag{8.7c}$$

where $P_{\text{tot}} = P_1 + P_2 + P_3 + \cdots + P_N$, R_{TN} are isolation resistances; Z_{Ad} is the arbitrary design impedance which can be determined by design situations, $R_1, R_2, \ldots, R_N$ are final output termination impedances, and $ZT_1, ZT_2, \ldots, ZT_N$ are the characteristic impedances of additional impedance transformers to transform R_{TN} into R_N.

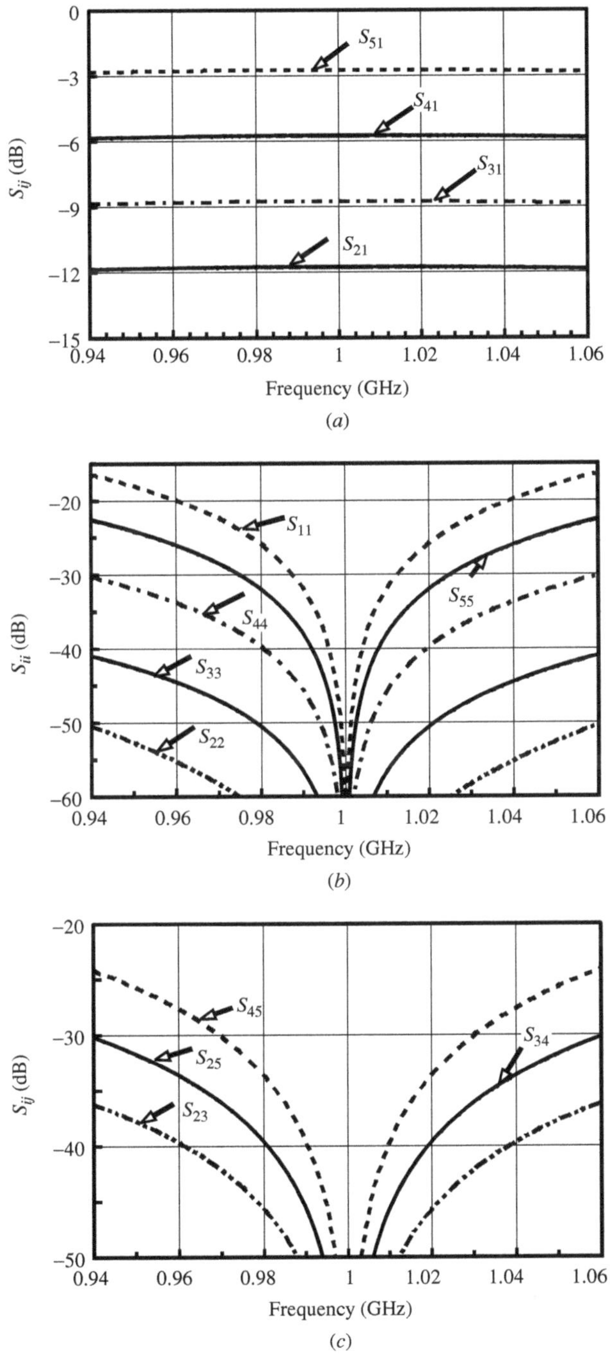

FIGURE 8.6 Simulation results of a four-way power divider with a power-split ratio of 1 : 2 : 4 : 8: (*a*) frequency responses of power division characteristics; (*b*) frequency responses of reflection coefficients; (*c*) frequency responses of isolation characteristics.

The amount of power in decibels delivered to each branch is calculated using

$$P_N = 10\ \log_{10} \frac{P_N}{P_{\text{tot}}}. \tag{8.8}$$

Based on the design formula, a four-way power divider with a power division ratio of $P_1 : P_2 : P_3 : P_4 = 1 : 2 : 4 : 8$, $Z_{Ad} = 2$, and $R_a = 25\ \Omega$ was designed at 1 GHz. The corresponding values were $Z_1 = 106.6\ \Omega$, $Z_2 = 53\ \Omega$, $Z_3 = 26.5\ \Omega$, and $Z_4 = 13.2\ \Omega$. It was simulated under ideal conditions, and simulation results are plotted in Fig. 8.6, where the power division responses are shown in Fig. 8.6(*a*), the reflection coefficients in Fig. 8.6(*b*), and the isolation characteristics in Fig. 8.6(*c*). Since the power division ratio is 1 : 2 : 4 : 8, the difference between them is equally 3 dB, and the value of S_{51} is −2.73 dB, based on (8.8). All the ports are perfectly matched, and perfect isolation between each port can be achieved as shown in Fig. 8.6(*b*) and (*c*), respectively.

To verify the method suggested above, a three-way power divider was designed at 3 GHz and fabricated with microstrip technology. The design data are given in Table 8.2. The power division ratio was $P_1 : P_2 : P_3 = 1 : 2 : 4$ and $R_a = 25\ \Omega$. In this case, $Z_{Ad} = 6$ and was determined so that the termination impedances, R_{T1}, R_{T2}, and R_{T3}, could have suitable values. The characteristic impedances of Z_1, Z_2, and Z_3 are 85.7 Ω, 42.9 Ω and 21.4 Ω, respectively, and the widths and lengths of microstrip lines on Teflon substrate ($\varepsilon_r = 2.88$, $h = 508\ \mu$m) are also given in Table 8.2. However, the width of the transmission line with Z_3 was too wide, so it needed to be realized in suitable transmission lines. Figure 8.7 shows an equivalent circuit where one transmission line can be realized with two transmission lines with twice the characteristic impedance. ZT_a, ZT_{T1}, and ZT_{T2} are the characteristic impedances of the quarter-wavelength impedance transformers to transform R_a, R_{T1}, and R_{T2} into equal impedances of 50 Ω. Port ④ terminated in R_{T3} is eliminated for easier measurements, so the number of remaining

TABLE 8.2 Fabrication Data for a Three-Way Power Divider[a]

$Z_{Ad} = 6$, $R_a = 25\ \Omega$, $P_1 : P_2 : P_3 = 1:2:4$	
$R_{T1} = 42\ \Omega$, $R_{T2} = 21\ \Omega$, $R_{T3} = 10.5\ \Omega$	
$\varepsilon_r = 2.88$, $h = 508\ \mu$m	
$Z_1 = 85.7\ \Omega$	$ZT_a = 35.35\ \Omega$
$w = 500.4\ \mu$m	$w = 2178\ \mu$m
$l = 16{,}820.3\ \mu$m	$l = 15{,}983\ \mu$m
$Z_2 = 42.9\ \Omega$	$ZT_{T1} = 45.82\ \Omega$
$w = 1653.5\ \mu$m	$w = 1499.4\ \mu$m
$l = 16{,}153.4\ \mu$m	$l = 16{,}213\ \mu$m
$Z_3 = 21.4\ \Omega$	$ZT_{T2} = 32.4\ \Omega$
$w = 4139\ \mu$m	$w = 2441\ \mu$m
$l = 15{,}603\ \mu$m	$l = 15{,}913\ \mu$m

[a] *w*, Center strip width; *l*, line length.

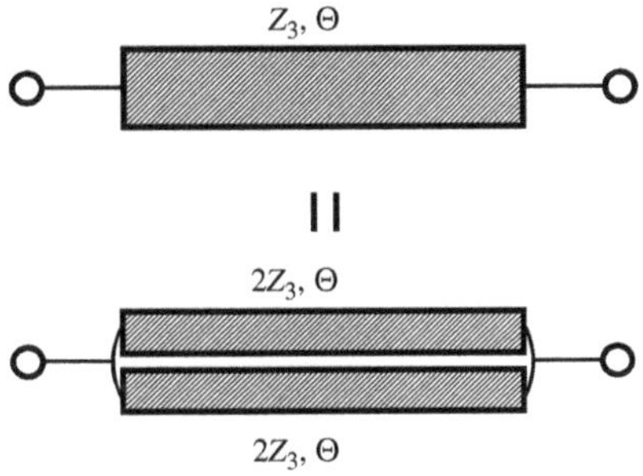

FIGURE 8.7 Fabrication of a transmission-line section with two other transmission-line sections.

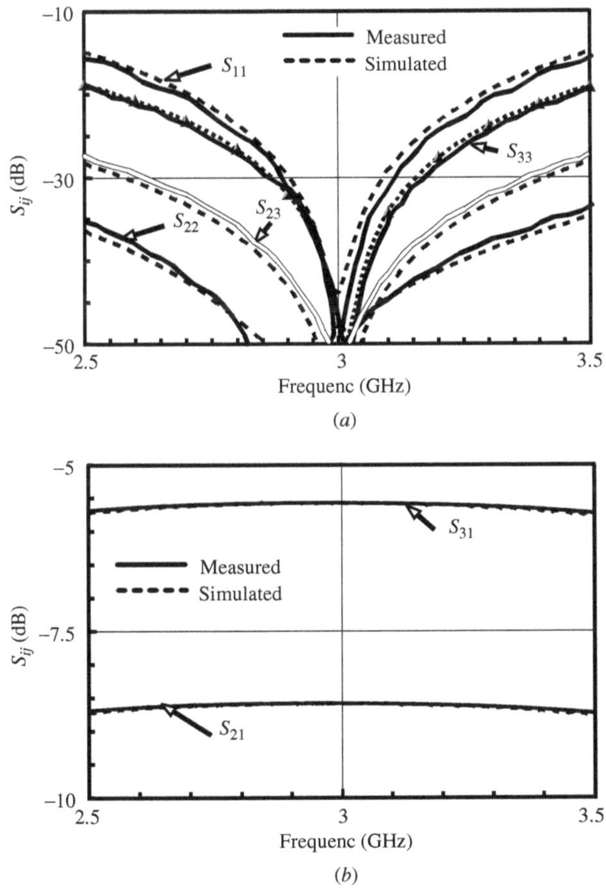

FIGURE 8.8 Results measured for a three-way power divider with a power-split ratio of 1 : 2 : 4: (*a*) frequency responses of reflection coefficients and isolation; (*b*) frequency responses of power division characteristics.

ports is three. The results measured are compared with the results simulated in Fig. 8.8, where matching and isolation results are given in Fig. 8.8(*a*), and the powers delivered to each port are shown in Fig. 8.8(*b*). All the results measured are in good agreement with those simulated.

EXERCISES

8.1 Determine Z_{Ad} so that a 100-Ω isolation resistor can be used when the three-port power divider in Fig. 8.1(*a*) is terminated in 50 Ω at port ①, 30 Ω at port ②, and 40 Ω at port ③.

8.2 Verify that one transmission line can be realized with two transmission lines, as shown in Fig. 8.7 using circuit parameters.

8.3 Design a three-way power divider with a power-split ratio of $1:2:4:8$ and termination impedances of 50 Ω, 40 Ω, 60 Ω, and 30 Ω at ports ①, ②, ③, and ④, respectively.

REFERENCES

1. H.-R. Ahn and I. Wolff, General Design Equations of Three-Port Power Dividers, Small-Sized Impedance Transformers, and Their Applications to Small-Sized Three-Port 3-dB Power Divider, *IEEE Trans. Microwave Theory Tech.*, Vol. 49, July 2001, pp. 1277–1288.
2. H.-R. Ahn, K. Lee, and N. H. Myung, General Design Equations of *N*-Way Arbitrary Power Dividers, *IEEE MTT-S Dig.*, June 2004, pp. 65–68.
3. E. J. Wilkinson, An *n*-Way Hybrid Power Divider, *IRE Trans. Microwave Theory Tech.*, Vol. 8, January 1960, pp. 116–118.
4. H.-R. Ahn, I.-S. Chang, and S.-W. Yun, Miniaturized 3-dB Ring Hybrid Terminated by Arbitrary Impedances, *IEEE Trans. Microwave Theory Tech.*, Vol. 42, December 1994, pp. 2216–2221.
5. H.-R. Ahn, I. Wolff and I.-S. Chang, Arbitrary Termination Impedances, Arbitrary Power Division and Small-Sized Ring Hybrids, *IEEE MTT-S Dig.*, June 1997, pp. 285–288.
6. H.-R. Ahn, I. Wolff, and I.-S. Chang, Arbitrary Termination Impedances, Arbitrary Power Division and Small-Sized Ring Hybrids, *IEEE Trans. Microwave Theory Tech.*, Vol. 44, December 1997, pp. 2241–2247.
7. H.-R. Ahn and I. Wolff, Three-Port 3-dB Power Divider Terminated by Different Impedances and Its Application to MMIC's, *IEEE Trans. Microwave Theory Tech.*, Vol. 47, June 1999, pp. 786–794.
8. L. I. Parad and R. L. Moynihan, Split-Tee Power Divider, *IRE Trans. Microwave Theory Tech.*, Vol. 8, January 1965, pp. 91–95.
9. F. Ardemagni, An Optimized *L*-Band Eight-Way Gysel Power Divider-Combiner, *IEEE Trans. Microwave Theory and Tech.*, Vol. 31, June 1983, pp. 491–495.

10. U. H. Gysel, A New N-Way Power Divider/Combiner Suitable for High Power Applications, *MTT-S Dig.*, 1975, pp. 116–118.

11. K. J. Russell, Microwave Power Combining Techniques, *IEEE Trans. Microwave Theory and Tech.*, Vol. 27, May 1979, pp. 472–478.

12. F. A. Alhargan, Circular and Annular Sector Planar Components of Arbitrary Angle for N-Way Power Divider/Combiners, *IEEE Trans. Microwave Theory and Tech.*, Vol. 42, July 2001, pp. 1617–1623.

13. H.-R. Ahn and I. Wolff, Asymmetric Four-Port and Branch-Line Hybrids, *IEEE Trans. Microwave Theory Tech.*, Vol. 48, September 2000, pp. 1585–1588.

14. H.-R. Ahn and I. Wolff, 3-dB Branch-Line Hybrids with Arbitrary Termination Impedance Values, *IEICE Trans. Electrons*, Vol. E82-C, No. 7, July 1999, pp. 1324–1326.

15. H.-R. Ahn and I. Wolff, Arbitrary Power Division Branch-Line Hybrid Terminated by Arbitrary Impedances, *IEE Electron. Lett.*, Vol. 35, No. 7, April 1999, pp. 572–273.

16. H.-R. Ahn and I. Wolff, 3-dB Branch-Line Hybrid Terminated by Arbitrary Impedances, *IEE Electron. Lett.*, Vol. 34, No. 11, May 1998, pp. 1109–1110.

17. H.-R. Ahn and I. Wolff, Asymmetric Four-Port Hybrids, Asymmetric 3-dB Branch-Line Hybrids, *Asia–Pacific Microwave Conf. Proc.*, Yokohama, Japan, December 1998, pp. 677–680.

18. H.-R. Ahn and I. Wolff, Asymmetric Four-Port and Branch-Line Hybrids, *IEEE Trans. Microwave Theory Tech.*, Vol. 48, September 2000, pp. 1585–1588.

19. H.-R. Ahn and I. Wolff, 3-dB Branch-Line Hybrids with Arbitrary Termination Impedance Values, *IEICE Trans. Electrons*, Vol. E82-C, No. 7, July 1999, pp. 1324–1326.

CHAPTER NINE

Asymmetric Ring-Hybrid Phase Shifters and Attenuators

9.1 INTRODUCTION

Phase shifters and attenuators are common two-port components. In this chapter, asymmetric ring-hybrid phase shifters and attenuators are introduced and discussed as an application of the asymmetric ring hybrid. Phase shifters have various applications in microwave equipment: as linearizers, antennas, and so on. Phase shifters using passive four-port devices such as branch-line hybrids [1], ring hybrids [2], parallel-coupled directional couplers (backward-wave couplers), and Lange couplers [3,4] can provide a constant phase shift across their two output ports. The 90° couplers, such as branch-line hybrids, parallel-coupled directional couplers, Lange couplers, and so on, have the advantage of using symmetric reflecting terminations. However, branch-line hybrids have narrow bandwidths [5–8], and parallel-coupled directional couplers and Lange couplers are not easy to fabricate. In addition to these disadvantages, they require two additional 90° delay lines to realize 180° phase shifters [9].

For 180° phase shifters, ring hybrids are good candidates. A phase shifter using a ring hybrid was proposed by White [2], but it had an additional 90° phase delay line to utilize two symmetric reflecting terminations. In this chapter, phase shifters and attenuators using asymmetric ring hybrids without a 90° phase delay line are presented. First, the scattering matrix of the asymmetric ring hybrid is derived from a two-port equivalent circuit proposed by Ahn et al. [10]. On the basis of the scattering matrix derived, that of an asymmetric ring-hybrid phase shifter is calculated, and why a 90° phase-delay line is not necessary is explained. A conventional ring-hybrid phase shifter and an asymmetric ring-hybrid phase shifter are simulated and compared with respect to a 180° phase

Asymmetric Passive Components in Microwave Integrated Circuits, by Hee-Ran Ahn

shift. The results will show that the performance of the asymmetric ring-hybrid phase shifter is better. While conventional phase shifters have used symmetric couplers together with symmetric reflecting terminations [2–4,11–13], an asymmetric coupler consists of an asymmetric ring hybrid and asymmetric reflecting terminations. This fact allows an additional function as an impedance transformer because of the arbitrary termination impedances. To analyze the phase shifters, the characteristic impedances of the two transmission lines for the reflecting terminations are normalized to the termination impedances R_b and R_d, and the normalized impedances NI_b and NI_d, allowing design freedom. Using this concept, there are many choices for designing the reflecting terminations and reducing their size. Normally, if the size of microwave components is smaller, the output performance is degraded in terms of bandwidth [14,15]. However, even though the reflecting terminations are smaller, the asymmetric ring-hybrid phase shifters produce the same performances with different values of NI_b and NI_d. This is a significant design advantage. On the basis of the concept described above, a uniplanar asymmetric ring-hybrid $-135°$ phase shifter terminated in 30 Ω and 60 Ω will be fabricated on an Al_2O_3 substrate at a center frequency of 3 GHz.

Additionally, asymmetric ring-hybrid attenuators are synthesized using the phase shifter concept. For a certain attenuation, two resistances are needed at the end of the extending transmission lines. To use commercial resistors in MICs and HICs, the normalized resistance ratios NR_{Lb} and NR_{Ld} are introduced for asymmetric ring-hybrid attenuators. Based on the analyses, 6-dB attenuators with three different phase shifts, $-170°$, $-160°$, and $-135°$, are simulated and subpoles considered. These subpoles may be used to increase the bandwidths of microwave components [18]. Also, a microstrip asymmetric ring-hybrid 4-dB attenuator with a 45° phase shift and termination impedances of 30 Ω and 60 Ω is fabricated on Al_2O_3 at a center frequency of 3 GHz. This attenuator has the additional function of an impedance transformer to transform a 30-Ω system to a 60-Ω system.

9.2 SCATTERING PARAMETERS OF ASYMMETRIC RING HYBRIDS

Figure 9.1 shows an asymmetric ring hybrid terminated in arbitrary impedances R_a, R_b, R_c, and R_d. As this asymmetric ring hybrid has no symmetry planes, the even- and odd-mode equivalent circuits cannot be constructed. To derive the scattering parameters, a two-port equivalent circuit is indispensable because derivation of the scattering parameters may be complicated using four-port calculations. Figure 9.2 shows a two-port equivalent circuit when the power is excited at port ① [10]. The admittance matrix $\mathbf{Y}$ of the network $[N]$ in Fig. 9.2 and the reference admittance matrix $\mathbf{y}$ between ports ① and ② are given as

$$\mathbf{Y} = \frac{1}{Z_1 j \sin\Theta_1} \begin{bmatrix} \dfrac{jZ_1}{Z_{in}} \sin\Theta_1 + \cos\Theta_1 & -1 \\ -1 & \cos\Theta_1 \end{bmatrix}, \tag{9.1}$$

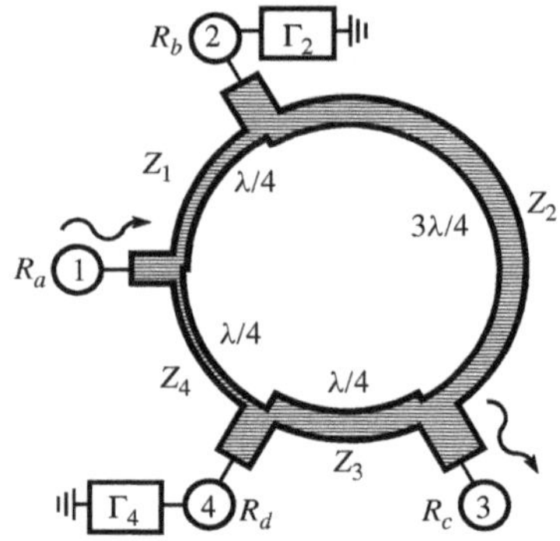

FIGURE 9.1 Asymmetric ring-hybrid phase shifter or attenuator.

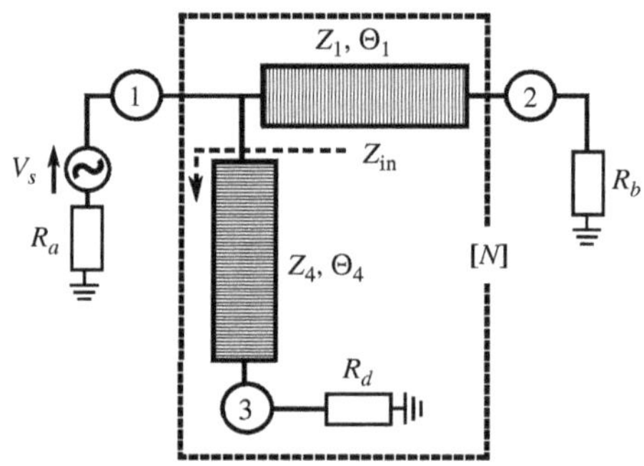

FIGURE 9.2 Two-port equivalent circuit of an asymmetric ring hybrid when power is fed into port ①.

where

$$Z_{\text{in}} = Z_4 \frac{R_d + jZ_4 \tan\Theta_4}{Z_4 + jR_d \tan\Theta_4},$$

$$y = \begin{bmatrix} \dfrac{1}{R_a} & 0 \\ 0 & \dfrac{1}{R_b} \end{bmatrix}.$$

Applying $\Theta_1 = \Theta_4 = \pi/2$ to (9.1) and performing additional calculations using the design equations, the scattering matrix $\boldsymbol{S}_{12}$ with respect to ports ① and ② is derived as

$$\boldsymbol{S}_{12} = \boldsymbol{k}\boldsymbol{S}^V\boldsymbol{k}^{-1*} = \begin{bmatrix} 0 & -j\sqrt{\dfrac{d_1^2}{d_1^2 + d_2^2}} \\ -j\sqrt{\dfrac{d_1^2}{d_1^2 + d_2^2}} & \dfrac{d_2^2}{d_1^2 + d_2^2} \end{bmatrix}, \tag{9.2}$$

where $\boldsymbol{kk}^* = \frac{1}{2}(\boldsymbol{y} + \boldsymbol{y}^*)$, $\boldsymbol{S}^V = -(\boldsymbol{Y} + \boldsymbol{y})^{-1}(\boldsymbol{Y} - \boldsymbol{y}^*)$, and the ratio of d_1 to d_2 is the scattering parameter ratio of S_{21} to S_{41}. A more precise process is described in the appendix in [18]. As shown in (9.2), the equivalent circuit in Fig. 9.2 is not perfectly matched at port ②. The purpose of the equivalent circuit is to derive S_{21} and S_{11}. On the basis of the scattering matrix $\boldsymbol{S}_{12}$ derived in (9.2), all scattering parameters are easily obtained. For example, the element S_{41} is calculated from the relation of $|S_{21}|^2 + |S_{41}|^2 = 1$ described by unitary property $\boldsymbol{SS}^{*T} = \boldsymbol{U}$, where $\boldsymbol{S}$ is a scattering matrix, $\boldsymbol{U}$ an identity matrix, and $\boldsymbol{S}^{*T}$ a complex conjugate and transverse scattering matrix.

Assuming that $\Theta_2 = 3\pi/4$ and $\Theta_3 = \pi/4$, the scattering parameters of the asymmetric ring hybrid using the unitary and reciprocity properties are therefore derived as

$$S_{\text{ring}} = \frac{1}{\sqrt{d_1^2 + d_2^2}} \begin{bmatrix} 0 & -jd_1 & 0 & -jd_2 \\ -jd_1 & 0 & jd_2 & 0 \\ 0 & jd_2 & 0 & -jd_1 \\ -jd_2 & 0 & -jd_1 & 0 \end{bmatrix}. \tag{9.3}$$

In the case of conventional ring hybrids [19–21], the relative output voltages at the output ports ① and ④ in Fig. 9.1 are given by $d_1/d_2 = Z_4/Z_1$. For perfect matchings, $(Z_0/Z_1)^2 + (Z_0/Z_2)^2 = 1$ is also required where the ring hybrid is terminated in the system impedances Z_0. However, for the asymmetric ring hybrid, the scattering parameter ratio is d_1 to d_2, which is no longer proportional to the characteristic admittance ratio of Z_4 to Z_1 [10]. It is a function of both the termination impedances R_b and R_d and characteristic impedances Z_4 and Z_1. Therefore, all transmission-line characteristic impedances differ from each other [10], and closed forms for the perfect matching and the arbitrary power divisions cannot be introduced in any asymmetric ring hybrids.

9.3 ASYMMETRIC RING-HYBRID PHASE SHIFTERS

If the asymmetric ring hybrid is terminated in reflection coefficients, Γ_2 and Γ_4 at ports ② and ④, respectively, as shown in Fig. 9.1, the scattering matrix of the asymmetric ring hybrid and the termination reflection matrix $\boldsymbol{\Gamma}_L$ are given as

$$\boldsymbol{S} = \begin{bmatrix} \boldsymbol{P}_{11} & \boldsymbol{P}_{12} \\ \boldsymbol{P}_{21} & \boldsymbol{P}_{22} \end{bmatrix}, \tag{9.4a}$$

$$\boldsymbol{\Gamma}_L = \begin{bmatrix} \Gamma_2 & 0 \\ 0 & \Gamma_4 \end{bmatrix}, \tag{9.4b}$$

where

$$\boldsymbol{P}_{11} = \begin{bmatrix} S_{11} & S_{13} \\ S_{31} & S_{33} \end{bmatrix} = 0,$$

$$\boldsymbol{P}_{12} = \begin{bmatrix} S_{12} & S_{14} \\ S_{32} & S_{34} \end{bmatrix} = \begin{bmatrix} -jd_1 & -jd_2 \\ jd_2 & -jd_1 \end{bmatrix},$$

$$\boldsymbol{P}_{21} = \begin{bmatrix} S_{21} & S_{23} \\ S_{41} & S_{43} \end{bmatrix} = \begin{bmatrix} -jd_1 & jd_2 \\ -jd_2 & -jd_1 \end{bmatrix},$$

$$\boldsymbol{P}_{22} = \begin{bmatrix} S_{22} & S_{24} \\ S_{42} & S_{44} \end{bmatrix} = 0.$$

The scattering matrix $\boldsymbol{S}'$ with respect to ports ① and ③ is calculated using a multiport concept, and it is given as

$$\boldsymbol{S}' = \boldsymbol{P}_{11} + \boldsymbol{P}_{12}\boldsymbol{\Gamma}_L[\boldsymbol{U} - \boldsymbol{P}_{22}\boldsymbol{\Gamma}_L]^{-1}\boldsymbol{P}_{21}, \tag{9.5a}$$

$$\mathbf{S}' = \frac{1}{d_1^2 + d_2^2} \begin{bmatrix} -(d_1^2\Gamma_2 + d_2^2\Gamma_4) & d_1 d_2(\Gamma_2 - \Gamma_4) \\ d_1 d_2(\Gamma_2 - \Gamma_4) & -(d_1^2\Gamma_4 + d_2^2\Gamma_2) \end{bmatrix}. \tag{9.5b}$$

The phase shifter is a two-port passive device whose basic function is to provide a change of RF signal phase with negligible (ideally, zero) attenuation. So for the perfect matching and zero attenuation conditions, the two results are obtained from (9.5) as

$$\Gamma_2 = -\frac{d_2^2}{d_1^2}\Gamma_4, \qquad S'_{12} = -\frac{d_2}{d_1}\Gamma_4, \tag{9.6a}$$

or

$$\Gamma_2 = -\frac{d_1^2}{d_2^2}\Gamma_4, \qquad S'_{12} = -\frac{d_1}{d_2}\Gamma_4. \tag{9.6b}$$

If the magnitude of Γ_4 is not unity from the equation $S'_{12} = -d_2/d_1\Gamma_4$ in (9.6*a*), the magnitude of the ratio of d_2 to d_1 should be greater than unity to satisfy $|S'_{21}| = 1$. In this case, the magnitude of Γ_2 in (9.6*a*) will be greater than unity from the equation $\Gamma_2 = (d_2/d_1)S'_{12}$. As this device is passive, the magnitude of Γ_2 should not be greater than unity. Due to that principle, if a ring hybrid is chosen for a phase shifter application, it should be a 3-dB ring hybrid (equal power division) and the magnitudes of Γ_2 and Γ_4 should ideally be unity.

The relation of $d_2 = d_1$ may also be derived from equating Γ_2 in (9.6*a*) to that of (9.6*b*). Applying $d_2 = d_1$ to (9.5), the scattering matrix contributed by two ports ① and ③ is

$$\boldsymbol{S}' = \frac{1}{2}\begin{bmatrix} -(\Gamma_2 + \Gamma_4) & \Gamma_2 - \Gamma_4 \\ \Gamma_2 - \Gamma_4 & -(\Gamma_2 + \Gamma_4) \end{bmatrix}. \tag{9.7}$$

Figure 9.3 shows a ring-hybrid phase shifter proposed by White [2]. It requires a $\lambda/4$ phase shift transmission line between port ④ and a diode. The diode operates as an electronic on–off switch when switched between a fixed forward bias and a reverse bias. Under forward bias, the diode offers very low impedance, thus approximating a short circuit (on-state), and under reverse bias it offers very high impedance, approximating an open circuit (off-state). Therefore, by switching the diodes between forward and reverse bias states, an open or shorted circuit is easily obtained. Because of the diode characteristics, if the bias condition of one

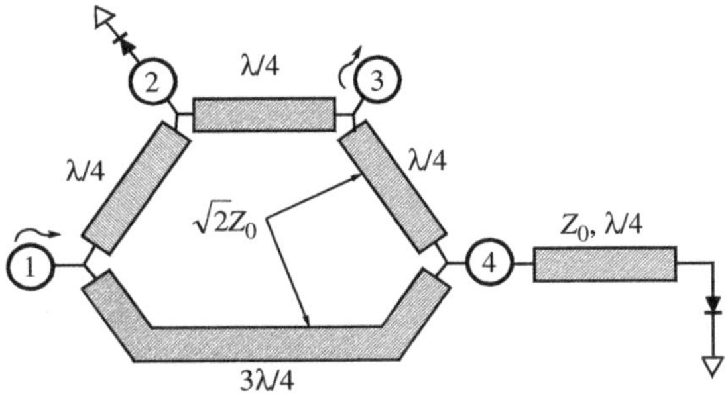

FIGURE 9.3 Conventional ring-hybrid phase shifter for symmetric reflecting terminations.

diode is different from that of the other diode, the perfect matching condition $\Gamma_2 = -\Gamma_4$ in (9.7) can be satisfied without a $\lambda/4$ phase shift transmission line.

Figure 9.4 shows the simulation results of two ring-hybrid phase shifters. "Aps" represents the asymmetric ring-hybrid phase shifter proposed in this chapter, and "Cps" indicates the conventional ring-hybrid phase shifter in Fig. 9.3. The asymmetric phase shifter is a ring-hybrid phase shifter with two different diode directions at ports ② and ④ (i.e., a forward bias for the diode at port ② and a backward bias at port ④ without an additional $\lambda/4$ transmission line). For these simulations, the two ring hybrids are terminated in equal impedances of 50 Ω, and a short or an open is used for forward- or reverse-bias operation of the diodes, respectively. These simulations were carried out under ideal conditions using a circuit simulator. The scattering parameters transmitted are plotted in Fig. 9.4(*a*), the return losses in Fig. 9.4(*b*), and the phase responses in Fig. 9.4(*c*). Two vertical dashed lines through Fig. 9.4(a)–(c) indicate where the return losses of the asymmetric phase shifter are −15 dB. When the two ring-hybrid phase shifters are compared in Fig. 9.4, both the insertion and return loss performances of "Aps" are better than those of "Cps." Most important, the phase responses of "Aps" are less dependent on frequencies than those of "Cps". Thus, the $\lambda/4$ phase shift transmission line in Fig. 9.3 results not only in greater size but also in more frequency dependencies.

Figure 9.5 shows four simple circuits to realize Γ_2 and Γ_4, where Z_b and Θ_d are the characteristic impedance and length of the transmission line connected at port ②, and Z_d and Θ_d are those at port ④. The subscripts s and o indicate short-circuited and open-circuited transmission lines, respectively. Thus, Γ_2, Z_b and Θ_b become Γ_{2s}, Z_{bs}, and Θ_{bs} when the transmission line is terminated in a short circuit, and Θ_{bo}, Z_{bo}, and Θ_{bo} when the transmission line is open-circuited in Fig. 9.5. From (9.7), the phase Θ of the asymmetric phase shifter under perfect matching conditions ($\Gamma_2 = -\Gamma_4$) is that of Γ_2, and it is derived as follows: In the case of $0 \leq \Theta_{bs} \leq 90^\circ$ in Fig. 9.5(*a*),

$$\Theta = \text{Ang}(\Gamma_2 = \Gamma_{2s}), \tag{9.8}$$

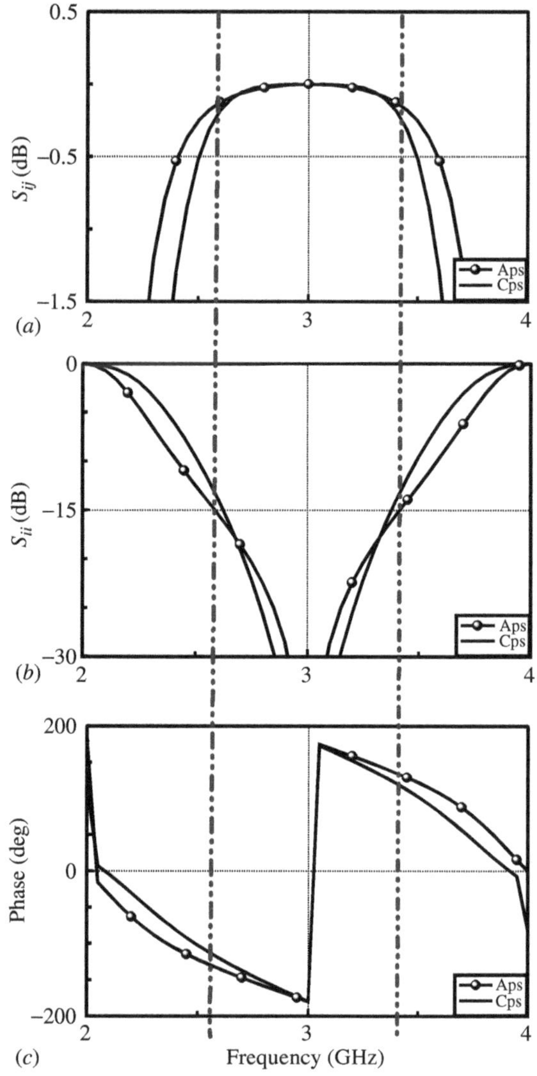

FIGURE 9.4 Simulation results of asymmetric and conventional ring-hybrid phase shifters: (*a*) insertion loss responses; (*b*) return loss responses; (*c*) phase frequency dependencies.

where

$$\Gamma_2 = \Gamma_{2s} = \frac{jNI_{bs}\tan\Theta_{bs} - 1}{jNI_{bs}\tan\Theta_{bs} + 1} \qquad \text{with} \quad NI_{bs} = \frac{Z_{bs}}{R_b};$$

and for $0 \leq \Theta_{bo} \leq 90^\circ$ in Fig. 9.5(*b*),

$$\Theta = \text{Ang}(\Gamma_2 = \Gamma_{2o}), \tag{9.9}$$

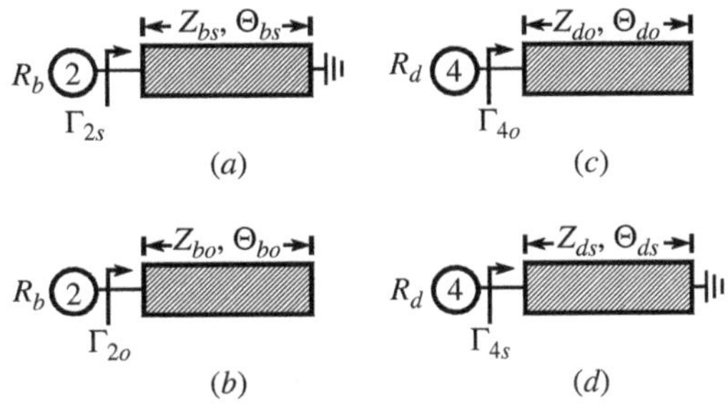

FIGURE 9.5 Simple structures to realize the reflection coefficients Γ_2 and Γ_4: (*a*) short-circuited transmission line for Γ_2; (*b*) open-circuited transmission line for Γ_2; (*c*) open-circuited transmission line for Γ_4; (*d*) short-circuited transmission line for Γ_4.

where

$$\Gamma_2 = \Gamma_{2o} = \frac{jNI_{bo} + \tan\Theta_{bo}}{jNI_{bo} - \tan\Theta_{bo}} \qquad \text{with} \quad NI_{bo} = \frac{Z_{bo}}{R_b}.$$

Figure 9.6 shows the response curves of Θ versus Θ_{bs}, Θ versus Θ_{bo}, and Θ versus NI_{bs} or NI_{bo} on the basis of (9.8) and (9.9). Variations in Θ using the circuit in Fig. 9.5(*a*) are shown in Fig. 9.6(*a*), while those using the circuit in Fig. 9.5(*b*) are shown in Fig. 9.6(*b*). Figure 9.6(*c*) shows the changes in Θ at several electrical lengths ($\Theta_{bs} = 22.5^\circ$, 45°, and 67.5°, and $\Theta_{bo} = 67.5^\circ$, 45°, and 22.5°) when NI_{bs} and NI_{bo} are varied. More detailed data for each response curve in Fig. 9.6 are given in Tables 9.1, 9.2, and 9.3, respectively. In the case of $NI_{bs} = NI_{bo} = 1$ in Fig. 9.6(*a*) and (*b*), the responses are linear; with NI_{bs} and NI_{bo} farther from 1, they become more nonlinear. Therefore, the phase Θ is dependent on the length of the transmission lines (Θ_{bs} and Θ_{bo}) and on termination impedance R_b at port ② together with the characteristic impedances Z_{bs} and Z_{bo} (i.e., NI_{bs} and NI_{bo}).

All response curves with NI_{bs} and those with $1/NI_{bs}$ are symmetric with respect to the line $NI_{bs} = 1$ in Fig. 9.6(*a*). Similar results with NI_{bo} can be seen in Fig. 9.6(*b*). Depending on which phase shifter is chosen, the load at port ② is determined referring to Fig. 9.6(*a*)–(*c*) and Tables 9.1 to 9.3. To realize the perfect matching condition $\Gamma_2 + \Gamma_4 = 0$, four pairs, (a)–(c), (a)–(d), (b)–(c), and (b)–(d), in Fig. 9.5 are possible, but the desirable choices are the pairs (a)–(c) for $0 \le \Theta_{bs} \le 90^\circ$ and (b)–(d) for $0 \le \Theta_{bo} \le 90^\circ$. The formulas for Z_d and Θ_d in Fig. 9.5(*c*) or (*d*) are as follows: For $0 \le \Theta_{bs} \le 90^\circ$,

$$NI_{do}\cot\Theta_{do} = \frac{1}{NI_{bs}\tan\Theta_{bs}}, \qquad \text{with} \quad NI_{do} = \frac{Z_{do}}{R_d}; \tag{9.10}$$

and for $0 \le \Theta_{bo} \le 90^\circ$,

$$NI_{ds}\cot\Theta_{ds} = \frac{1}{IN_{bo}\tan\Theta_{bo}}, \qquad \text{with} \quad NI_{ds} = \frac{Z_{ds}}{R_d}. \tag{9.11}$$

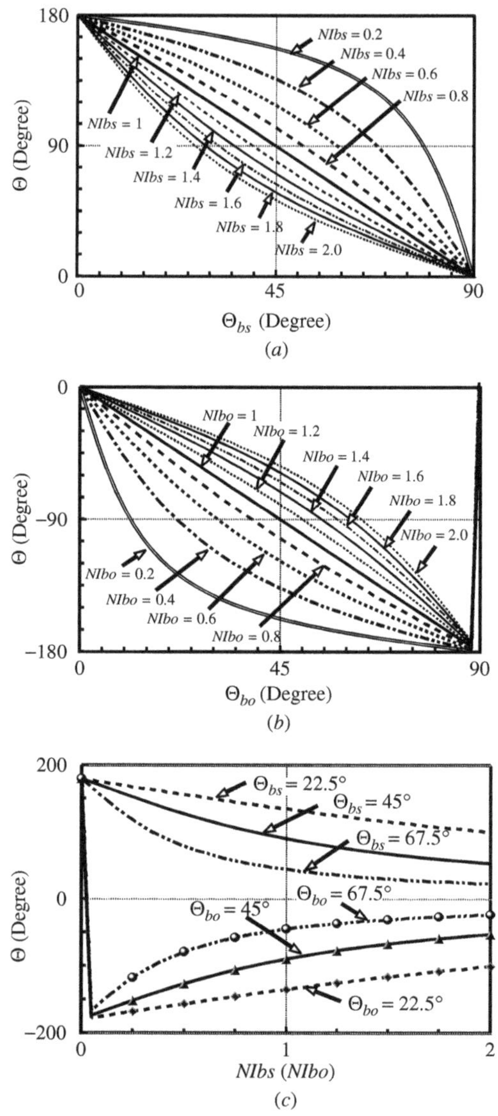

FIGURE 9.6 Phase responses depending on NI_b and Θ_b: (*a*) depending on Θ_{bs} and NI_{bs}; (*b*) depending on Θ_{bo} and NI_{bo}; (*c*) depending on NI_b and Θ_b. (From Ref. 18 with permission from IEEE.)

For $NI_{bs} \cdot NI_{do} = 1$ in (9.10), Θ_{bs} is always equal to Θ_{do}, and $\Theta_{bo} = \Theta_{ds}$ coincides with $NI_{bo} \cdot NI_{ds} = 1$ in (9.11). In (9.8)–(9.11), NI_{do}, NI_{bs}, NI_{bo}, and NI_{ds} are arbitrary positive constants that can be determined depending on the design situations and the arbitrary termination impedances, R_b and R_d. Therefore, there are many design choices for asymmetric ring-hybrid phase shifters.

TABLE 9.1 Phase Responses (deg) That Depend on NI_{bs} and Θ_{bs}

Θ_{bs} (deg) \ NI_{bs}	0.2	0.4	0.6	0.8	1	1.2	1.4	1.6	1.8	2.0
180	0	0	0	0	0	0	0	0	0	0
157.5	44.8	26.4	18.3	13.96	11.3	9.41	8.09	7.09	6.31	5.68
135.0	64.2	46.0	34.6	27.4	22.5	19.0	16.5	14.5	12.96	11.7
112.5	73.3	59.1	48.1	39.9	33.6	29.1	25.5	22.7	20.4	18.5
90.0	78.7	68.2	59.0	51.3	45	39.8	35.5	32.0	29.1	26.6
67.5	82.4	75.0	68.2	61.9	56.3	51.3	46.9	43.1	39.7	36.8
45.0	85.3	80.6	76.0	71.7	67.5	63.6	59.9	56.5	53.3	50.4
22.5	87.7	85.5	83.2	80.96	78.75	76.58	74.4	72.3	70.3	68.3
0	90	90	90	90	90	90	90	90	90	90

TABLE 9.2 Phase Responses (deg) That Depend on NI_{bo} and Θ_{bo}

Θ_{bo} (deg) \ NI_{bo}	0.2	0.4	0.6	0.8	1	1.2	1.4	1.6	1.8	2.0
0	0	0	0	0	0	0	0	0	0	0
−22.5	2.28	4.55	6.8	9.0	11.3	13.4	15.6	17.7	19.7	21.7
−45	4.74	9.41	13.96	18.33	22.5	26.43	30.10	33.53	36.71	39.64
−67.5	7.61	14.96	21.85	28.13	33.75	38.72	43.09	46.91	50.26	53.19
−90.0	11.3	21.80	30.96	38.66	45	50.19	54.46	57.99	60.95	63.44
−112.5	16.7	30.91	41.92	50.13	56.25	60.89	64.49	67.33	69.63	71.53
−135.0	25.8	44.00	55.38	62.63	67.5	70.96	73.52	75.49	77.04	78.30
−157.5	45.2	63.56	71.66	76.04	78.75	80.59	81.91	82.91	83.69	84.32
−180	90	90	90	90	90	90	90	90	90	90

There are many ways to design an asymmetric ring-hybrid phase shifter, but one of them is as follows:

1. The termination impedances R_a, R_b, R_c, and R_d are determined based on the design situation. When $R_b = R_d$, the relations of $Z_1 = Z_4$ and $Z_2 = Z_3$ always hold regardless of any value of R_a and R_c. If $R_b = R_d$, R_b and R_d should probably be chosen to be $\sqrt{R_a R_c}$.
2. The load at port ② for the phase shift required based on (9.8) and (9.9) is determined referring to Fig. 9.6(*a*) and (*b*) and Table 9.1 or 9.2.
3. The circuit at port ④ in Fig. 9.5(*c*) or (*d*) is designed for perfect matching. Since the response curves with NI_{bs} and those with $1/NI_{bs}$ are symmetric with respect to the line $NI_{bs} = 1$ in Fig. 9.6(*a*), to hold the relation $NI_{do} \cdot NI_{bs} = 1$ is important when determining the circuit in Fig. 9.5(*c*). In the

TABLE 9.3 Phase Changes (deg) in Θ with $\Theta_{bs} = \Theta_{bo} = 22.5°$, 45°, and 67.5° Versus NI_{bs} and NI_{bo}

Θ_{bs} (deg) \ NI_{bs}	1/1.9	1/1.6	1/1.3	1	1.3	1.6	1.9
22.5	155.4	150.97	144.65	135	123.4	112.93	103.59
45	124.48	115.99	104.86	90	75.14	64.01	55.51
67.5	76.41	67.07	56.60	45	35.35	29.03	24.60
Θ_{bo} (deg) \ NI_{bo}	1/1.9	1/1.6	1/1.3	1	1.3	1.6	1.9
67.5	−155.40	−150.97	−144.65	−135	−123.40	−112.93	−103.59
45	−55.52	−64.01	−75.14	−90	−104.86	−115.99	−124.48
22.5	−24.60	−29.03	−35.35	−45	−56.60	−67.07	−76.41

case of NI_{bo} and $1/NI_{bo}$, the relation $NI_{bo} \cdot NI_{ds} = 1$ is also important for the circuit in Fig. 9.5(*d*).

9.3.1 Uniplanar Asymmetric Ring-Hybrid −135° Phase Shifter

If a uniplanar phase shifter with termination impedances of 30 Ω and 60 Ω is needed for a phase shift of −135°, let $R_b = R_d$ be $\sqrt{30 \times 60}\ \Omega = 42.4\ \Omega$. Then $Z_1 = Z_4 = 50.5\ \Omega$ and $Z_2 = Z_3 = 71.4\ \Omega$. The circuit to realize Γ_2 for the −135° phase shifter is that in Fig. 9.5(*a*) based on the results in Fig. 9.6(*a*) and Table 9.1. As the value of NI_{bs} increases, the length of Θ_{bs} decreases. Therefore, several cases are given in Table 9.4 together with the data for the ring hybrid. Any case with changes in NI_{bs} produces the same output performances; experimental data for the case of $NI_{bs} = 1.2$ are given in Table 9.5. This uniplanar phase shifter was realized in CPW (coplanar waveguide) technology on Al_2O_3 substrate ($\varepsilon_r = 9.9$, $h = 635\ \mu m$) and designed at a center frequency of 3 GHz. The layout for the measurement is depicted in Fig. 9.7. It is terminated in 30 Ω and 60 Ω, so two quarter-wavelength transformer lines with characteristic impedances Z_{01} and Z_{03} are needed. Measured and simulated results are compared in Fig. 9.8.

9.4 ASYMMETRIC RING-HYBRID ATTENUATOR WITH PHASE SHIFTS

If the magnitude of Γ_2 in (9.7) is less than unity, the ring-hybrid phase shifter in Fig. 9.1 may be an attenuator. Simple circuits to realize Γ_2 and Γ_4 for an asymmetric ring-hybrid attenuator are depicted in Fig. 9.9. Each transmission line with the characteristic impedance Z_{ba} or Z_{da} is terminated in R_{Lb} or R_{Ld} instead of the short or open circuit in Fig. 9.5. Assuming that $R_b = Z_{ba}$ and

TABLE 9.4 Several Choices for the Design of an Asymmetric Ring-Hybrid 135° Phase Shifter

Ring Hybrid

Termination impedance (Ω)	R_a	R_b	R_c	R_d
	30	42.4	60	42.4
Characteristic impedance (Ω)	Z_1	Z_2	Z_3	Z_4
	50.5	71.4	71.4	50.5

135° Phase Shift

NI_{bs}	$Z_{bs}(\Omega)$	Θ_{bs} (deg)	NI_{do}	$Z_{do}(\Omega)$	Θ_{do} (deg)
2	84.4	11.7	0.5	21.2	11.7
1.8	76.4	12.96	0.55	23.6	12.96
1.6	67.9	14.5	0.63	26.5	14.5
1.4	59.4	16.5	0.71	30.3	16.5
1.2	50.9	19.0	0.83	35.4	19.0
1	42.2	22.5	1	42.4	22.5

TABLE 9.5 Fabrication Data for a Uniplanar Asymmetric Ring-Hybrid 135° Phase Shifter for $NI_{bs} = 1.2$[a]

Ring hybrid; $Z_{01} = 38.7\ \Omega$, $Z_{03} = 54.8\ \Omega$
Phase shift for Γ_{2s} and Γ_{4o}; Z_{bs} and Z_{do}

Z_{01}	Z_{03}	$Z_1 = Z_4$	Z_2
$w = 583\ \mu m$ $g = 100\ \mu m$ $l = 11,059\ \mu m$	$w = 334\ \mu m$ $g = 200\ \mu m$ $l = 10,994\ \mu m$	$w = 324\ \mu m$ $g = 150\ \mu m$ $l = 10,950\ \mu m$	$w = 125\ \mu m$ $g = 200\ \mu m$ $l = 32,635\ \mu m$
Z_3	Z_{bs}	Z_{do}	Z_1 50.5 Ω
$w = 125\ \mu m$ $g = 200\ \mu m$ $l = 10,878\ \mu m$	$w = 444\ \mu m$ $g = 200\ \mu m$ $l = 2,337\ \mu m$	$w = 896\ \mu m$ $g = 100\ \mu m$ $l = 2,375\ \mu m$	Z_2 71.5 Ω

[a] w, Center strip width; g, gap width; l, line length.

$R_d = Z_{da}$ for easy analyses, the reflection coefficients Γ_{2a} and Γ_{4a} are calculated as

$$\Gamma_2 = \Gamma_{2a} = \frac{(NR_{Lb} - 1) - j(NR_{Lb} - 1)\tan\Theta_{ba}}{(NR_{Lb} + 1) + j(NR_{Lb} + 1)\tan\Theta_{ba}}, \tag{9.12}$$

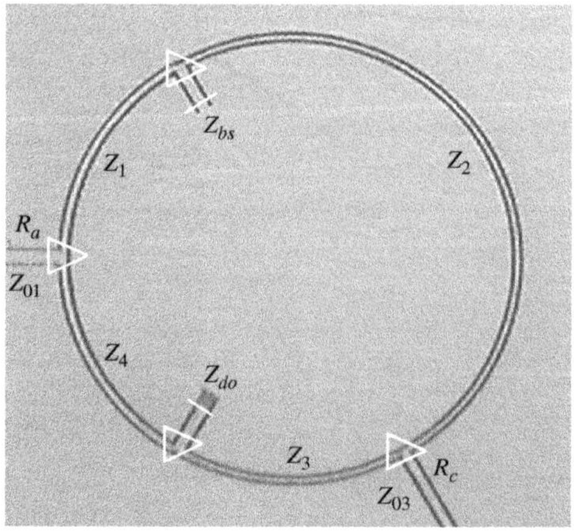

FIGURE 9.7 Layout of a uniplanar asymmetric ring-hybrid $-135°$ phase shifter terminated in 30 Ω and 60 Ω. (From Ref. 18 with permission from IEEE.)

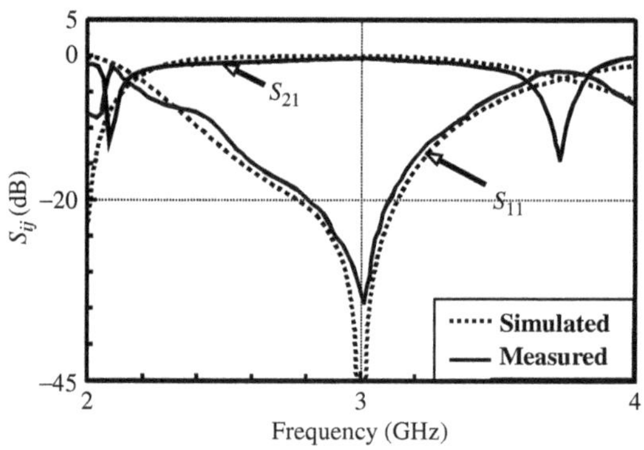

FIGURE 9.8 Measured and simulated results for the uniplanar asymmetric $-135°$ phase shifter.

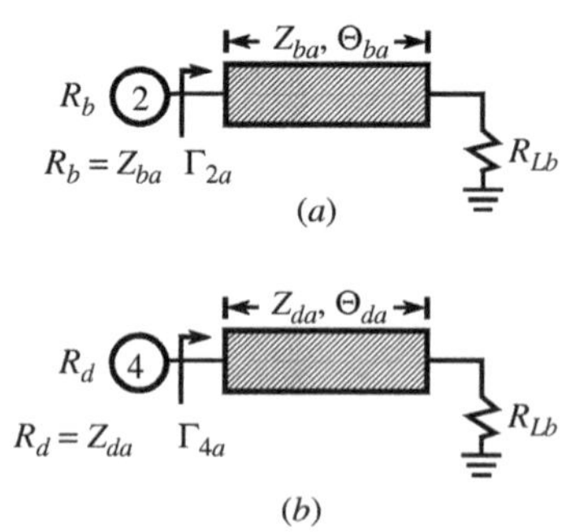

FIGURE 9.9 Simple circuits for the reflection coefficients (*a*) Γ_{2a} and (*b*) Γ_{4a}.

where $NR_{Lb} = R_{Lb}/Z_{ba}$ and $R_b = Z_{ba}$, and

$$\Gamma_4 = \Gamma_{4a} = \frac{(NR_{Ld} - 1) - j(NR_{Ld} - 1)\tan\Theta_{da}}{(NR_{Ld} + 1) + j(NR_{Ld} + 1)\tan\Theta_{da}}, \tag{9.13}$$

where $NR_{Ld} = R_{Ld}/Z_{da}$ and $R_d = Z_{da}$. Therefore, the magnitudes of Γ_2 and Γ_4 are

$$\text{Mag}(\Gamma_{2a}) = \left|\frac{NR_{Ld} - 1}{NR_{Ld} + 1}\right|, \tag{9.14}$$

$$\text{Mag}(\Gamma_{4a}) = \left|\frac{NR_{Lb} - 1}{NR_{Lb} + 1}\right|. \tag{9.15}$$

Also the angle of $\Gamma_2 = \Gamma_{2a}$ results in the following: For $0 < NR_{Lb} < 1$,

$$\text{Ang}(\Gamma_{2a}) = \Theta \qquad \text{with} \quad NI_{bs} = 1; \tag{9.16}$$

and for $NR_{Lb} > 1$,

$$\text{Ang}(\Gamma_{2a}) = \Theta \qquad \text{with} \quad NI_{bo} = 1. \tag{9.17}$$

On the basis of (9.14) and (9.15), the magnitude responses of T_{2a} or T_{4a} are plotted in Fig. 9.10, where in region I, $0 < NR_{Lb} < 1$ or $0 < NR_{Ld} < 1$, and in region II, $NR_{Lb} > 1$ or $NR_{Ld} > 1$. There are two values of NR_{Lb} to realize the asymmetric attenuator, one in region I and the other in region II. However, the output signals are out of phase by 180° for the two values, and one is always the inverse of the other one. The same situation holds with the two values of NR_{Ld}. Values of NR_{Lb} depending on attenuations are given in Table 9.6.

On the basis of Table 9.6, three 6-dB attenuators with phase shifts −170°, −160°, and −135° were simulated and the termination impedances for all were 40 Ω and 60 Ω. The phase shift is a negative value of Θ, so Fig. 9.6(*a*) shows cases where Θ = 170°, 160°, and 135° with $NI_{bs} = 1$ and $0 < NR_{Lb} < 1$ from

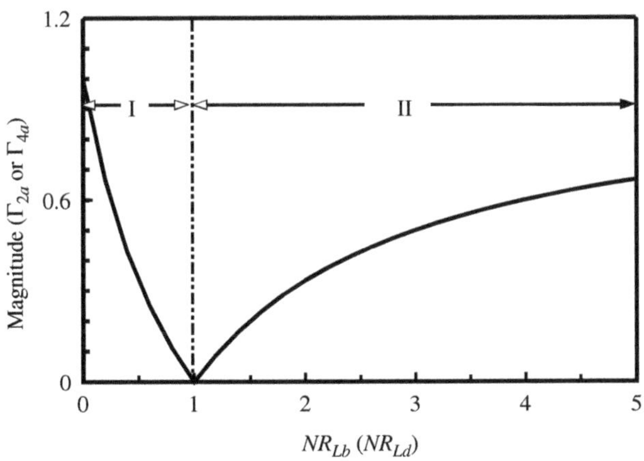

FIGURE 9.10 Magnitude of Γ_{2a} or Γ_{4a} depending on NR_{Lb} or NR_{Ld}.

TABLE 9.6 Values of NR_{Lb} in Region I of Fig. 9.10 Versus Attenuation

Attenuation (dB)	NR_{Lb}	Attenuation (dB)	NR_{Lb}
1	0.0575	8	0.43050
2	0.11462	9	0.47621
3	0.17099	10	0.51949
4	0.22627	11	0.56025
5	0.28012	12	0.59847
6	0.33227	13	0.63415
7	0.38247	14	0.66732

(9.16). Therefore, NR_{Lb} for the 6-dB attenuation for the phase shifts is 0.3327 in Table 9.6. From the relation of $NR_{Lb} = R_{Lb}/R_b$ in (9.12), if a commercially available 15-Ω system is used for the R_{Lb}, the termination impedance R_b will be 45.14 Ω. The value of NI_{Ld} is the inverse of NR_{Lb}, or 1/0.33227, to satisfy the perfect matching condition $\Gamma_2 + \Gamma_4 = 0$. When a commercial 150-Ω system is utilized for the R_{Ld}, the termination impedance R_d will be 49.84 Ω.

For the 6-dB attenuators, the lengths of Θ_{ba} for the $-170°$, $-160°$, and $-135°$ phase shifts are 5°, 10°, and 22.5°, respectively, from (9.16), Fig. 9.6(*a*), and Table 9.1 with $NI_{bs} = 1$. These 6-dB attenuators are terminated in 40 Ω and 60 Ω. Therefore, two quarter-wavelength transformer lines to transform the termination impedances into 50 Ω's are needed. Figure 9.11 shows the simulation results of the three attenuators with phase shifts $-170°$, $-160°$, and $-135°$. Their insertion loss and matching characteristics are plotted in Fig. 9.11(*a*) and their phase responses in Fig. 9.11(*b*). Simulated insertion losses are all −6 dB, and simulated reflection coefficients are −157.4 dB, −156.4 dB, and −154.4 dB at 3 GHz. The phase responses of the three attenuators are −10°, −20°, and −45° at a center frequency of 3 GHz in Fig. 9.11(*b*), because two 90° impedance transformers are connected at the input and output ports. The subpoles in Fig. 9.11(*a*) are poles that appear when Θ_{ba} is not zero. The longer Θ_{ba} is, the farther the subpole appears from the center frequency of 3 GHz. The Θ_{ba} values for the $-170°$, $-160°$, and $-135°$ phase shifters are 5°, 10°, and 22.5°, and the subpoles appear at 2.87 GHz, 2.75 GHz, and 2.48 GHz, respectively. How far the subpole appears from the design center frequency is not linear and depends on several factors: the length of Θ_{ba}, the input and output termination impedances, the resistances R_{Lb} and R_{Ld}, and so on. Therefore, this concept may be used to increase bandwidths.

9.4.1 Microstrip Asymmetric Ring-Hybrid 4-dB Attenuator with 45° Phase Shift

On the basis of design equations (9.14)–(9.17) and Table 9.6, an asymmetric ring-hybrid 4-dB attenuator with a 45° phase shift [20] was designed at a center frequency of 3 GHz and fabricated in microstrip technology on Al_2O_3 substrate

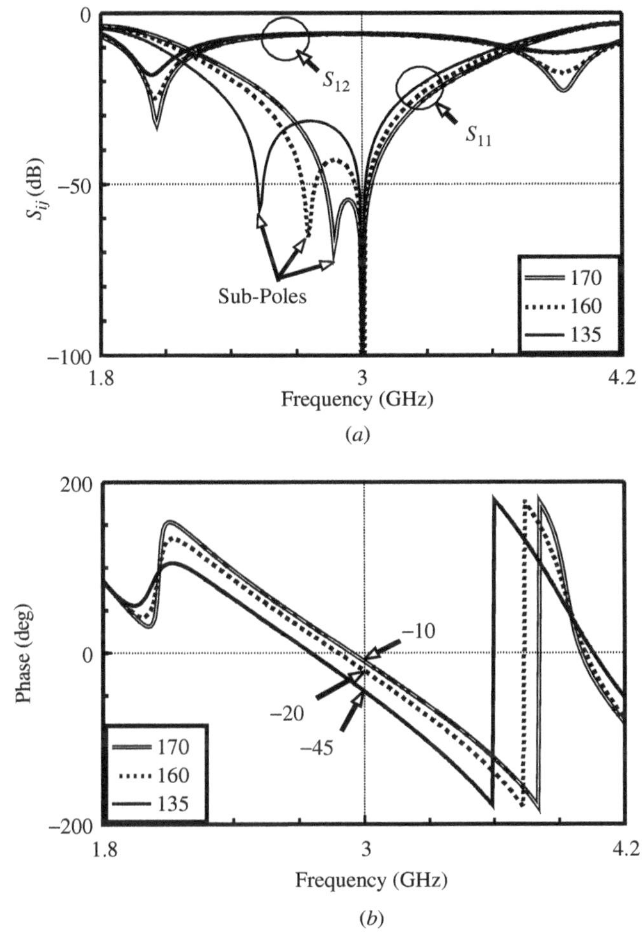

FIGURE 9.11 Asymmetric ring-hybrid 6-dB attenuators with three phase shifts: -170°, -160° and -135°. (*a*) Insertion and return losses; (*b*) phase responses.

($\varepsilon_r = 9.9$, $h = 635$ μm). The NR_{Lb} value for a 4-dB attenuation is the inverse of the value in Table 9.6, because NR_{Lb} for a 45° phase shift is located in region II in Fig. 9.10. In this case, $NR_{Lb} = 1/0.22627$, or 4.4195. If a commercially available 221 Ω is used for R_{Lb}, $R_b = 50.00$ Ω and NR_{Ld} is 0.22627, to satisfy the perfect matching condition. When a commercial 12-Ω resistor is chosen for R_{Ld}, R_d is 53.0 Ω. If the microstrip attenuator is terminated in $R_a = 30$ Ω and $R_c = 60$ Ω, the characteristic impedances necessary for the transmission lines in Fig. 9.1 are $Z_1 = 54.65$ Ω, $Z_2 = 77.5$ Ω, $Z_3 = 79.75$ Ω, and $Z_4 = 56.4$ Ω. The lengths of Θ_{ba} and Θ_{da} are both 22.5°. Necessary data for the fabrication are given in Table 9.7, where Z_{01} and Z_{03} are also the characteristic impedances of two 90° impedance transformers to transform 30 Ω and 60 Ω to 50 Ω's. The attenuator package is illustrated in Fig. 9.12(*a*), and the attenuator itself

TABLE 9.7 Fabrication Data for a Microstrip Asymmetric Ring Hybrid with a 4-dB Attenuator and a 45° Phase Shift[a]

$-45°$ phase shift with 4 dB attenuation $\rightarrow NR_{Lb} = 4.4195 \rightarrow R_{Lb} = 221\ \Omega$
$\rightarrow R_b = 50.0\ \Omega$, $NR_{Ld} = 0.2263 \rightarrow R_{Ld} = 12\ \Omega \rightarrow R_d = 53.0\ \Omega$

Ring Hybrid and Phase Shift Parts

$Z_1 = 54.7\ \Omega$	$Z_2 = 77.5\ \Omega$	$Z_3 = 79.8\ \Omega$
$w = 501\ \mu$	$w = 202.3\ \mu$	$w = 185\ \mu$
$l = 9736\ \mu$	$l = 29996\ \mu$	$l = 10018\ \mu$
$Z_4 = 56.4\ \Omega$	$Z_{ba} = 49.8\ \Omega$	$Z_{da} = 53.0\ \Omega$
$w = 468\ \mu$	$w = 613\ \mu$	$w = 537\ \mu$
$l = 9562\ \mu$	$l = 2414\ \mu$	$l = 2427\ \mu$
$Z_{01} = 38.7\ \Omega$	$Z_{02} = 54.8\ \Omega$	
$w = 988\ \mu$	$w = 499\ \mu$	
$l = 9433\ \mu$	$l = 9738\ \mu$	

[a] w, Center conductor; l, length of microstrip line

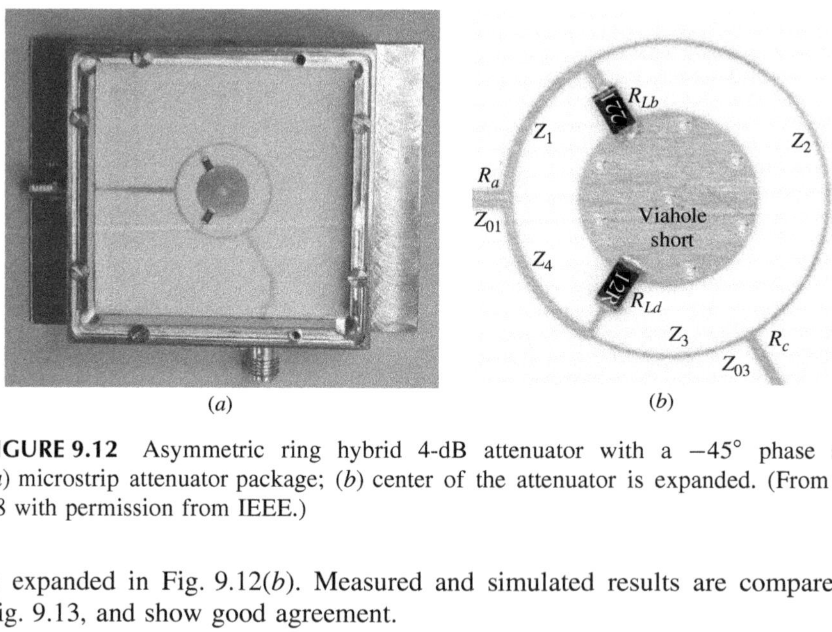

FIGURE 9.12 Asymmetric ring hybrid 4-dB attenuator with a $-45°$ phase shift: (*a*) microstrip attenuator package; (*b*) center of the attenuator is expanded. (From Ref. 18 with permission from IEEE.)

is expanded in Fig. 9.12(*b*). Measured and simulated results are compared in Fig. 9.13, and show good agreement.

EXERCISES

9.1 Derive (9.2) for the network in Fig. 9.2.

9.2 Design a 45° phase shifter using the asymmetric ring hybrid in Fig. 9.1 terminated in $R_a = 50\ \Omega$, $R_b = 40\ \Omega$, $R_c = 60\ \Omega$, and $R_d = 30\ \Omega$.

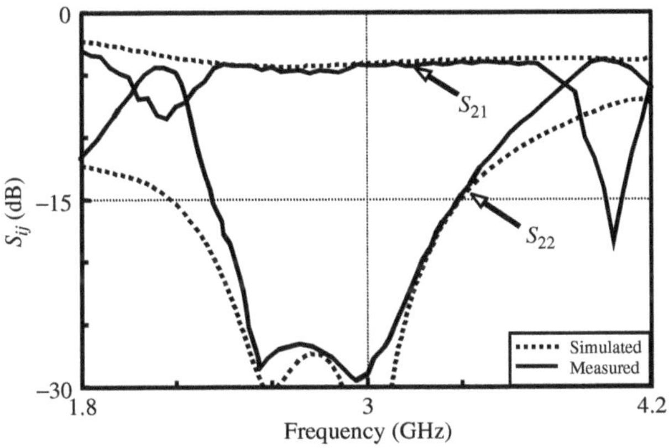

FIGURE 9.13 Measured and simulated results for the asymmetric ring-hybrid 4-dB attenuator with a −45° phase shift.

9.3 Derive and plot the magnitude of Γ_{2a} or Γ_{4a} for $R_b \neq Z_{ba}$ and $R_d \neq Z_{da}$ in (9.13)–(9.15).

9.4 Find a relation between the subpoles and Θ_{ba} in Fig. 9.9.

9.5 Design a 3-dB attenuator with a 45° phase shift.

REFERENCES

1. C.-L. Chen, W. E. Courtney, L. J. Mahoney, M. J. Manfra, A. Chu, and H. A. Atwater, A Low-Loss Ku-Band Monolithic Analog Phase Shifter, *IEEE Trans. Microwave Theory Tech.*, Vol. 35, March 1987, pp. 315–319.
2. J. F. White, Diode Phase Shifters for Array Antennas, *IEEE Trans. Microwave Theory Tech.*, Vol. 22, June 1974, pp. 658–674.
3. S. Lucyszyn and I. D. Robertson, Analog Reflection Topology Building Blocks for Adaptive Microwave Signal Processing Applications, *IEEE Trans. Microwave Theory Tech.*, Vol. 43, March 1995, pp. 601–611.
4. C. Andricos, I. J. Bahl, and E. L. Griffin C-band 6 Bit GaAs Monolithic Phase Shifter, *IEEE Trans. Microwave Theory Tech.*, Vol. 33, December 1985, pp. 1591–1596.
5. H.-R. Ahn and I. Wolff, Asymmetric Four-Port and Branch-Line Hybrids, *IEEE Trans. Microwave Theory Tech.*, Vol. 48, September 2000, pp. 1585–1588.
6. H.-R. Ahn and I. Wolff, 3-dB Branch-Line Hybrids with Arbitrary Termination Impedance Values, *IEICE Trans. Electrons*, Vol. E82-C, No. 7, July 1999, pp. 1324–1326.
7. H.-R. Ahn and I. Wolff, Arbitrary Power Division Branch-Line Hybrid Terminated by Arbitrary Impedances, *IEE Electron. Lett.*, Vol. 35, No. 7, April 1999, pp. 572–273.
8. H.-R. Ahn and I. Wolff, 3-dB Branch-Line Hybrid Terminated by Arbitrary Impedances, *IEE Electron. Lett.*, Vol. 34, No. 11, May 1998, pp. 1109–1110.
9. Y. Yahara, Y. Kadowaki, H. Hoshika, and K. Shirahata, Broad-Band 180° Phase Shift Section in X-Band, *IEEE Trans. Microwave Theory Tech.*, Vol. 23, March 1975, pp. 307–309.

10. H.-R. Ahn, I. Wolff, and I.-S. Chang, Arbitrary Termination Impedances, Arbitrary Power Division and Small-Sized Ring Hybrids, *IEEE Trans. Microwave Theory Tech.*, Vol. 44, December 1997, pp. 2241–2247.
11. C. E. Free and C. S. Aitchison, Improved Analysis and Design of Coupled-Line Phase Shifters, *IEEE Trans. Microwave Theory Tech.*, Vol. 43, September 1995, pp. 2126–2131.
12. V. P. Meschanov, I. V. Metelnikova, V. D. Tupikin, and G. G. Chumaevskaya, A New Structure of Microwave Ultrawide-Band Differential Phase Shifter, *IEEE Trans. Microwave Theory Tech.*, Vol. 42, May 1994, pp. 762–765.
13. B. M. Schiffman, A New Class of Broad-Band Microwave 90-Degree Phase Shifters, *IRE Trans. Microwave Theory Tech.*, Vol. 6, April 1958, pp. 232–237.
14. H.-R. Ahn, I.-S. Chang, and S.-W. Yun, Miniaturized 3-dB Ring Hybrid Terminated by Arbitrary Impedances, *IEEE Trans. Microwave Theory Tech.*, Vol. 42, December 1994, pp. 2216–2221.
15. M. A. Hamid and M. M. Yunik, On the Design of Stepped Transmission-Line Transformers, *IEEE Trans. Microwave Theory Tech.*, Vol. 15, September 1967, pp. 528–529.
16. H.-R. Ahn and I. Wolff, Novel Ring Filter as a Wide-Band 180° Transmission-Line, *EUMC Proc.*, Vol. III, October 1999, pp. 95–98.
17. T. Wang and K. We, Size-Reduction and Band-Broadening Design Technique of Uniplanar Hybrid Ring Coupler Using Phase Inverter for M(H)MIC's, *IEEE Trans. Microwave Theory Tech.*, Vol. 47, February 1999, pp. 198–206.
18. H.-R. Ahn and I. Wolff, Asymmetric Ring Hybrid Phase-Shifters and Attenuators, *IEEE Trans. Microwave Theory Tech.*, Vol. 50, April 2002, pp. 1146–1155.
19. C. Y. Pon, Hybrid-Ring Directional Coupler for Arbitrary Power Divisions, *IRE Trans. Microwave Theory Tech.*, Vol. 9, November 1961, pp. 529–535.
20. F. Mikucki and A. K. Agrawal, A Broad-Band Printed Circuit Hybrid Ring Power Divider, *IEEE Trans. Microwave Theory Tech.*, Vol. 37, January 1989, pp. 112–117.
21. A. K. Agrawal and G. F. Mikucki, A Printed Circuit Hybrid Ring Directional Coupler for Arbitrary Power Division, *IEEE Trans Microwave Theory Tech.*, Vol. 34, December 1986, pp. 1401–1407.

CHAPTER TEN

Ring Filters and Their Use in a New Measurement Technique for Inherent Ring-Resonance Frequency

10.1 INTRODUCTION

It is well known that ring resonators have low radiation loss, high Q factors, and two orthogonal resonant modes. Because of these special properties, they have been used widely for the measurement of dielectric constants and dispersion relations [1,2], bandpass filters, and duplexers [3,4]. Microstrip open- and closed-ring resonators have been discussed intensively [5,6], and mixers, oscillators, and tuning filters have been realized based on circuit theory concepts [7]. However, these circuits suffer from high insertion loss due to gap discontinuities and inaccurate analyses of gap capacitances, which cannot be neglected. To overcome this disadvantage, the ring filter has been introduced as a 180° phase shifter [8].

In this chapter, the ring filter is analyzed systematically and its wideband characteristics explained. Each ring filter consists of a ring and two short stubs that are connected at the 90° and 270° points of the ring. Quite different from conventional ring-based circuits, the feeding lines are coupled directly to the ring to alleviate the high insertion loss caused by gaps in conventional ring-based circuits. Since the termination impedance where each short stub is connected has an effect on design, it is defined as a hypothetical port and the relations between bandwidths, termination impedances, and the characteristic impedance of the short stub are

Asymmetric Passive Components in Microwave Integrated Circuits, by Hee-Ran Ahn

studied. Then ring filters that have optimal wideband performances are tested. The results measured are shown to be in good agreement with the results predicted. The bandwidth measured shows more than 100% fractional bandwidth with return losses less than −15 dB, which can be obtained in conventional filter design techniques only when more than five stages are used [9].

It is very important in the design of ring-based circuits to know the exact resonance frequency [10], so there have been many trials to determine it: using gap coupling [11], a simple cavity model with magnetic sidewalls [12,13], a planar waveguide model [14,15], and so on [16]. However, even though several of the methods proposed are simple ones [1,2,11], the ring-resonance frequency determined by the method may shift due to the gap discontinuities, which may be very harmful in narrowband filter designs. Also, a number of methods require much time and somewhat complicated mathematical programs based on field theory analyses [12–16].

In this chapter, a new and simple method to determine an inherent ring-resonance frequency is introduced as another use of ring filters. A ring filter consists of a ring and two short stubs, but in another sense it appears as two filters connected in parallel, an up-filter and a down-filter. Therefore, the power excited at the input is divided just like a three-port power divider or ring hybrid [17–22], flows into the up- and the down-filters, and combines at the output. If the two powers are out of phase at the output, the input power cannot be delivered to the output. Two input powers out of phase by 180° can be obtained by making the two stubs differ in length. The phase of the power transmitted (S_{21}) changes abruptly at the vicinity of the inherent ring-resonance frequency, which indicates that all the excited power is reflected under certain conditions. Therefore, if the frequency is determined, it will be the exact inherent ring-resonance frequency. In the lossless case, the return loss reaches 0 dB at the inherent ring-resonance frequency but does not reach 0 dB if there are losses (e.g., conductor loss, dielectric loss). Therefore, the return losses at the inherent ring-resonance frequency are investigated depending on the differences between the two short stubs. By the method explained above, exact gap capacitances for coupling in conventional ring-based circuits may also be found, and the inherent ring-resonance frequency of several common forms of the ring may be determined. To verify the method suggested in this chapter, two different ring filters are tested, and the results measured are shown to be in good agreement with those predicted.

10.2 RING FILTERS

10.2.1 Analyses of Ring Filters

A ring filter is depicted in Fig. 10.1(*a*) and its up-filter in Fig. 10.1(*b*). It is terminated in Z_1 and Z_2 and consists of a ring, two short stubs, and a resistor. Theoretically, the resistor is not needed, but it is indispensable for better performance during fabrication. The two short stubs are located at the 90° and 270° points of the ring, which may be considered as hypothetical ports whose

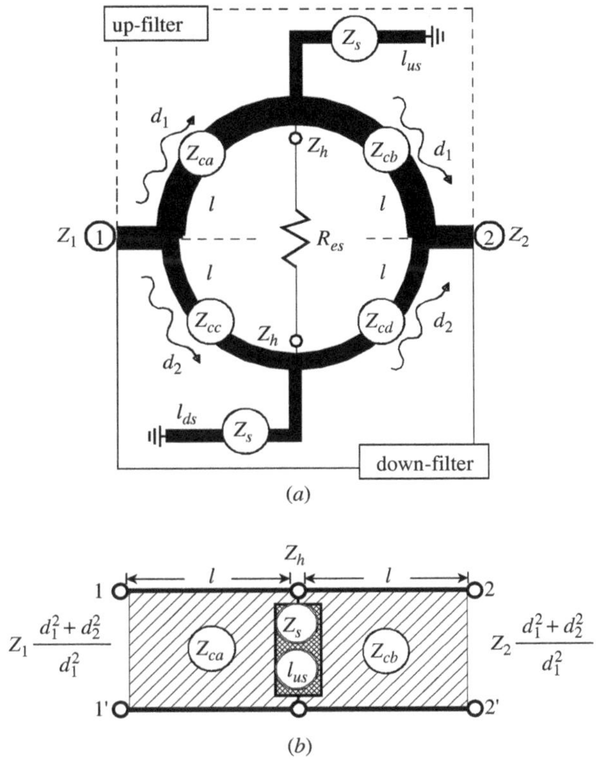

FIGURE 10.1 (*a*) Ring filter; (*b*) up-filter.

termination impedances are Z_h. The Z_h value is needed to design the ring filters and may be chosen arbitrarily when $Z_1 = Z_2$, and $Z_h = (Z_1 + Z_2)/2$ or $\sqrt{Z_1 Z_2}$ when $Z_1 \neq Z_2$. The four transmission-line sections forming the ring have equal lengths l, and their characteristic impedances are Z_{ca}, Z_{cb}, Z_{cc}, and Z_{cd}, as shown in Fig. 10.1(*a*). If the two short stubs do not exist in Fig. 10.1(*a*), all the power excited at port ① will be delivered to port ② at all frequencies, and perfect matching occurs at multiples of the design center frequency. Therefore, each stub is necessary to reject the power at even multiples of the design center frequency and to achieve filter characteristics as a resonator. To reduce high insertion loss caused by gaps in conventional ring-based circuits, two feeding lines are coupled directly to the ring. Since the lengths of the two short stubs are about 90°, the power excited at port ① is divided depending on the power division ratio of d_1 to d_2 indicated in Fig. 10.1(*a*), and the divided power is combined at port ②. Thus, it can be understood that the up- and down-filters shown in Fig. 10.1(*a*) are connected in parallel, and the termination impedances of the up-filter may be derived as explained in Fig. 10.1(*b*) [20,21]. Z_s, l_{us}, and l_{ds} are the characteristic impedance and lengths of the short stubs of the up- and down-filters.

The *ABCD* parameters of the up-filter in Fig. 10.1(*b*) are

$$A_{\mathrm{up}} = \cosh^2\gamma l + \frac{Z_{ca}}{2Z_s}\sinh 2\gamma l \coth\gamma l_{us} + \frac{Z_{ca}}{Z_{cb}}\sinh^2\gamma l, \tag{10.1a}$$

$$B_{\mathrm{up}} = \frac{Z_{ca} + Z_{cb}}{2}\sinh 2\gamma l + \frac{Z_{ca}Z_{cb}}{Z_s}\sinh^2\gamma l \coth\gamma l_{us}, \tag{10.1b}$$

$$C_{\mathrm{up}} = \frac{\sinh 2\gamma l}{2Z_{ca}} + \frac{1}{Z_s}\cosh^2\gamma l \coth\gamma l_{us} + \frac{\sinh 2\gamma l}{2Z_{cb}}, \tag{10.1c}$$

$$D_{\mathrm{up}} = \frac{Z_{cb}\sinh^2\gamma l}{Z_{ca}} + \frac{Z_{cb}}{2Z_s}\coth\gamma l_{us}\sinh 2\gamma l + \cosh^2\gamma l, \tag{10.1d}$$

where $\gamma = \alpha + j\beta$ (α and β are attenuation and phase constants), and

$$Z_{ca} = \sqrt{Z_1 Z_h \frac{d_1^2 + d_2^2}{d_1^2}}, \tag{10.2a}$$

$$Z_{cb} = \sqrt{Z_2 Z_h \frac{d_1^2 + d_2^2}{d_1^2}}. \tag{10.2b}$$

In the same way, the *ABCD* parameters of the down-filter are derived as a function of Z_{cc}, Z_{cd}, Z_s, l, and l_{ds}, where $Z_{cc} = \sqrt{Z_1 Z_h (d_1^2 + d_2^2)/d_2^2}$ and $Z_{cd} = \sqrt{Z_2 Z_h (d_1^2 + d_2^2)/d_2^2}$. Using the relation between the *ABCD* and admittance parameters, the admittance parameters contributed by ports ① and ② may be derived as

$$Y_{11} = \frac{D_{\mathrm{up}}}{B_{\mathrm{up}}} + \frac{D_{do}}{B_{do}}, \tag{10.3a}$$

$$Y_{12} = Y_{21} = -\left(\frac{1}{B_{\mathrm{up}}} + \frac{1}{B_{\mathrm{do}}}\right), \tag{10.3b}$$

$$Y_{22} = \frac{A_{\mathrm{up}}}{B_{\mathrm{up}}} + \frac{A_{do}}{B_{do}}, \tag{10.3c}$$

where A_{do}, B_{do}, C_{do}, D_{do} are *ABCD* parameters of the down filter.

If the port ② is terminated in Z_2 as shown in Fig. 10.1(*a*), the reflection coefficient at port ① Γ is

$$\Gamma = \frac{1 - Y_{\mathrm{in}} Z_1}{1 + Y_{\mathrm{in}} Z_1}, \tag{10.4}$$

where

$$Y_{\mathrm{in}} = \frac{Y_{22} + Y_2}{Y_{11}(Y_{22} + Y_2) - Y_{12}Y_{21}} \quad \text{and} \quad Y_2 = \frac{1}{Z_2}.$$

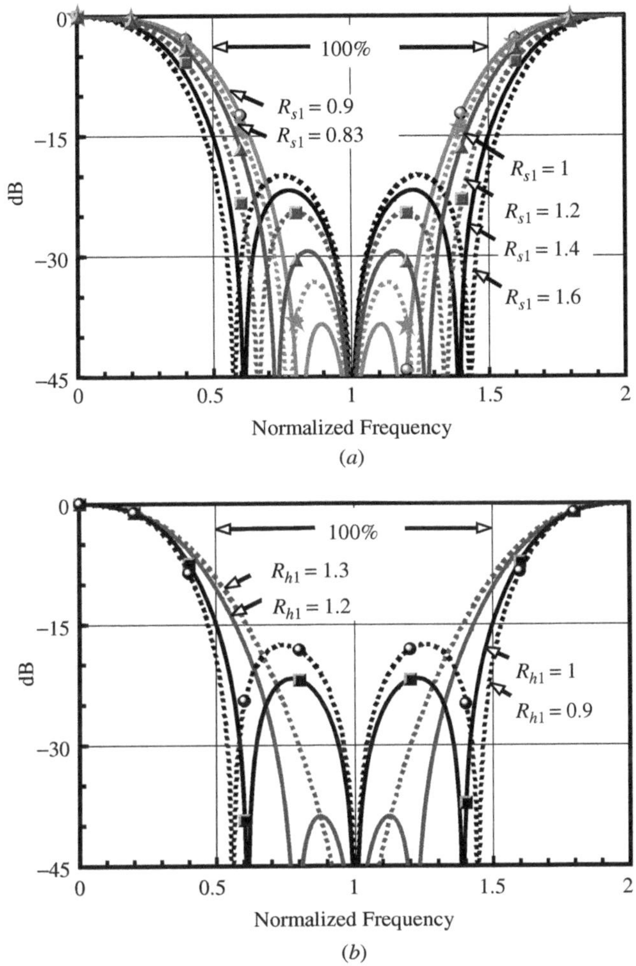

FIGURE 10.2 Calculated reflection coefficients: (*a*) depending on R_{s1} with $Z_1 = Z_2 = Z_h$; (*b*) depending on R_{h1} with $R_{s1} = 1.4$.

Based on (10.1)–(10.4), the frequency responses of the reflection coefficient Γ were calculated by Matlab 6.1. For simplicity, $\alpha = 0$ (lossless), no discontinuity effect, and $Z_1 = Z_2$, $d_1 = d_2$, $\beta_o l = \beta_o l_{us} = \beta_o l_{ds} = 90^\circ$ were assumed, where β_o is a propagation constant at the design center frequency. Two types of calculated results are plotted in Fig. 10.2, where R_{s1} is the ratio of Z_s to Z_1 and R_{h1} is that of Z_h to Z_1 in Fig. 10.1(*a*). Figure 10.2(*a*) shows reflection coefficients depending on R_{s1} in the case of $Z_1 = Z_2 = Z_h$. Figure 10.2(*a*) indicates that the higher values of Z_s have wider bandwidths and that the fractional bandwidth, the frequency range where Γ is less than -15 dB, is about 100% with $R_{s1} = 1.4$. Those for various values of R_{h1} are plotted in Fig. 10.2(*b*), where $R_{s1} = 1.4$ or $Z_s = 1.4Z_1$. Figure 10.2(*b*) shows that the bandwidths become smaller as R_{h1} increases. Up to $R_{h1} = 1.2$, three poles exist, which is a prime cause of the wideband characteristics of ring filters. However, the poles do not appear at $R_{h1} = 1.3$. Therefore, if Z_s and Z_h are chosen properly, ring filters with fractional bandwidths greater than 100% may be designed.

10.2.2 Measurements

Based on this study, a ring filter was fabricated on a substrate with $\varepsilon_r = 2.88$, $h = 508$ μm, and $\tan\delta = 0.0064$ at 3 GHz in microstrip technology. In this case, the termination impedances of Z_1, Z_2, and Z_h were each 50 Ω, the characteristic impedance of the short stub was 85 Ω, $Z_{ca} = Z_{cb} = Z_{cc} = Z_{cd} = 70.71$ Ω, and $l = l_{us} = l_{ds} = 90^\circ$. The design center frequency is 3 GHz, the length and width of the transmission-line section are 16,630.5 μm and 731 μm, and those of the short stub are 16,810 μm and 509 μm. A photo of the ring filter is shown in Fig. 10.3(*a*), and the insertion and return losses measured are compared with those predicted in Fig. 10.3(*b*) and (*c*), respectively. In practice, the phases of S_{21} of the up- and down-filters are not always the same because the two short stubs could not be realized equally. That causes a sudden phase inversion, or transmission zero, so the resistance between the two Z_h ports in Fig. 10.1 is needed to prevent it. The smaller the resistance, the better. The reason for the transmission zero is explained in the next section.

A circuit for the prevention of transmission zero (POTZ) consists of two resistors, as shown in Fig. 10.3(*a*). The output performances of the ring filter will be better with a lower resistance, so two 10-Ω chip resistors were used for a sufficient distance between the two transmission lines. The insertion and return losses predicted in Fig. 10.3(*b*) and (*c*) were carried out using an EM simulator, Empire (IMST, Kamp-Lintfort, Germany) that is based on finite-difference time-domain analyses. The deviation between the measured and predicted values at the edge of the passband in Fig. 10.4(*c*) is due to the stray capacitance and inductance caused by soldering the two resistors on the transmission lines. Except for those points, the results measured agree well with those predicted.

10.3 NEW MEASUREMENT TECHNIQUE FOR INHERENT RING-RESONANCE FREQUENCY

10.3.1 Lossless Case

In Fig. 10.1(*a*), if the two powers of the up- and down-filters are out of phase at port ②, the excited power at port ① cannot be delivered to port ②. If a certain condition is satisfied between the two short stubs, a frequency exists where all the excited power is reflected. That is the inherent ring-resonance frequency. To derive the condition, losslessness ($\alpha = 0$) and lack of discontinuity effect are assumed, and values of l, l_{us} and l_{ds} are set as

$$\beta_o l = \frac{\pi}{2}, \tag{10.5a}$$

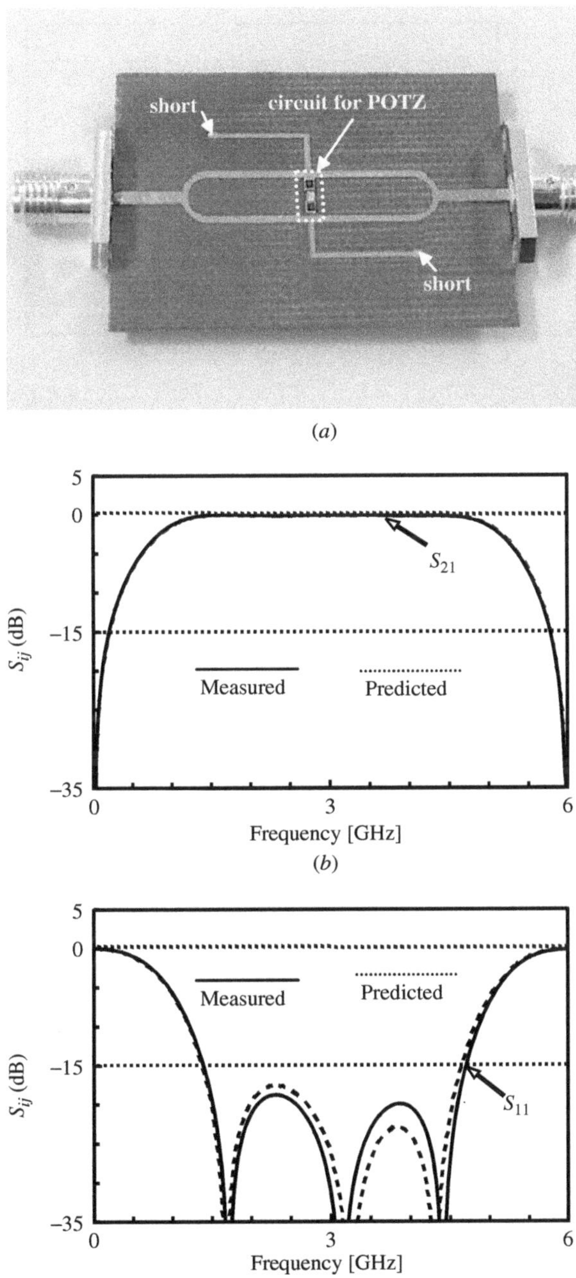

FIGURE 10.3 Ring filter for measurements: (*a*) photo; (*b*) measured and predicted insertion losses; (*c*) measured and predicted return losses.

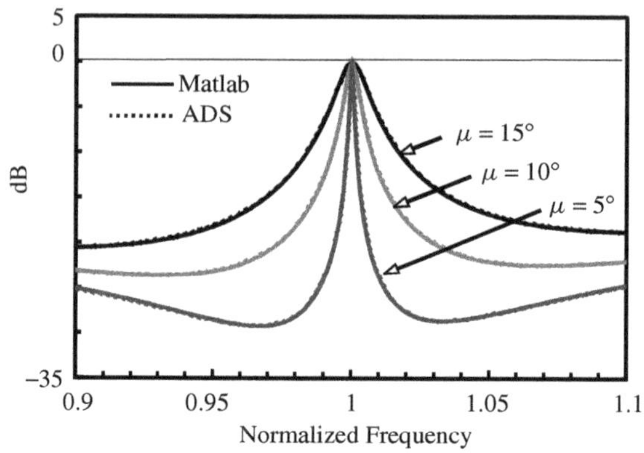

(*a*)

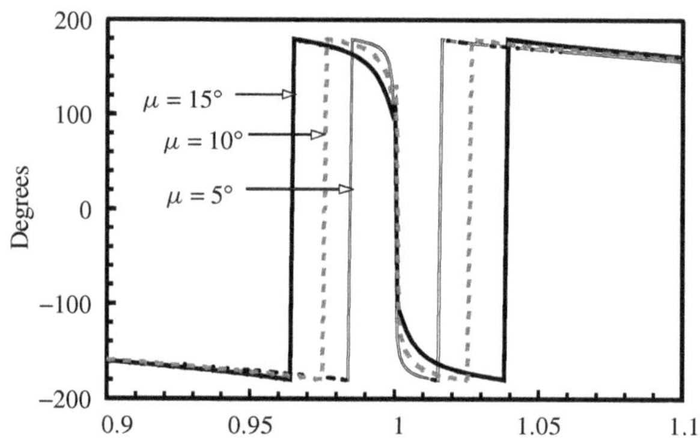

(*b*)

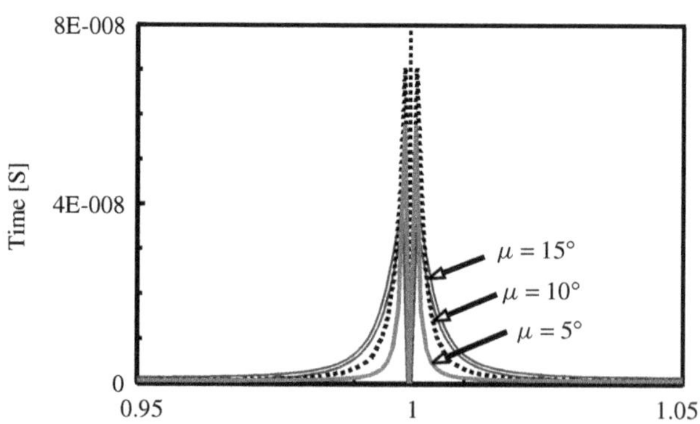

(*c*)

FIGURE 10.4 Frequency responses: (*a*) return losses; (*b*) phase responses; (*c*) group delays.

$$\beta_o l_{us} = \frac{\pi}{2} + \mu, \tag{10.5b}$$

$$\beta_o l_{ds} = \frac{\pi}{2} + \nu, \tag{10.5c}$$

where β_o is a propagation constant at a design center frequency. Substituting the propagation constant in (10.5) for γ in (10.1)–(10.3), the admittance parameters of the ring filter in Fig. 10.1(a) are calculated as

$$Y_{11} = -j\frac{Z_s}{Z_{ca}{}^2}\cot\mu - j\frac{Z_s}{Z_{cc}{}^2}\cot\nu, \tag{10.6a}$$

$$Y_{12} = Y_{21} = -\frac{jZ_s}{Z_{ca}Z_{cb}}\cot\mu - \frac{jZ_s}{Z_{cc}Z_{cd}}\cot\nu, \tag{10.6b}$$

$$Y_{22} = -j\frac{Z_s}{Z_{cb}{}^2}\cot\mu - j\frac{Z_s}{Z_{cd}{}^2}\cot\nu, \tag{10.6c}$$

which are derived at the design center frequency.

For an equal power division $d_1 = d_2$, (10.6) is simplified as

$$Y = -jZ_s\frac{\sin(\mu+\nu)}{\sin\mu\sin\nu}\begin{bmatrix} \dfrac{1}{Z_{ca}{}^2} & \dfrac{1}{Z_{ca}Z_{cb}} \\ \dfrac{1}{Z_{ca}Z_{cb}} & \dfrac{1}{Z_{cb}{}^2} \end{bmatrix}. \tag{10.7}$$

The values $\mu \neq 0$, $\nu \neq 0$, and $\mu + \nu = 0$ in (10.7) result in

$$Y = \begin{bmatrix} 0 & 0 \\ 0 & 0 \end{bmatrix}, \tag{10.8}$$

which implies that matching and power transfer cannot occur when $\mu + \nu = 0$, $\mu \neq 0$, and $\nu \neq 0$. This fact may be used to determine an inherent ring-resonance frequency. Frequency responses of the return losses, phase responses of the insertion losses, and group delays depending on μ are plotted in Fig. 10.4(a), (b) and (c), respectively. They are all satisfied with the conditions $\mu + \nu = 0$ and $\mu \neq 0$. The return losses were under ideal conditions calculated based on (10.1)–(10.4) using Matlab 6.1 and compared with those obtained using a commercial program, ADS 2002. The comparisons are plotted in Fig. 10.4(a), where the solid lines are those found using Matlab 6.1 and the dashed lines are those found using ADS 2002. The two groups of results are almost identical, and all the excited power is reflected at a frequency that is the inherent ring-resonance frequency. Figure 10.4(b) shows sudden phase inversions, where all are satisfied with the conditions $\mu + \nu = 0$ and $\mu \neq 0$. Due to the sudden phase inversions, high values of the group delay in Fig. 10.4(c) occur, which is consistent with (10.8). Therefore, the conditions $\mu + \nu = 0$ and $\mu \neq 0$ are sufficient and necessary to determine the inherent ring-resonance frequency.

10.3.2 Loss Case

If loss is taken into consideration for microstrip lines, the complex propagation constant is

$$\gamma = \alpha + j\beta = j\omega\sqrt{\mu\varepsilon'[1 - j\tan\delta_e(f)]}, \tag{10.9}$$

where $\omega = 2\pi f$ and $\tan\delta_e(f)$ is the effective loss tangent, including dielectric, conductor, and other losses at a given frequency f.

Since $\tan\delta_e(f) << 1$, (10.5) may be formulated as

$$\gamma l = \frac{\pi}{2}\frac{f}{f_0}\cdot\frac{\tan\delta_e(f)}{2} + j\left(\frac{\pi}{2}\frac{f}{f_0}\right), \tag{10.10a}$$

$$\gamma l_{us} = \left(\frac{\pi}{2} + \mu\right)\frac{f}{f_0}\cdot\frac{\tan\delta_e(f)}{2} + j\left(\frac{\pi}{2} + \mu\right)\frac{f}{f_0}, \tag{10.10b}$$

$$\gamma l_{ds} = \left(\frac{\pi}{2} + \nu\right)\frac{f}{f_0}\cdot\frac{\tan\delta_e(f)}{2} + j\left(\frac{\pi}{2} + \nu\right)\frac{f}{f_0}. \tag{10.10c}$$

Based on (10.1)–(10.4) and (10.9)–(10.10), frequency responses of the ring filters with loss were calculated under the conditions $\mu + \nu = 0$ and $\mu \neq 0$, and the reflection coefficients calculated are plotted in Fig. 10.5, where $Z_1 = Z_2 = Z_h = 50\ \Omega$; $d_1 = d_2$; Z_{ca}, Z_{cb}, Z_{cc}, Z_{cd}, and Z_s in Fig. 10.1(a) are all 70.71 Ω; and the X-axis is normalized to 3 GHz. The value of $\tan\delta_e(f)$ was taken to be as 0.0064 at 3 GHz for these calculations. From the results in Fig. 10.5, the amount of reflected power at the inherent ring-resonance frequency increases as μ grows. Substituting equations (10.10) into (10.1) and (10.3) results in the

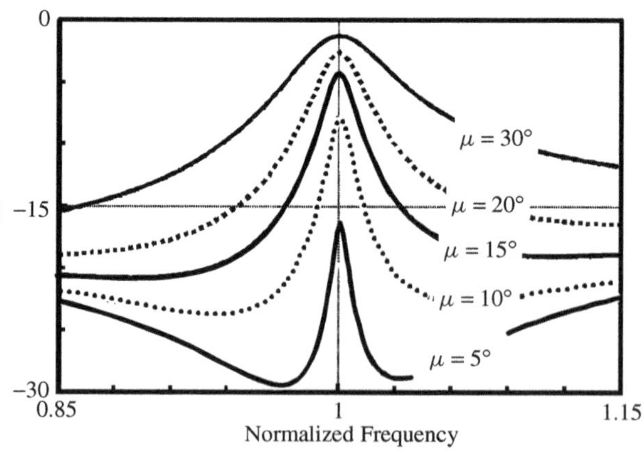

FIGURE 10.5 Frequency responses of reflection coefficients, including dielectric and conductor losses.

admittance parameters with loss at $f = f_0$:

$$Y_{11} = \frac{\dfrac{Z_{cb}}{Z_{ca}} \cosh^2 \alpha l + K_u \dfrac{Z_{cb}}{2Z_m} \sinh 2\alpha l + \sinh^2 \alpha l}{\dfrac{1}{2}(Z_{cb} + Z_{ca}) \sinh 2\alpha l + K_u \dfrac{Z_{ca} Z_{cb}}{Z_s} \cosh^2 \alpha l} + \frac{\dfrac{Z_{cd}}{Z_{cc}} \cosh^2 \alpha l + K_d \dfrac{Z_{cd}}{2Z_s} \sinh 2\alpha l + \sinh^2 \alpha l}{\dfrac{1}{2}(Z_{cc} + Z_{cd}) \sinh 2\alpha l + K_d \dfrac{Z_{cs} Z_{cd}}{Z_s} \cosh^2 \alpha l}, \tag{10.11a}$$

$$Y_{12} = Y_{21} = \frac{1}{\dfrac{1}{2}(Z_{cb} + Z_{ca}) \sinh 2\alpha l + K_u \dfrac{Z_{ca} Z_{cb}}{Z_s} \cosh^2 \alpha l} + \frac{1}{\dfrac{1}{2}(Z_{cc} + Z_{cd}) \sinh 2\alpha l + K_d \dfrac{Z_{cs} Z_{cd}}{Z_s} \cosh^2 \alpha l} \tag{10.11b}$$

$$Y_{22} = \frac{\sinh^2 \alpha l + K_u \dfrac{Z_{ca}}{2Z_s} \sinh 2\alpha l + \dfrac{Z_{ca}}{Z_{cb}} \cosh^2 \alpha l}{\dfrac{1}{2}(Z_{cb} + Z_{ca}) \sinh 2\alpha l + K_u \dfrac{Z_{ca} Z_{cb}}{Z_s} \cosh^2 \alpha l} + \frac{\sinh^2 \alpha l + K_d \dfrac{Z_{cc}}{2Z_s} \sinh 2\alpha l - \dfrac{Z_{cc}}{Z_{cd}} \cosh^2 \alpha l}{\dfrac{1}{2}(Z_{cc} + Z_{cd}) \sinh 2\alpha l + K_d \dfrac{Z_{cs} Z_{cd}}{Z_s} \cosh^2 \alpha l}, \tag{10.11c}$$

where $\alpha l = (\pi/4) \tan \delta_e(f_0)$,

$$K_u = \frac{e^{2(\alpha l_{us} + j\mu)} + 1}{e^{2(\alpha l_{us} + j\mu)} - 1} \quad \text{and} \quad K_d = \frac{e^{2(\alpha l_{ds} - j\mu)} + 1}{e^{2(\alpha l_{ds} - j\mu)} - 1}$$

with

$$\alpha l_{us} = \left(\frac{\pi}{2} + \mu\right) \frac{\tan \delta_e(f_0)}{2} \quad \text{and} \quad \alpha l_{ds} = \left(\frac{\pi}{2} - \mu\right) \frac{\tan \delta_e(f_0)}{2}.$$

To calculate attenuations due to conductor loss α_c, dielectric loss α_d, and so on, for most microstrip lines, many factors must be considered: for example, relative dielectric constant, dielectric loss, conductivity, characteristic impedance, and width of the microstrip line [23]. Moreover, none of them can be ensured to be correct, which causes errors. However, if the return loss of the ring filter at the inherent ring-resonance frequency is known or measured, $\tan \delta_e(f)$ may be calculated correctly at a given frequency using (10.11). To verify the method to determine the inherent ring-resonance frequency, two ring filters with $\mu + \nu = 0$ and $\mu \neq 0$ were fabricated on a substrate ($\varepsilon_r = 2.88$, $h = 508$ μm) in microstrip technology. They were designed at 3 GHz, $\tan \delta_e(f)$ was taken as 0.0064, and the loss due to feeding lines is ignored. A device with $\mu = 20^\circ$ is shown in Fig. 10.6, and the results measured are compared with those predicted in Fig. 10.7. Measured return loss with $\mu = 20^\circ$ is -2.26 dB and the simulated return loss is -2.28 dB at 3 GHz. In the case of $\mu = 15^\circ$, the measured return loss is -3.4 dB and the simulated return loss is -3.71 dB. Due to conductor and dielectric losses, the values measured show not 0 dB but -2.26 dB or -3.4 dB at the inherent ring resonance frequency, 3 GHz.

In the design of ring-based circuits, it is necessary to determine the ring-resonance frequency of various forms of the ring. Figure 10.8 shows

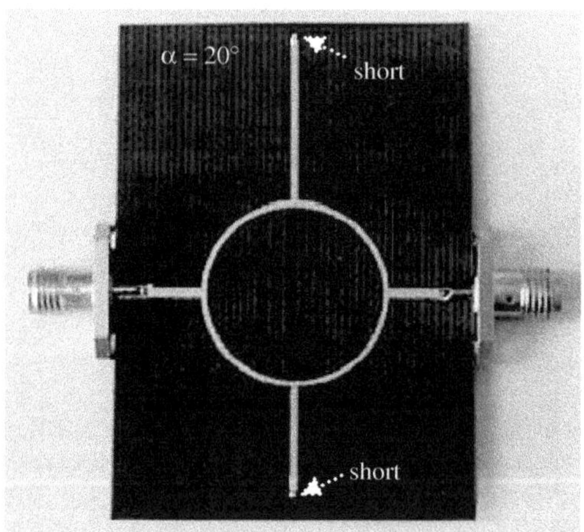

FIGURE 10.6 Ring filter photo with $\mu + \nu = 0°$, $\mu = 20°$.

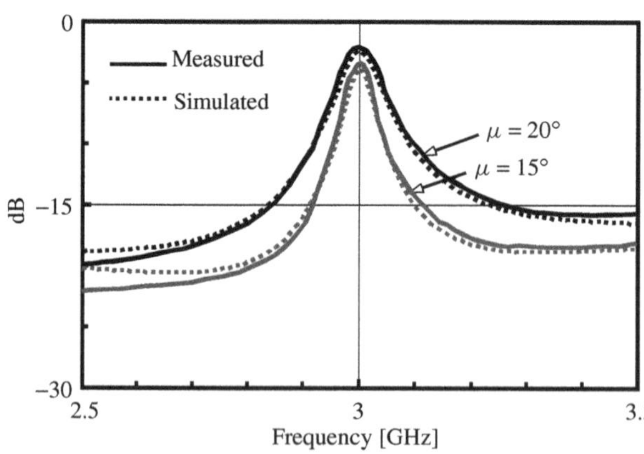

FIGURE 10.7 Measured and simulated return losses.

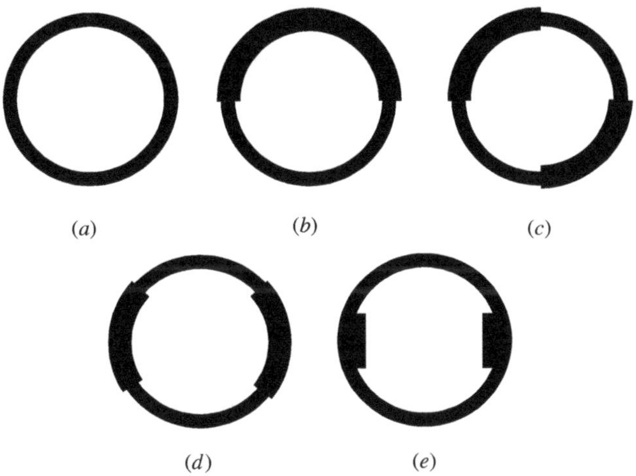

FIGURE 10.8 Various forms of ring: (*a*) simple form; (*b*) different-power-division form; (*c*) isolation form; (*d*) CVT form; (*e*) CCT form.

various forms that may be produced by combining the Z_1, Z_2, d_1, and d_2 in Fig. 10.1(*a*), or by using small impedance transformers, CVTs (constant-VSWR impedance transformers), and CCTs (constant-conductance impedance transformers) [17,18], to be introduced in the next chapter. When $Z_1 = Z_2$ and $d_1 = d_2$ in Fig. 10.1(*a*), the simple form shown in Fig. 10.8(*a*), which is very popular, is produced. The form depicted in Fig. 10.8(*b*) is for $d_1 \neq d_2$. When the power division ratio is d_1 to d_2 at port ① and the combined power ratio is d_2 to d_1 at port ②, the excited power at port ① cannot be delivered to port ①, as shown in Fig. 10.1(*c*). This fact may be used for isolation characteristics, as ascertained by Ahn et al. [20]. The CVT and CCT forms in Fig. 10.8(*d*) and (*e*) are from CVT and CCT ring filters [18]. All the ring-resonance frequencies may be determined by the method suggested above, and that in Fig. 10.8(*c*) may be determined without the two stubs.

10.4 CONCLUSIONS

It was suggested that a ring filter be designed as a wideband filter consisting of a ring and two short stubs connected at the 90° and 270° points of the ring. The termination impedance at the hypothetical ports and the characteristic impedances of the short stubs are related to the bandwidth of the ring filter, so their optimal values are needed for the wideband characteristics. Such a ring filter was designed, compared, and tested. Its measured bandwidth was more than 100%, which can be achieved with more than five stages in conventional filter design technique. As an application of the ring filter, a technique was suggested for measurement of the inherent resonance frequency. By knowing the exact inherent ring-resonance frequency, the capacitances due to gap coupling in conventional ring-based circuits may be found exactly, and several different forms of the ring that may be utilized in the design of ring-based circuits were proposed. They can be produced by combinations of various termination impedances, arbitrary power division ratios, and CVTs or CCTs in the ring filters, and their inherent ring-resonance frequency may be determined by the method proposed in this chapter.

EXERCISES

10.1 Derive the *ABCD* parameters of a down-filter.

10.2 Explain why there is just one pole with $R_{h1} = 1.3$ in Fig. 10.2.

10.3 Derive (10.6).

10.4 Using the ring filter, explain the method used to determine all losses, including radiation loss.

10.5 Derive (10.11).

REFERENCES

1. I. Wolff and N. Knoppik, Microstrip Ring Resonator and Dispersion Measurement on Microstrip Lines, *Electron. Lett.*, Vol. 7, No. 26, December 1971, pp. 779–781.
2. I. Wolff, Microstrip Bandpass Filter Degenerate Modes of a Microstrip Ring Resonator, *Electron. Lett.*, Vol. 8, No. 12, June 1972, pp. 302–303.
3. H. Yabuki, M. Sagawa, M. Matsuo, and M. Makimoto, Stripline Dual-Mode Ring Resonators and Their Application to Microwave Devices, *IEEE Trans. Microwave Theory Tech.*, Vol. 44, May 1996, pp. 723–729.
4. K. Chang, *Microwave Ring Circuits and Antennas*, Wiley, New York, 1996, pp. 42–124.
5. I. Wolff and V. Tripathi, The Microstrip Open-Ring Resonator, *IEEE Trans. Microwave Theory Tech.*, Vol. 32, January 1984, pp. 102–107.
6. V. Tripathi and I. Wolff, Perturbation Analysis and Design Equations for Open- and Closed-Ring Microstrip Resonators, *IEEE Trans. Microwave Theory Tech.*, Vol. 32, April 1984, pp. 409–410.
7. K. Chang, S. Martine, M. Fuchen, and J. L. Kline, On the Study of Microstrip Ring and Varactor-Tuned Ring Circuits, *IEEE Trans. Microwave Theory Tech.*, Vol. 35, December 1987, pp. 1288–1295.
8. H.-R. Ahn and I. Wolff, Novel Ring Filter as a Wide-Band 180° Transmission-Line, *EUMC Proc.*, Vol. III, October 1999, pp. 95–98.
9. G. Matthaei, L. Young, and E. M. T. Jones, *Microwave Filters, Impedance-Matching Networks, and Coupling Structures*, Artech House, Dedham, MA, 1980, pp. 521–562.
10. L. Zhu and K. Wu, A Joint Field/Circuit Model of Line-to-Ring Coupling Structures and Its Application to the Design of Microstrip Dual-Mode Filters and Ring

Resonator Circuits, *IEEE Trans. Microwave Theory Tech.*, Vol. 47, October 1999, pp. 1938–1948.

11. P. Troughton, Measurement Technique in Microstrip, *Electron. Lett.*, Vol. 5, No. 2, January 1969, pp. 25–26.
12. M. Mao, S. Jones, and G. D. Vendelin, Millimeter-Wave Integrated Circuits, *IEEE Trans. Microwave Theory Tech.*, Vol. 16, July 1968, pp. 455–461.
13. Y. S. Wu and F. J. Rosenbaum, Mode-Chart for Microstrip Ring Resonators, *IEEE Trans. Microwave Theory Tech.*, Vol. 21, July 1973, pp. 487–489.
14. R. P. Ovens, Curvature Effects in Microstrip Ring Resonators, *Electron. Lett.*, Vol. 12, July 1976, pp. 356–357.
15. A. M. Khilla, Computer-Aided Design for Microstrip Mode-Chart for Microstrip Ring Resonators, *Proc. 11th EUMC*, 1981, pp. 1–6.
16. F. Tefiku and E. Yamashita, An Efficient Method for the Determination of Resonant Frequencies of Shielded Circular Disk and Ring Resonators, *IEEE Trans. Microwave Theory Tech.*, Vol. 41, February 1993, pp. 343–346.
17. H.-R. Ahn and I. Wolff, General Design Equations of Three-Port Power Dividers, Small-Sized Impedance Transformers, and Their Applications to Small-Sized Three-Port 3-dB Power Divider, *IEEE Trans. Microwave Theory Tech.*, Vol. 49, July 2001, pp. 1277–1288.
18. H.-R. Ahn, N.-H. Myung, and I. Wolff, Small-Sized Wideband CVT- and CCT-Ring Filters, *IEEE MTT-S Dig.*, June 2003, pp. 1607–1610.
19. H.-R. Ahn, I.-S. Chang, and S.-W. Yun, Miniaturized 3-dB Ring Hybrid Terminated by Arbitrary Impedances, *IEEE Trans. Microwave Theory Tech.*, Vol. 42, December 1994, pp. 2216–2241.
20. H.-R. Ahn, I. Wolff, and I.-S. Chang, Arbitrary Termination Impedances, Arbitrary Power Division and Small-Sized Ring Hybrids, *IEEE Trans. Microwave Theory Tech.*, Vol. 44, December 1997, pp. 2241–2247.
21. H.-R. Ahn and I. Wolff, Three-Port 3-dB Power Divider Terminated by Different Impedances and Its Application to MMIC's, *IEEE Trans. Microwave Theory Tech.*, Vol. 47, June 1999, pp. 786–794.
22. H.-R. Ahn and I. Wolff, Asymmetric Ring Hybrid Phase-Shifters and Attenuators, *IEEE Trans. Microwave Theory Tech.* Vol. 50, April 2002, pp. 1146–1155.
23. D. M. Pozar, *Microwave Engineering*, Addison-Wesley, Reading MA, 1990, pp. 185–186.

CHAPTER ELEVEN

Small Impedance Transformers, CVTs and CCTs, and Their Applications to Small Power Dividers and Ring Filters

11.1 SMALL TRANSMISSION-LINE IMPEDANCE TRANSFORMERS

The most important design tool in microwave passive and active components is the concept of impedance matching. All values can be plotted in a reflection coefficient plane (Smith chart), and those of passive component loads are $|\Gamma| \leq 1$, where Γ is a complex reflection coefficient. Therefore, the Smith chart has been used extensively in the design of matching circuits [1]. However, although the plane of the Smith chart expresses reflection coefficients of loads, the coordinates written on the Smith chart are load oriented and the numbers on it are not linear. Therefore, reading errors may easily be committed.

Impedance transformers have been used for various applications, so there have been many trials to reduce their size. One of these utilizes a technique to match a complex load to a real impedance (admittance) with only one transformer line. This matching technique was proposed graphically [2] and theoretically [3]. However, in Ref. 3 the formula for the length of the impedance transformer is not correct, and the graphical method [2] requires two impedance (admittance) normalizations, a process in which reading errors are easily committed [4,5]. Also, it was suggested that the graphical method [2] be considered only if the complex load were located on the left-half plane in the Smith chart.

In this chapter, formulas for transformer lines to match a complex load to a real reference load are derived mathematically, and two types of small

Asymmetric Passive Components in Microwave Integrated Circuits, by Hee-Ran Ahn

transmission-line impedance transformers are constructed using the formulas derived. These small impedance transformers will be very useful in miniaturized microwave circuits, and the design equations derived will be utilized for other general circuit matching. As applications of the small impedance transformers, small three-port 3-dB power dividers, 45° three-port power dividers, and small ring filters are introduced and discussed.

11.2 MATHEMATICAL APPROACH FOR CVTs AND CCTs

A transmission-line transformer to match a complex load Z_L to a real reference impedance Z_r is depicted in the lower part of the Smith chart shown in Fig. 11.1, and a graphical procedure to find a transformer line of characteristic impedance Z_0' and length d is illustrated on the Smith chart, where z_L is a complex load normalized to Z_r, $\sqrt{r_1}$ is a characteristic impedance Z_0' normalized to Z_r, C is the center of the circle passing through the two points z_L and the center of the Smith chart, and z_L' is the complex load Z_L renormalized to Z_0'.

By the relation $\Gamma = (z_L - 1)/(z_L + 1)$, real and imaginary parts of the reflection coefficient, Γ_r and Γ_i, are derived as

$$\Gamma_r(r, x) = \frac{(r^2 - 1) + x^2}{(1 + r)^2 + x^2}, \tag{11.1}$$

$$\Gamma_i(r, x) = \frac{2x}{(1 + r)^2 + x^2}, \tag{11.2}$$

where r and x are the real and imaginary parts of z_L, respectively. Since C is the center of the circle passing through z_L and the center of the Smith chart, the

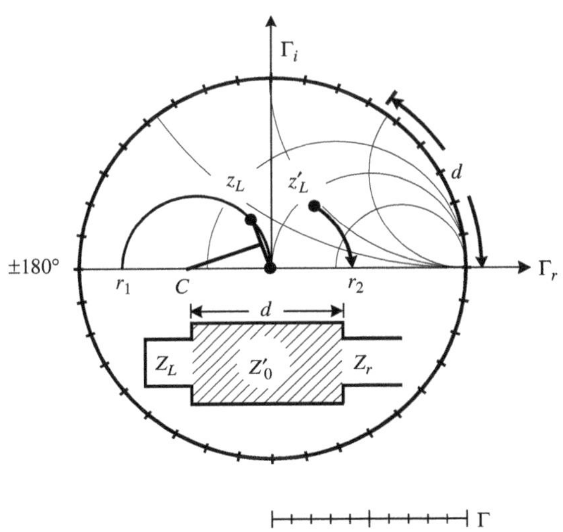

FIGURE 11.1 Smith chart to design a transformer line to transform a complex load Z_L into a reference real load Z_r.

perpendicular bisector of the line connecting the center of the Smith chart with z_L is projected onto the real axis, point C. Assuming that the value at r_1 is two times that at C, the reflection coefficient at r_1 is

$$\Gamma_r(r,x)|r_1 = \frac{[\Gamma_r(r,x)|_{z_L}]^2 + [\Gamma_i(r,x)|_{z_L}]^2}{\Gamma_r(r,x)|_{z_L}}, \tag{11.3}$$

where $\Gamma_r(r,x)|_{z_L}$ and $\Gamma_i(r,x)|_{z_L}$ are the real and imaginary parts of the reflection coefficient at z_L, respectively. A formula for the impedance Z_0' based on (11.1) and (11.3) yields

$$\frac{Z_0'}{Z_r} = \sqrt{\frac{1+\Gamma_r(r,x)|r_1}{1-\Gamma_r(r,x)|r_1}}. \tag{11.4}$$

The length d of the transformer line in Fig. 11.1 is obtained from the point z_L' and the complex load renormalized to Z_0', and they are as follows:
For $\Gamma_r(r,x)|r_1 \le 0$,

$$\tan 2\Theta_0 = \frac{\Gamma_i(r,x)|_{Z_L'}}{\Gamma_r(r,x)|_{Z_L'}}, \tag{11.5}$$

For $\Gamma_r(r,x)|_{r_1} \ge 0$ and $\Gamma_i(r,x)|_{Z_L'} \ge 0$,

$$\tan\left[2\left(\Theta_0 - \frac{\pi}{2}\right)\right] = \frac{\Gamma_i(r,x)|_{Z_L'}}{\Gamma_r(r,x)|_{Z_L'}}, \tag{11.6a}$$

For $\Gamma_r(r,x)|_{r_1} \ge 0$ and $\Gamma_i(r,x)|_{Z_L'} \le 0$,

$$\tan\left[2\left(\Theta_0 + \frac{\pi}{2}\right)\right] = \frac{\Gamma_i(r,x)|_{Z_L'}}{\Gamma_r(r,x)|_{Z_L'}}, \tag{11.6b}$$

where $0 \le \Theta_0 \le \pi$ and $\Theta_0 = \beta d$ with an assumption of $\alpha = 0$.

11.2.1 CVTs and CCTs

Impedance transformers of the 2 : 1 quarter-wave length type are often utilized for 3-dB power dividers (ring hybrids and three-port power dividers with equal power division). A 2 : 1 quarter-wavelength impedance transformer is plotted from 0.5 to 1 on an admittance Smith chart as demonstrated by the solid half-circle in Fig. 11.2(a), and a constant-VSWR circle passing through the point 0.5 is illustrated as a dashed half-circle. To reduce its size, the load admittance 0.5 needs to be changed from a real value to a complex value, and there are several ways to accomplish this. One of them is by moving up along a constant-VSWR circle passing through 0.5, and another is along a constant $g = 0.5$ circle, as indicated by the arrows in Fig. 11.2(a). Other cases are also available, but for easy construction of the transformers, only these two are discussed. The circuit resulting from the first method is shown in Fig. 11.2(b), that from the second in Fig. 11.2(c). The transformer in Fig. 11.2(d) is a conventional reduced-size

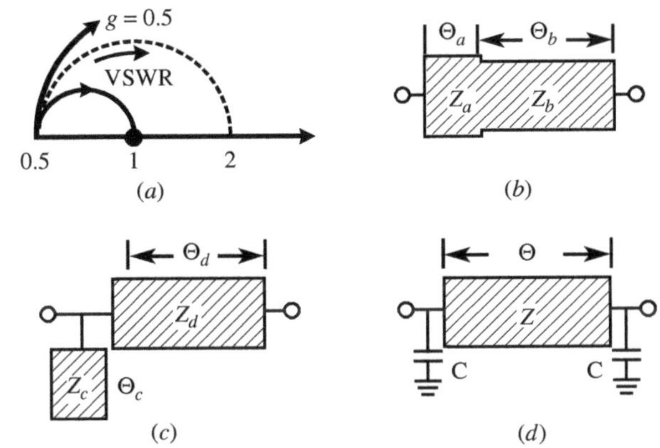

FIGURE 11.2 Impedance transformers to transform a real impedance into a real reference impedance: (*a*) part of a Smith chart (for *IR* = 2); (*b*) CVT; (*c*) CCT; (*d*) conventional reduced-size impedance transformer.

transformer equivalent to a quarter-wavelength transformer [6]. From the background outlined above, the transmission-line transformer in Fig. 11.2(*b*) is called a constant-VSWR transmission-line transformer (CVT) and that in Fig. 11.2(*c*) a constant-conductance transmission-line transformer (CCT). Here $\Theta_a + \Theta_b$ in the CVT and Θ_d in the CCT of Fig. 11.2 are much shorter than a quarter wavelength.

Figure 11.3(*a*) and (*b*) show the impedance Z_b and the electrical length Θ_b of the CVT in Fig. 11.2(*b*), depending on Θ_a variations. Figure 11.3(*c*) and (*d*) indicate the impedance Z_d and the electrical length Θ_d of the CCT in Fig. 11.2(*c*), depending on Θ_c variations. They were produced based on (11.4) and (11.5). IR 2.0, IR 1.8, IR 1.6, IR 1.4, and IR 1.2 in the graphs indicate impedance transformation ratios. For these simulations, two impedances Z_a and Z_c were set to 50 Ω, and the reference impedance Z_r was also set to 50 Ω. Figure 11.3(*a*) and (*c*) show that the impedance Z_b or Z_d increases with an increase in the transformation ratio (*IR*). The dependencies of Θ_b on Θ_a in Fig. 11.3(*b*) are almost linear in the CVT case; however, those of Θ_d on Θ_c in Fig. 11.3(*d*) are linear only when the *IR* is greater than 1.6. The total length of the CVT is $\Theta_a + \Theta_b$, but that of the CCT is only Θ_d, because the transmission line with Z_c in Fig. 11.2 is an open-circuited stub, which is easily replaced by a capacitor.

The transmission-line sections of the ring hybrid and three-port power divider with an equal power division are representative impedance transformers and transform 100 Ω into 50 Ω when terminated in equal impedances 50 Ω. In this case, the characteristic impedance and electrical length of the transmission-line section are well known: 70.71 Ω and 90°, respectively. This is case IR 2 on the graphs. When $\Theta_a = 0°$ or $\Theta_c = 0°$ with IR 2 in Fig. 11.3, the characteristic impedance of Z_b or Z_d is 70.7 Ω from Fig. 11.3(*a*) or (*c*), and its length is 90° from Fig. 1.3(*b*) or (*d*). In this way, if the impedance transformation ratio is known, the impedance and length of the transformers can be known with the variations in complex loads. Detailed data are tabulated in Tables 11.1 to 11.4. A

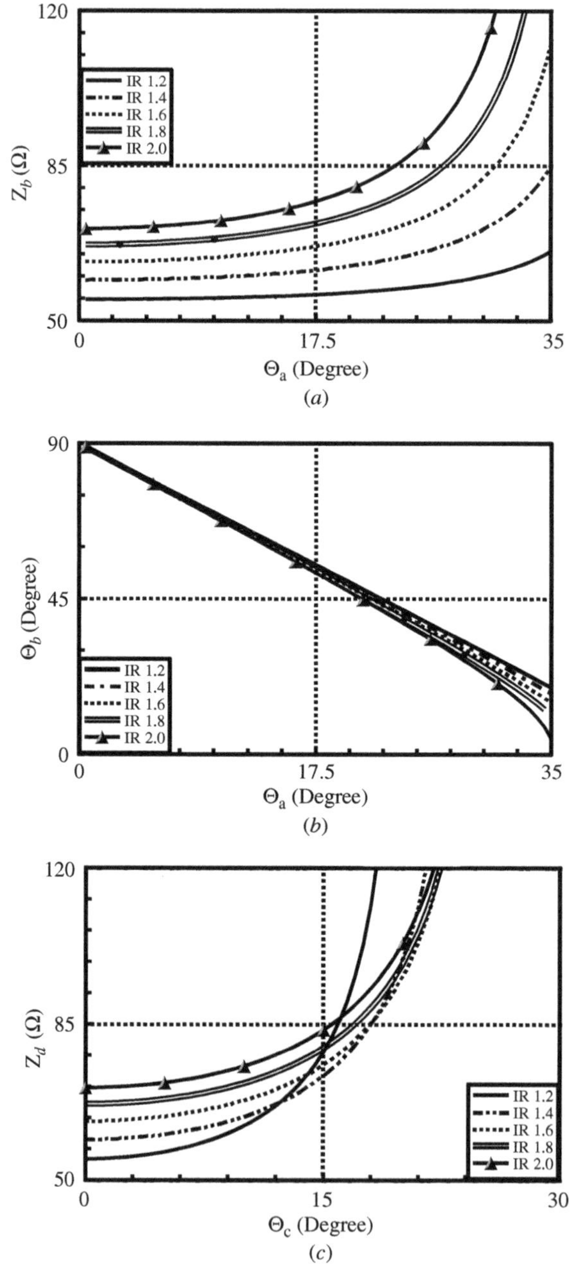

FIGURE 11.3 Design graphs for CVT and CCT, depending on impedance transformation ratios: (*a*) impedance Z_b depending on Θ_a; (*b*) length Θ_b depending on Θ_a; (*c*) impedance Z_d depending on Θ_c; (*d*) length Θ_b depending on Θ_c. (From Ref. 20 with permission from IEEE.)

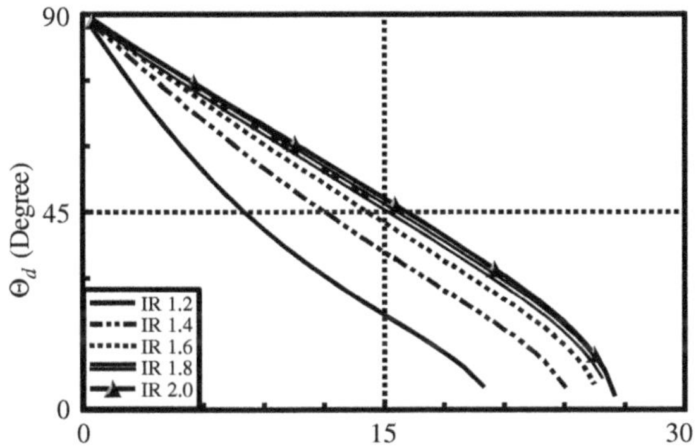

(*d*)

FIGURE 11.3 (*continued*)

TABLE 11.1 Impedance Z_b (Ω) That Depends on Θ_a

	Z_b (Z_a = 50 Ω, Z_r = 50 Ω) for:				
Θ_a (deg)	$IR = 1.2$	$IR = 1.4$	$IR = 1.6$	$IR = 1.8$	$IR = 2.0$
0	54.77	59.16	63.25	67.08	70.71
5	54.85	59.32	63.48	67.40	71.12
10	55.10	59.82	64.25	68.44	72.45
15	55.55	60.76	65.70	70.44	75.03
20	56.33	62.37	68.23	73.99	79.70
25	57.65	65.19	72.80	80.63	88.80
30	60.09	70.71	82.37	95.74	111.80
35	65.65	85.16	113.45	166.95	440.86

TABLE 11.2 Length Θ_b (deg) That Depends on Θ_a

	Θ_b (Z_a = 50 Ω, Z_r = 50 Ω) for:				
Θ_a (deg)	$IR = 1.2$	$IR = 1.4$	$IR = 1.6$	$IR = 1.8$	$IR = 2.0$
0	90	90	90	90	90
5	79.96	79.86	79.72	79.56	79.39
10	69.91	69.70	69.42	69.09	68.72
15	59.86	59.53	59.08	58.55	57.97
20	49.80	49.31	48.65	47.88	47.01
25	39.72	39.02	38.07	36.93	35.66
30	29.58	28.56	27.12	25.35	23.28
35	19.33	17.63	15.03	11.31	4.66

TABLE 11.3 Impedance $Z_d(\Omega)$ That Depends on Θ_c

	$Z_d(Z_c = 50\ \Omega, Z_r = 50\ \Omega)$ for:				
Θ_c (deg)	$IR = 1.2$	$IR = 1.4$	$IR = 1.6$	$IR = 1.8$	$IR = 2.0$
0	54.77	59.16	63.25	67.08	70.71
5	56.35	60.30	64.30	68.15	71.81
10	62.17	64.26	67.91	71.75	75.57
15	78.81	73.48	75.94	79.66	83.75
20	254.87	99.88	95.92	98.54	103.131
25			253.30	194.17	195.94

TABLE 11.4 Length Θ_d (deg) That Depends on Θ_c

	$\Theta_d(Z_c = 50\ \Omega, Z_r = 50\ \Omega)$ for:				
Θ_c (deg)	$IR = 1.2$	$IR = 1.4$	$IR = 1.6$	$IR = 1.8$	$IR = 2.0$
0	90	90	90	90	90
5	59.39	69.73	73.30	74.98	75.89
10	37.24	51.58	57.44	60.35	61.94
15	21.54	35.96	42.66	46.16	48.09
20	5.133	21.45	28.24	31.78	33.66
25			9.70	13.79	15.30

careful study of Tables 11.3 and 11.4 for $IR = 1.6$ shows that $Z_d = 95.92\ \Omega$ and $\Theta_d = 28.24^\circ$ when $\Theta_c = 20^\circ$. This indicates that a 1.6 : 1 impedance transformer can be realized with a transmission line of 28.24° and a capacitor.

Several CVTs and CCTs whose impedance transformation ratio is 1.6 : 1 were simulated on the basis of the Tables 11.1 to 11.4. For the simulations, Z_L and Z_r in Fig. 11.1 were set to 80 Ω and 50 Ω, respectively, and the impedance transformers were designed at a center frequency of 3 GHz. Figure 11.4(*a*) shows the simulated reflection coefficients of CVTs and CCTs where the number after CVT or CCT represents the electrical length of Θ_a or Θ_c in Fig. 11.2(*b*) or (*c*), respectively. The reflection coefficients of CVTs and CCTs are in the range −73.36 to −94.259 dB at 3 GHz. The frequency responses of CVT 20 and CCT 10 are about the same, and that of CCT 20 is the worst in terms of bandwidth. In Fig. 11.4(*b*), CCT 20 (the worst case) is compared with conventional impedance transformers in Fig. 11.2(*d*). For the conventional transformers, $Z = 133.7\ \Omega$ and $C = 0.739$ pF when $\Theta = 28.24^\circ$, $Z = 64.22\ \Omega$ and $C = 0.146$ pF when $\Theta = 80^\circ$. "Con" on the graph indicates conventional transformers and the next number is the electrical length of the transmission-line sections. Figure 11.4(*b*) shows that the frequency responses of the CCT 20 are about the same as those of Con 80. Thus, the length of the conventional impedance transformer in Fig. 11.2(*d*) must be at least 80° in order for the performance of the conventional transformer to be equal to that of CCT 20.

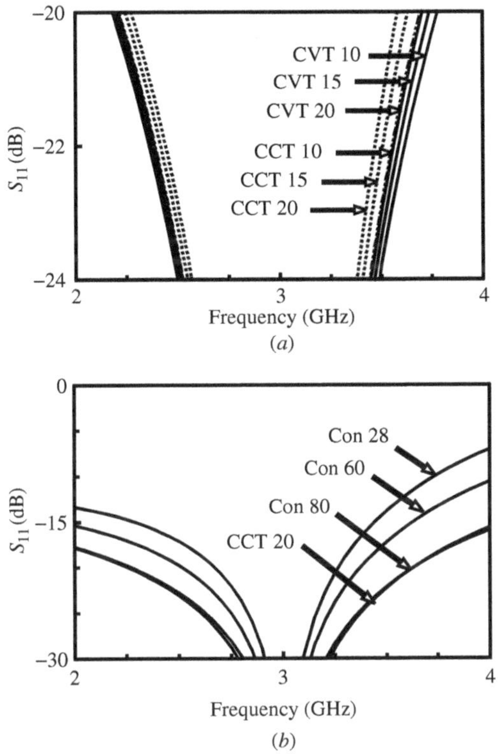

FIGURE 11.4 Simulated reflection coefficients of 1.6 : 1 impedance transformers. (*a*) reflection coefficients of CVTs and CCTs ($Z_r = 50\ \Omega$); (*b*) comparison of one CCT and conventional reduced-size impedance transformers.

11.2.2 Microstrip CVTs and CCTs (1.6 : 1 Impedance Transformers)

On the basis of design equations (11.4)–(11.5) and Tables 11.1 to 11.4, four 1.6 : 1 impedance transformers were fabricated on Al_2O_3 substrate ($\varepsilon_r = 9.9$, $h = 635\ \mu m$) in microstrip technology. The layouts for the impedance transformers are depicted in Fig. 11.5, and the results measured are plotted in Fig. 11.6. The design center frequency is 3 GHz and the transformers are terminated in 80 Ω and 50 Ω. Therefore, a quarter-wavelength impedance transformer to transform 80 Ω to 50 Ω is needed for the measurements; it is labeled Ipt. in Fig. 11.5. CVT 20, CVT 30, CCT 15, and CCT 20 are illustrated in Fig. 11.5(*a*)–(*d*), respectively. From Table 11.2, the total electrical length of the CVT 20 is 68.65° and that of CVT 30 is 57.12°, and from Table 11.4, that of CCT 15 is 42.66° and that of CCT 20 is 28.24°. The results in Fig. 11.6 are very similar to those in Fig. 11.4(*a*). Since the reference impedance Z_r is 50 Ω in these small impedance transformers, the characteristic impedances of Z_a and Z_c are all 50 Ω. The data for the fabrication of 1.6 : 1 microstrip impedance transformers are given in Table 11.5.

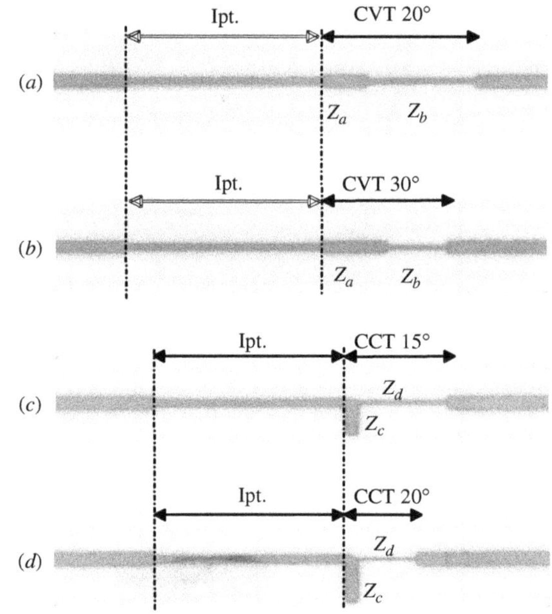

FIGURE 11.5 Layouts for the microstrip small 1.6 : 1 impedance transformers: (*a*) CVT 20°; (*b*) CVT 30°; (*c*) CCT 15°; (*d*) CCT 15°.

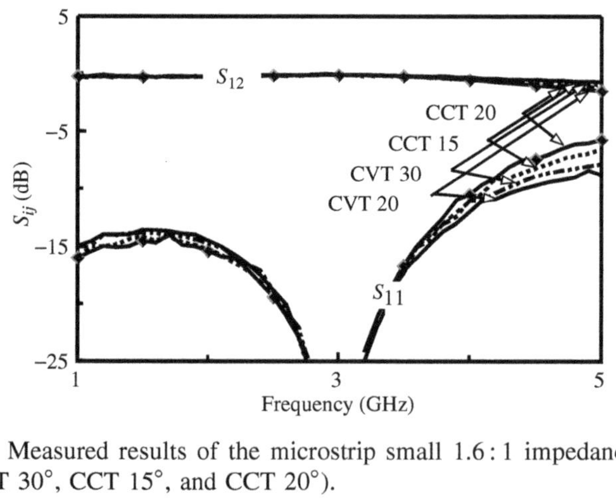

FIGURE 11.6 Measured results of the microstrip small 1.6 : 1 impedance transformers (CVT 20°, CVT 30°, CCT 15°, and CCT 20°).

11.2.3 Bounded Length of CVTs and CCTs

In the case of the impedance transformation ratio $IR = 2$ in Fig. 11.2(*b*) and (*c*), the locus of Θ_a or Θ_c starts from a point ($g = 0.5$, $b = 0$) on the real axis as shown in Fig. 11.2(*a*), where g is a normalized conductance and b is a

TABLE 11.5 Fabrication Data for Microstrip 1.6 : 1 Impedance Transformers[a]

Impedance Transformer		CVT 20°	CVT 30°
$w = 356$ μm, $l = 9853$ μm	Z_a	$w = 608$ μm	$w = 608$ μm
	Θ_a	$l = 2147$ μm	$l = 3212$ μm
	Z_b	$w = 292$ μm	$w = 167$ μm
	Θ_b	$l = 5357$ μm	$l = 3025$ μm
		CCT 15°	CCT 20°
$w = 356$ μm, $l = 9853$ μm	Z_c	$w = 607$ μm	$w = 607$ μm
	Θ_c	$l = 1610$ μm	$l = 2147$ μm
	Z_d	$w = 215$ μm	$w = 98$ μm
	Θ_d	$l = 4733$ μm	$l = 3179$ μm

[a] w, Width of microstrip line, l, length of microstrip line.

normalized susceptance. In the general case, the locus of Θ_a or Θ_c starts from a point $g = 1/IR$, $b = 0$ as the impedance transformation ratio IR is varied. When $g = 1/IR$ and $b = 0$ on the Smith chart, the real part of the reflection coefficient of the point is $(IR - 1)/(1 + IR)$, based on (11.1). In addition to that point, all the constant-conductance circles pass through a point (1, 0) on the reflection coefficient plane in Fig. 11.1. Therefore, the constant $g = 1/IR$ circle is expressed as

$$\left(\Gamma_r - \frac{1}{IR+1}\right)^2 + \Gamma_i^2 = \left(\frac{IR}{IR+1}\right)^2, \tag{11.7}$$

and the constant-VSWR circle related with IR is

$$\Gamma_r^2 + \Gamma_i^2 = \left(\frac{1-IR}{IR+1}\right)^2. \tag{11.8}$$

The CVT and CCT are passive components. Therefore, the point r_1 in Fig. 11.1 must not be outside the circle $|\Gamma| = 1$. For each locus of Θ_a or Θ_c to exist inside the circle $|\Gamma| = 1$, two conditions must be satisfied and result in

$$\left(\Gamma_r + \tfrac{1}{2}\right)^2 + {\Gamma_i}^2 \leq \left(\tfrac{1}{2}\right)^2, \tag{11.9}$$

$$\left(\Gamma_r - \tfrac{1}{2}\right)^2 + \Gamma_i^2 \leq \left(\tfrac{1}{2}\right)^2. \tag{11.10}$$

The possible ranges of Θ_a and Θ_c, a family of circles, a constant-VSWR circle, a constant-conductance circle, and two circles from (11.9) and (11.10) labeled Im+ and Im− are illustrated in Fig. 11.7(*a*). The family circles are phase angles between normalized voltages and currents with 0° to 40° [1], and the center of the Smith chart in Fig. 11.7(*a*) is expanded in Fig. 11.7(*b*), where the changed

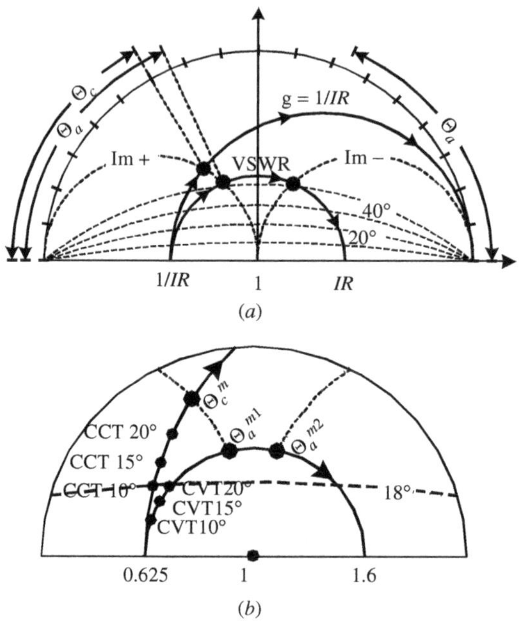

FIGURE 11.7 Bounded lengths of Θ_a and Θ_c: (*a*) half Smith chart where a family of phase angle circles is drawn, illustrating bounded lengths; (*b*) loads of the CVTs and CCTs with $IR = 1.6$ and bounded lengths Θ_a^{m1}, Θ_a^{m2}, and Θ_c^{m}.

complex loads of CVTs and CCTs designed in Fig. 11.4(*a*) are marked with dots. Here, the changed complex load of the CCT 20 is located on the highest phase angle circle, whereas that of the CVT 10 is on the lowest phase angle circle and those of the CVT 20 and the CCT 10 are nearby on the same phase angle circle. If the changed complex load is located on the lower phase angle circle, the bandwidth is wider, as can be seen in Fig. 11.4(*a*). Therefore, the phase angle circles are related to the *Q*-factor, and the bandwidth of the CVT 10 shows the best performance among the small transformers in Fig. 11.4(*a*). Depending on where the changed complex loads of the CVTs or CCTs are located, this means that the relative bandwidths of the CVTs and CCTs may be determined from Fig. 11.7(*b*). These values agree with the results of Fig. 11.4(*a*).

The constant-VSWR circle intersects the two circles Im+ and Im− at two points, Θ_a^{m1} and Θ_a^{m2}, in Fig. 11.7(*b*). Θ_a^{m1} is a maximum length of Θ_a in one region and Θ_a^{m2} is a minimum length in another region. The formulas for Θ_a^{m1} and Θ_a^{m2} in relation to the impedance transformation ratio *IR* are

$$\tan(\pi - 2\Theta_a^{m1}) = \frac{\Gamma_i^{m1}}{\Gamma_r^{m1}}, \tag{11.11}$$

$$\tan(\pi - 2\Theta_a^{m2}) = \frac{\Gamma_i^{m2}}{\Gamma_r^{m2}} = -\frac{\Gamma_i^{m1}}{\Gamma_r^{m1}}, \tag{11.12}$$

TABLE 11.6 Bounded Lengths (deg) of CVTs and CCTs

IR	Θ_a^{m1}	Θ_a^{m2}	Θ_c^{m}
1.2	42.392	47.608	20.439
1.4	40.203	49.797	24.311
1.6	38.329	51.671	25.833
1.8	36.699	53.300	26.423
2.0	35.264	53.736	26.565

where $\Gamma_r^{m1} = -(1 - IR)^2/(IR + 1)^2$ and $\Gamma_i^{m1} = 2(IR - 1)\sqrt{IR}/(IR + 1)^2$ from (11.8)–(11.10). The constant $g = 1/IR$ circle intersects the circle Im+ at only one point, Θ_c^m, in Fig. 11.7(*b*) and the imaginary part Γ_i^m for Θ_c^m is derived as $2\sqrt{IR - 1}/(IR + 3)$ from the two circles in (11.7) and (11.9). The susceptance b for Γ_i^m is $\sqrt{IR - 1}/IR$ from (11.2), and the transmission line with Z_c is the open-circuited stub in Fig. 11.2(*c*). Therefore, the length Θ_c^m is

$$\tan \Theta_c^m = \frac{\sqrt{IR - 1}}{IR}. \qquad (11.13)$$

From Fig. 11.7, the ranges where Θ_a exists are $0° \leq \Theta_a \leq \Theta_a^{m1}$ and $\Theta_a^{m2} \leq \Theta_a \leq 90°$. Therefore, the CVT cannot be realized in the range between Θ_a^{m1} and Θ_a^{m2}, and the values given in Tables 11.1 and 11.2 and in Fig. 11.4(*a*) are all $0 \leq \Theta_a \leq \Theta_a^{m1}$. If Θ_a is greater than Θ_a^{m2}, the changed complex load of the CVT is located on the right half-plane of the Smith chart, and the length Θ_b is greater than 90° from (11.6*a*). Therefore, CVTs with $\Theta_a^{m2} \leq \Theta_a \leq 90°$ may be realizable but not recommendable. In the case of the CCTs, Θ_c is in the range $0 \leq \Theta_c \leq \Theta_c^m$, as shown in Fig. 11.7(*b*). All data for Θ_a^{m1}, Θ_a^{m2}, and Θ_c^m are written in Table 11.6 as the impedance transformation ratio IR is varied.

11.2.4 Phase Responses of CVTs and CCTs

The conventional impedance transformers to transform a real impedance into another real impedance are all $\lambda/4$ long, so they have only 90° phase shifts.

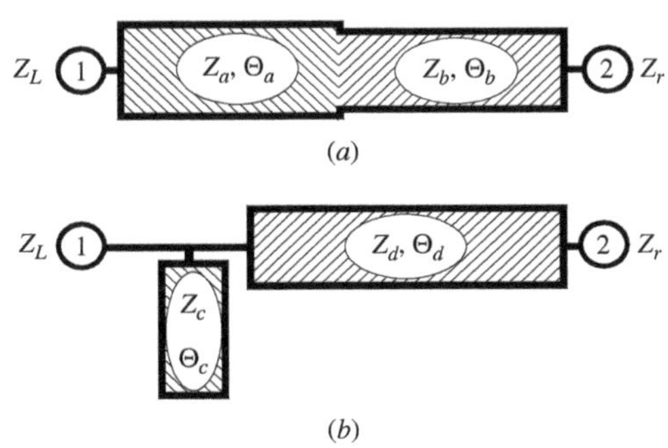

FIGURE 11.8 (*a*) CVT; (*b*) CCT.

However, the CVTs and CCTs can have arbitrary phase shifts of less than 90°, depending on the characteristic impedances and electrical lengths of the transmission lines in Fig. 11.2. Figure 11.8 shows the CVT and CCT that transform a real impedance Z_L into another real impedance, Z_r.

The *ABCD* parameters of the CVT in Fig. 11.8(*a*) are

$$\begin{bmatrix} A & B \\ C & D \end{bmatrix}_{\text{CVT}} = \begin{bmatrix} \cos\Theta_a & jZ_a \sin\Theta_a \\ j\dfrac{\sin\Theta_a}{Z_a} & \cos\Theta_a \end{bmatrix} \begin{bmatrix} \cos\Theta_b & jZ_b \sin\Theta_b \\ j\dfrac{\sin\Theta_b}{Z_b} & \cos\Theta_b \end{bmatrix}$$

$$= \begin{bmatrix} \cos\Theta_a \cos\Theta_b - \dfrac{Z_a}{Z_b}\sin\Theta_a \sin\Theta_b & j(Z_b \cos\Theta_a \sin\Theta_b + Z_a \sin\Theta_a \cos\Theta_b) \\ j\left(\dfrac{\sin\Theta_a \cos\Theta_b}{Z_a} + \dfrac{\cos\Theta_a \sin\Theta_b}{Z_b}\right) & \cos\Theta_a \cos\Theta_b - \dfrac{Z_b}{Z_a}\sin\Theta_a \sin\Theta_b \end{bmatrix}, \tag{11.14}$$

which can, with reference to Table 2.1, be converted into an admittance matrix. The scattering parameter S_{12} terminated in two different impedances, Z_L and Z_r, in Fig. 11.8(*a*) is calculated [7] as

$$S_{12} = \frac{1}{\Delta_y} Y_{12}(y_{22} + y_{22}^*)\sqrt{\frac{\text{Re}(y_{11})}{\text{Re}(y_{22})}}, \tag{11.15}$$

where $\Delta_y = Y_{11}Y_{22} + Y_{11}y_{22} - Y_{12}Y_{21} + Y_{22}y_{11} + y_{11}y_{22}$, $y_{11} = 1/Z_L$, $y_{22} = 1/Z_r$, and $\boldsymbol{Y}$ is an admittance matrix of the CVT.

In the configuration of the CCT in Fig. 11.8(*b*), the admittance parameters are easier than the *ABCD* parameters to obtain the scattering parameter S_{12}. They are

$$\begin{bmatrix} Y_{11} & Y_{12} \\ Y_{21} & Y_{22} \end{bmatrix}_{\text{CCT}} = \begin{bmatrix} j\left(\dfrac{\cot\Theta_c}{Z_c} + \dfrac{\cot\Theta_d}{Z_d}\right) & \dfrac{j\sin\Theta_d}{Z_d} \\ \dfrac{j\sin\Theta_d}{Z_d} & \dfrac{j\cot\Theta_d}{Z_d} \end{bmatrix}. \tag{11.16}$$

The scattering parameter S_{12} of CCT terminated in two different impedances, Z_L and Z_r, is calculated similarly.

Based on (11.14)–(11.16), the phases of $S_{21} = S_{12}$ in CVTs and CCTs are as plotted in Fig. 11.9. The phases of CVTs against Θ_a are described in Fig. 11.9(*a*), those of CCTs versus Θ_c in Fig. 11.9(*b*). *IR* is an impedance transformer ratio, Z_L/Z_r. The bounded lengths of Θ_a in Fig. 11.8(*a*) are between 35° and 42°, depending on *IR*, but the phase responses are identical and linear. The bounded length of Θ_c in Fig. 11.8(*b*) is only 20.439° for *IR* = 1.2. The simulated results show that the phases of the CVTs and CCTs are less than 90° and indicate

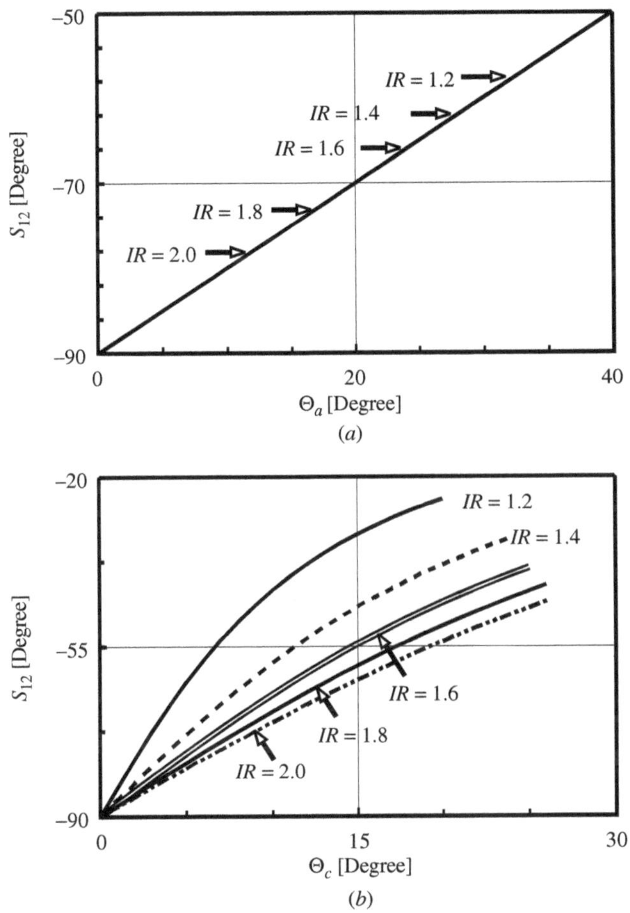

FIGURE 11.9 Phase responses of $S_{12} = S_{21}$: (*a*) CVT; (*b*) CCT.

that these two types of small impedance transformers may be utilized not only as impedance transformers but also as phase shifters that have less than 90° phase shifts.

11.3 CVT3PDs AND CCT3PDs

CVTs and CCTs are good candidates for reducing the size of three-port power dividers. Applying CVTs and CCTs to conventional three-port 3-dB power dividers results circuits, including isolation circuits, as shown in Fig. 11.10. The three-port 3-dB power divider based on a CVT (CVT3PD) is shown in Fig. 11.10(*a*) and that based on a CCT (CVT3PD) is shown in Fig. 11.10(*b*), where Z_c' is $Z_c/2$ and IC is the isolation circuit between ports ② and ③.

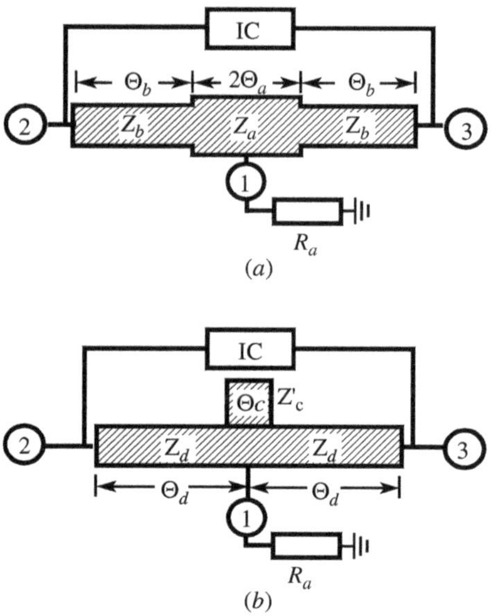

FIGURE 11.10 Two two-port small three-port 3-dB power dividers: (*a*) CVT3PD; (*b*) CCT3PD. (From Ref. 20 with permission from IEEE.)

11.3.1 Isolation Circuits of CVT3PDs and CVT3PDs

The CVT3PD in Fig. 11.10(*a*) is symmetric, so it can be analyzed using the easy-to-use method of even- and odd-mode excitations. The even-mode admittance y_1 [8] of the in-phase circuit contributed by two ports, ② and ③, without an IC (isolation circuit) is

$$y_1 = Y_{o/c} = Y_b \frac{Y_{\text{CVToc}} + jY_b \tan\Theta_b}{Y_b + jY_{\text{CVToc}} \tan\Theta_b}, \tag{11.17}$$

where $Y_a = 1/Z_a$, $Y_b = 1/Z_b$, and

$$Y_{\text{CVToc}} = Y_a \frac{1 + j2R_aY_a \tan\Theta_a}{2R_aY_a + j\tan\Theta_a},$$

and the odd-mode admittance y_2 of the out-of-phase circuit is

$$y_2 = Y_{s/c} = Y_b \frac{Y_{\text{CVTsc}} + jY_b \tan\Theta_b}{Y_b + jY_{\text{CVTsc}} \tan\Theta_b}, \tag{11.18}$$

where $Y_{\text{CVTsc}} = -jY_a \cot\Theta_a$.

For perfect isolation between ports ② and ③ in Fig. 11.10(*a*), the admittance parameter Y_{12} of all the circuits, including ICs, must be zero [see (7.6)], where the subscript of Y_{12} is related to a two-port circuit involving ports ② and ③.

Since the IC is connected in parallel with the rest of the circuit, the impedance Z_{IC} of the IC is $2/(y_1 - y_2)$ and results in

$$Z_{\mathrm{IC}} = 2Z_b \left(\frac{Y_{\mathrm{CVToc}} + jY_b \tan\Theta_b}{Y_b + jY_{\mathrm{CVToc}} \tan\Theta_b} - \frac{Y_{\mathrm{CVTsc}} + jY_b \tan\Theta_b}{Y_b + jY_{\mathrm{CVTsc}} \tan\Theta_b} \right)^{-1}. \tag{11.19}$$

The impedance Z_{IC} of the CCT3PD in Fig. 11.10(*b*) is derived similarly as

$$Z_{\mathrm{IC}} = 2\left(Y_d \frac{Y_{\mathrm{CCToc}} + jY_d \tan\Theta_d}{Y_d + jY_{\mathrm{CCToc}} \tan\Theta_d} + jY_d \cot\Theta_d \right)^{-1}, \tag{11.20}$$

where $Y_{\mathrm{CCToc}} = 1/(2R_a) + j\tan\Theta_c/Z_c$, $Y_d = 1/Z_d$.

When $\Theta_a = 0^\circ$, Θ_b is 90° from the results in Table 11.2, and the corresponding Z_{IC} is Z_b^2/R_a in (11.19). For a conventional three-port 3-dB power divider, assuming that the termination impedances are all 50 Ω, the impedance transformation ratio is 2 and the characteristic impedance of the transformer is $50\sqrt{2}$ Ω. With $Z_b = 50\sqrt{2}$ Ω and $R_a = 50$ Ω, the isolation resistance of Z_b^2/R_a becomes 100 Ω, a well-known isolation resistance [9], and coincides with $\Theta_c = 0^\circ$ and $\Theta_d = 90^\circ$ in (11.20).

Two perfect isolation impedances were simulated based on (11.19) and (11.20), and the results are plotted in Fig. 11.11 as Θ_a and Θ_c are varied. The perfect isolation impedance for the CVT3PD is plotted in Fig. 11.11(*a*) and that for CCT3PD in Fig. 11.11(*b*). The left axis of the two graphs in Fig. 11.11 indicates the real part of the IC, or resistance, and the right axis indicates the imaginary part of the IC, or capacitance. The capacitance is calculated at a frequency of 3 GHz, and R_a is set to 50 Ω.

For a 3-dB power divider terminated by equal impedances, the impedance transformation ratios, Θ_a and Θ_c, are both 2. In Fig. 11.11(*a*) and (*b*), if Θ_a and Θ_c are both 0°, the isolation impedances are both 100 Ω, and when Θ_a and Θ_c both start to deviate from 0°, capacitance starts to appear. If Θ_a is greater than Θ_a^{m1} ($\Theta_a^{m1} = 35.264^\circ$ from Table 11.6) in Fig. 11.11(*a*), the characteristic impedance Z_b is imaginary, as can be calculated from (11.4), but positive resistance still exists, due to the termination impedance R_a and the transmission line with characteristic impedance Z_a. If Θ_a is greater than 45°, the absolute value of the negative resistance due to the imaginary characteristic impedance Z_b becomes greater than that of the positive resistance due to the termination impedance R_a and the transmission line with characteristic impedance Z_a. Therefore, the resulting value is negative up to Θ_a^{m2}. When $\Theta_a = \Theta_a^{m2}$, the characteristic impedance of Z_b is 0 Ω and the resulting resistance of the isolation circuit is 0 Ω. Afterward, positive resistance and negative capacitance appear. In this case, the total length of $\Theta_a + \Theta_b$ is greater than 90° and positive resistance, together with inductance (i.e., positive resistance and negative capacitance) appear. When Θ_a is close to 90°, the characteristic impedance of Z_b is $50/\sqrt{2}$ Ω and the resulting isolation circuit is only a positive resistance of 50 Ω. This case shows that a 2 : 1 impedance transformer may be realized with two transmission lines, each of

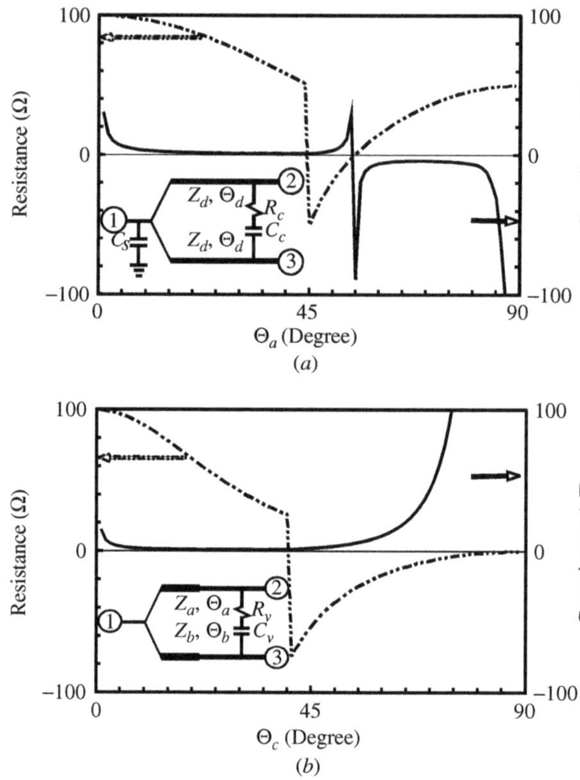

FIGURE 11.11 Simulation results of isolation circuits of the (*a*) CVT3PD and (*b*) CCT3PD.

which is 90° long and whose characteristic impedances are 50 Ω and $50/\sqrt{2}$ Ω when the reference impedance $Z_r = 50$ Ω. The total length of the CVT3PD with $\Theta_a^{m2} \leq \Theta_a \leq 90°$ is much longer than that with $0 \leq \Theta_a \leq \Theta_a^{m1}$, but the CVT3PD with $0 \leq \Theta_a \leq \Theta_a^{m1}$ may be used when transmission lines with high characteristic impedance are not easily realized. On the other hand, since the available region of Θ_c for the CCT3PD is only $0 \leq \Theta_c \leq \Theta_c^m$, the positive resistance is bounded in one region in Fig. 11.11(*b*), and when Θ_c is equal to Θ_c^m, the length of Z_d is 0° and the complex impedance of Z_{IC} is $50 - j50$ Ω.

In conclusion, the two graphs in Fig. 11.11 show that the isolation circuit must consist not only of resistance but also of inductance or capacitance for perfect isolation if the electrical length of transmission lines is not λ/4 in three-port power dividers.

11.3.2 Design of CVT3PDs and CCT3PDs

On the basis of the isolation impedances derived, four of many possible designs are given in Table 11.7. These four examples are practically realizable. Two

TABLE 11.7 Design Data for CVT3PD and CCT3PD

CVT3PD terminated in equal impedance 50 Ω at 3 GHz:
$Z_a = 50\ \Omega$, $\Theta_a = 29°$, $Z_b = 104.8\ \Omega$, $\Theta_b = 25.9°$ $Z_{IC} = 76.5 - j42.4\ \Omega \rightarrow R_v = 76.5\ \Omega$, $C_v = 1.25$ pF
$Z_a = 50\ \Omega$, $\Theta_a = 20°$, $Z_b = 79.7\ \Omega$, $\Theta_b = 47°$ $Z_{IC} = 88.30 - j32.14\ \Omega \rightarrow R_v = 88.30\ \Omega$, $C_v = 1.65$ pF
CCT3PD terminated in equal impedance 50 Ω at 3 GHz:
$Z_c = 50\ \Omega$, $\Theta_c = 25°$, $Z_d = 195.9\ \Omega$, $\Theta_d = 15.3°$, $C_s = 0.99$ pF $Z_{IC} = 53.5 - j49.9\ \Omega \rightarrow R_c = 53.5\ \Omega$, $C_c = 1.06$ pF
$Z_c = 50\ \Omega$, $\Theta_c = 20°$, $Z_d = 103.1\ \Omega$, $\Theta_d = 33.66°$, $C_s = 0.77$ pF $Z_{IC} = 65.36 - j47.58\ \Omega \rightarrow R_c = 65.36\ \Omega$, $C_c = 1.11$ pF

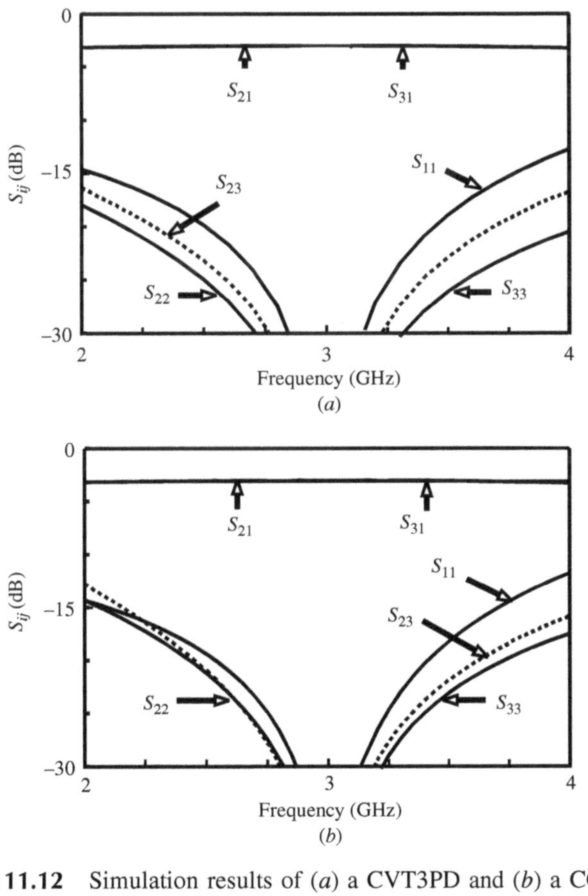

FIGURE 11.12 Simulation results of (*a*) a CVT3PD and (*b*) a CCT3PD.

of them were simulated at a center frequency of 3 GHz, and the results are plotted in Fig. 11.12. The results of the first CVT3PD in Table 11.7 are plotted in Fig. 11.12(*a*), and those of the first CCT3PD in Table 11.7 are given in Fig. 11.12(*b*). The design data for the CVT3PD are, as given in Table 11.7, $Z_a = 50\ \Omega$, $\Theta_a = 29^\circ$, $Z_b = 104.8\ \Omega$, $\Theta_b = 25.9^\circ$, $R_v = 76.5\ \Omega$, and $C_v = 1.25$ pF [see the circuit inserted in Fig. 11.11(*a*)] and those of the CVT3PD are $Z_c = 50\ \Omega$, $\Theta_c = 25^\circ$, $Z_d = 195.9\ \Omega$, $\Theta_d = 15.3^\circ$, $C_s = 0.99$ pF, $R_c = 53.5\ \Omega$, and $C_c = 1.06$ pF [see the circuit inserted in Fig. 11.11(*b*)]. In the two cases, determination of the arbitrary design impedance Z_{Ad} mentioned in Section 8.2 is not necessary, but the Z_{Ad} is 50 Ω and represents only one case in (8.1) and (8.2) because both of them are terminated in equal impedances 50 Ω. The impedance transformation ratios IR_{12} or IR_{13} are both 2. Therefore, the results for S_{11} are, in terms of the bandwidth, worse than those for S_{22} or S_{33}. The performance of S_{23} may be improved if a that has a larger bandwidth is used. The frequency responses of the CVT3PD are $|S_{12}| = |S_{13}| = -3.01$ dB, $|S_{11}| = -87.072$ dB, $|S_{23}| = -70.113$, and $|S_{22}| = |S_{33}| = -70.085$ dB, and those of the CCT3PD are $|S_{12}| = |S_{13}| = -3.01$ dB, $|S_{11}| = -58.945$ dB, $|S_{23}| = -70.261$, and $|S_{22}| = |S_{33}| = -57.447$ at a center frequency of 3 GHz. Only four examples with equal power division were discussed, designed, and simulated, but using the general design equations, the CCT3PD and CVT3PD concepts may be extended to all possible cases of three-port power dividers. Therefore, any small three-port power divider may be obtained.

11.4 ASYMMETRIC THREE-PORT 45° POWER DIVIDER TERMINATED IN ARBITRARY IMPEDANCES

CVTs and CCTs can be used for asymmetric three-port 45° power dividers terminated in arbitrary impedances. As demonstrated earlier, CVTs and CCTs may be used as phase shifters with arbitrary phase shifts of less than 90°. Using one CCT with a 45° phase shift, a three-port 45° power divider terminated in arbitrary impedances (asymmetric three-port 45° power divider) can be constructed. Three-port 45° power dividers are very important for suppressing harmonics in active mixers and amplifiers [10]. The length of a CCT with a 45° phase shift can be less than 45°, and asymmetric 45° three-port power dividers may be terminated in arbitrary impedances. Due to these distinct points, the total size of MICs (microwave integrated circuits) or MMICs (monolithic microwave integrated circuits) can be reduced by including asymmetric four-, three-, and two-port components [11]–[18].

Figure 11.13 shows an asymmetric three-port 45° power divider terminated in arbitrary impedances, R_a, R_b, and R_c. It consists of a Wilkinson-type three-port power divider and a CCT or CVT for the 45° phase shift. The S_{21} value of the CVT or CCT is unity in magnitude and arbitrary in phase. Therefore, the equivalent circuit of CVT or CCT may be considered as a transmission line with a certain length and arbitrary termination impedances.

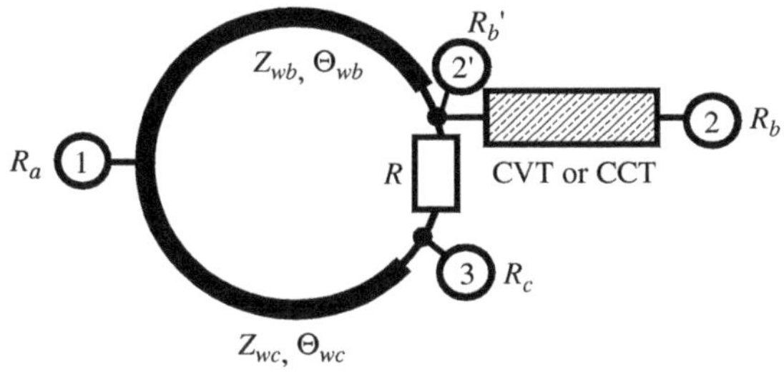

FIGURE 11.13 Asymmetric three-port 45° power divider terminated in arbitrary impedances R_a, R_b, and R_c.

The design equations for asymmetric 45° power dividers are derived for two cases.

1. For $R_b \neq R_c$ and $\Theta_{wb} = \Theta_{wc} = 90°$,

$$Z_{wb} = Z_{wc} = \sqrt{2R_aR_c}, \tag{11.21a}$$

$$R = 2R_c, \tag{11.21b}$$

$$IR = \frac{R_b}{R_c} \quad \text{or} \quad IR = \frac{R_c}{R_b}. \tag{11.21c}$$

2. For $R_b = R_c$ and $\Theta_{wb} = \Theta_{wc} = 90°$,

$$Z_{wb} = Z_{wc} = \sqrt{2R_a\sqrt{R_b'R_c}} \quad \text{or} \quad \sqrt{(R_b' + R_c)R_a}, \tag{11.22a}$$

$$R = 2\sqrt{R_b'R_c} \quad \text{or} \quad (R_b' + R_c) \tag{11.22b}$$

$$IR = \frac{R_b'}{R_c} \quad \text{or} \quad IR = \frac{R_c}{R_b'}, \tag{11.22c}$$

where R_b' is an assumed termination impedance at port ②' and may be chosen arbitrarily.

11.4.1 Asymmetric 45° Power Divider Terminated in 30 Ω, 60 Ω, and 50 Ω

If a 45° power divider is terminated in 30 Ω, 60 Ω, and 50 Ω in Fig. 11.13, R_b is not equal to R_c. Therefore, the design equations in (11.22) can be used and give $Z_{wb} = 54.77$ Ω, $R = 100$ Ω, and $IR = 1.2$. From the simulation results in Fig. 11.9, a 45° phase shift is possible for a CCT, and its data are $\Theta_c = 9.45°$, $Z_d = 61.21$ Ω, and $\Theta_d = 39.28°$. The stub with $\Theta_c = 9.45$ is open-circuited in Fig. 11.8 and can be realized with a capacitor. Thus, the main length of the CCT is only 39.28°.

With the data calculated, the asymmetric 45° power divider was, under ideal conditions, simulated at a center frequency of 3 GHz, and the results are plotted

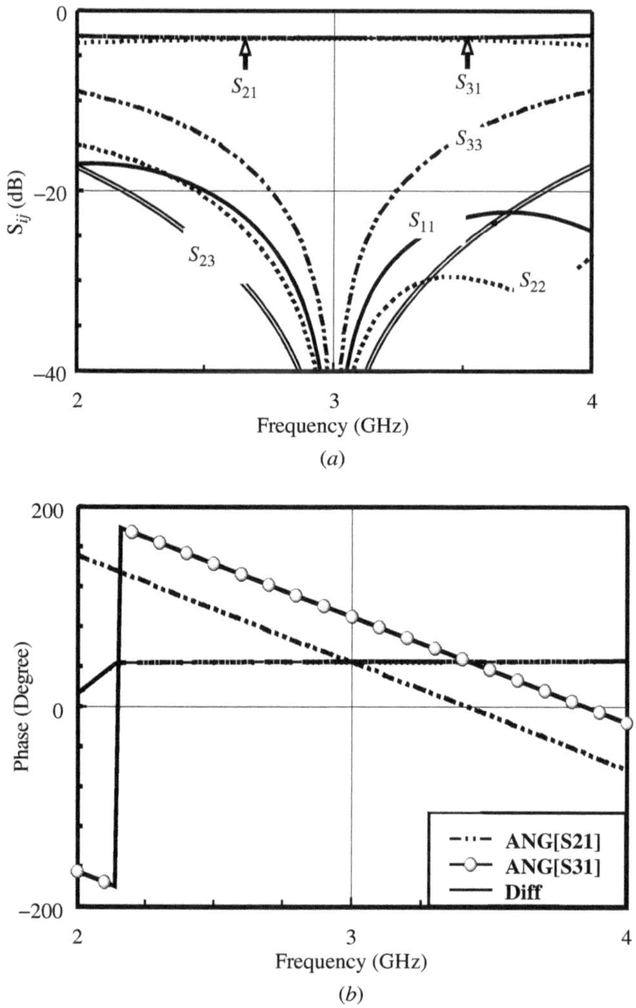

FIGURE 11.14 Simulation results of an asymmetric three-port 45° power divider terminated in 30 Ω, 60 Ω, and 50 Ω: (*a*) scattering parameters; (*b*) phase responses of S_{21} and S_{31}.

in Fig. 11.14. The simulated power division, isolation, and matching results are illustrated in Fig. 11.14(*a*), and the phase responses of the power divisions and the phase difference between the two outputs are given in Fig. 11.14(*b*). The results show that the phase difference between two outputs is within $45° \pm 1°$ in the frequency range 2.28 to 3.94 GHz and $|S_{21}| = -3.01$ dB, $|S_{31}| = -3.01$ dB, $|S_{32}| = -156.17$ dB, $|S_{11}| = -89.403$ dB, $|S_{22}| = -83.39$ dB, and $|S_{33}| = -154.73$ dB at a center frequency of 3 GHz. Based on the data, a uniplanar three-port 45° power divider was fabricated, and its layout is depicted in Fig. 11.15. Since it is not terminated in 50 Ω, the 90° impedance transformers are needed for the

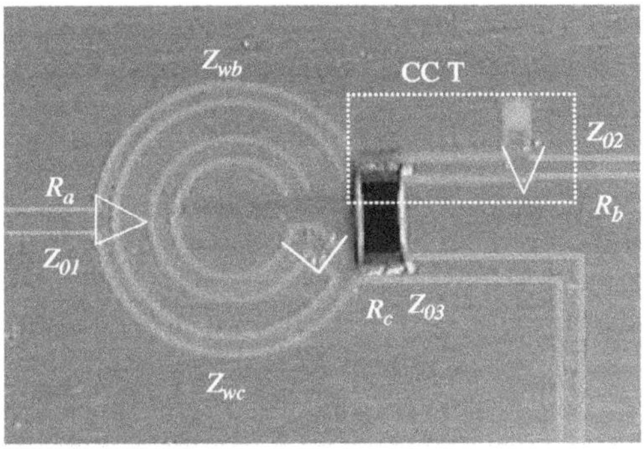

FIGURE 11.15 Layout of an asymmetric three-port 45° power divider terminated in 30 Ω, 60 Ω, and 50 Ω.

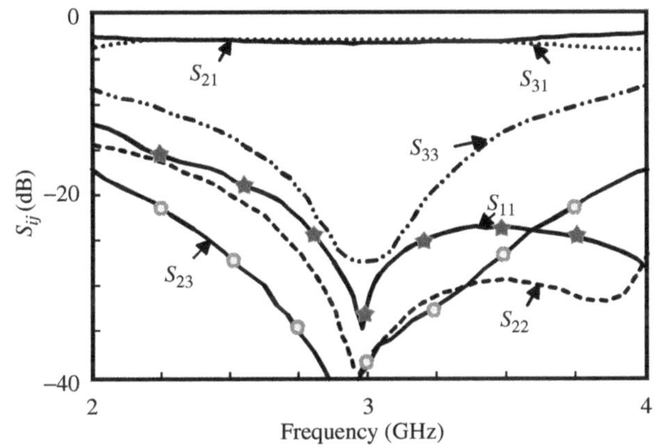

FIGURE 11.16 Measurements of the asymmetric three-port 45° power divider terminated in 30 Ω, 60 Ω, and 50 Ω.

measurements, and the Z_{01}, Z_{02}, and Z_{03} in Fig. 11.15 are the impedance transformers. Figure 11.16 compares the scattering parameters measured (solid lines) with the parameters simulated (all nonsolid lines), showing that they are in good agreement.

11.5 CVT AND CCT RING FILTERS

As other applications of CVTs and CCTs, small ring filters are introduced. To reduce the size of ring filters, four 90° transmission lines can be replaced by CVTs and CCTs [19,20], and the two types of small ring filters will be called CVT and CCT ring filters. The length of one transmission-line section can be reduced by 34%, and two cases 300° in circumference are simulated and compared. The results show more than 100% fractional bandwidths with return losses of less than −15 dB, which can be obtained in conventional filter design techniques only in more than five stages. To verify the performances, a CCT ring filter is measured and shows good agreement with the simulation.

11.5.1 Analyses of Ring Filters

In the ring filter depicted in Fig. 11.17, the feeding lines are connected directly to the ring. The circumference of the ring is one wavelength, and the two short stubs are each λ/4 long. The power excited at port ① is divided depending on the power ratio, d_1 to d_2, and the divided power is combined at port ②. If the two short stubs do not exist, all the power excited at port ① will be delivered to port ② at all frequencies, and perfect matching will occur at multiples of a design center frequency. Therefore, each stub is necessary to reject the power at even multiples of the design center frequency and to achieve a filter characteristic as a resonator.

All three impedance transformers in Fig. 11.18 transform a real impedance IR into unity, where IR is an impedance transformation ratio. A quarter-wavelength impedance transformer with its characteristic impedance $\sqrt{IR}$ is shown in Fig. 11.18(*a*), and CVT and CCT are shown in Fig. 11.18(*b*) and (*c*), respectively, where $Z_{L-\mathrm{CVT}}$ and $Z_{L-\mathrm{CCT}}$ are the input impedances looking into the transmission line connected in a load IR.

$Z_{L-\mathrm{CVT}}$ and $Z_{L-\mathrm{CCT}}$ in Fig. 11.18 are given as

$$Z_{L-\mathrm{CVT}} = Z_a \frac{IR + jZ_a \tan \Theta_a}{Z_a + jIR \tan \Theta_a}, \tag{11.23a}$$

$$Z_{L-\mathrm{CCT}} = \frac{Z_c IR}{jIR \tan \Theta_c + Z_c} \tag{11.23b}$$

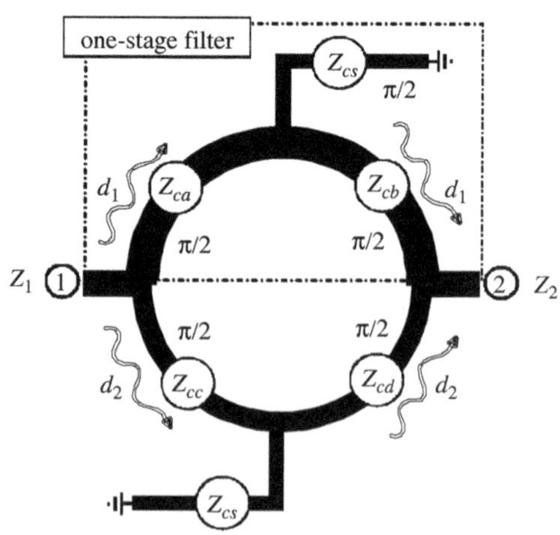

FIGURE 11.17 Conventional ring filter.

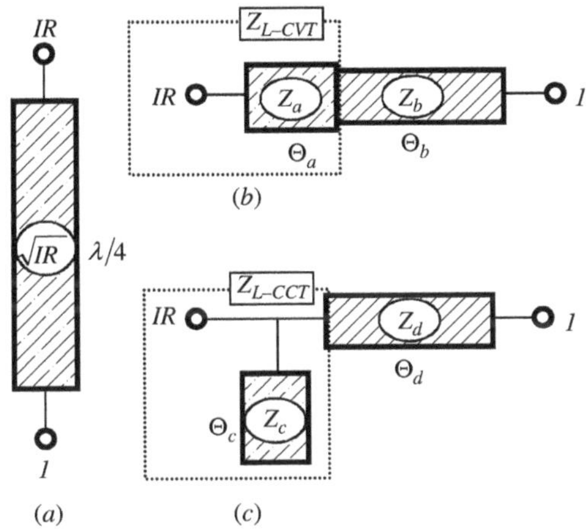

FIGURE 11.18 Three types of impedance transformers: (*a*) 90° impedance transformer; (*b*) CVT; (*c*) CCT.

The input impedances in (11.23) indicate that the real and imaginary parts of $Z_{L-\mathrm{CVT}}$ change at the same time with Z_a variations, but that only the imaginary part of $Z_{L-\mathrm{CCT}}$ varies with changes in Z_c. Thus, Z_c can have other values, including unity.

The 90° transmission-line sections with Z_{ca}, Z_{cb}, Z_{cc}, and Z_{cd} in Fig. 11.17 are all impedance transformers to transform a real impedance into another

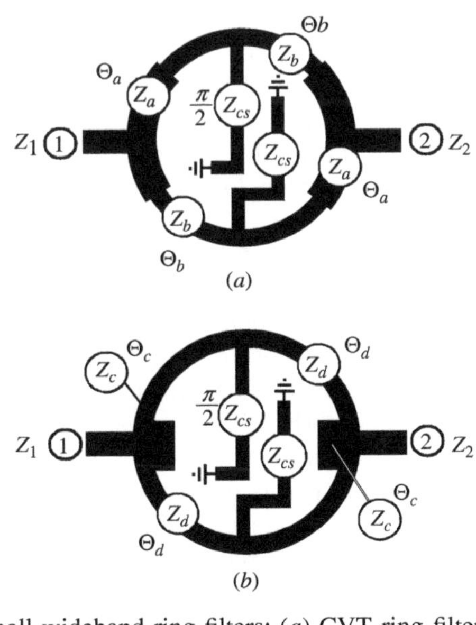

FIGURE 11.19 Small wideband ring filters: (*a*) CVT ring filter; (*b*) CCT ring filter.

TABLE 11.8 Design Data and Simulation Results for Conventional, CVT, and CCT Ring Filters

Conventional Ring Filter	CVT Ring Filter	CCT Ring Filter
$Z_1 = Z_2 = 50\ \Omega$	$Z_0 = 50\ \Omega$	$Z_0 = 50\ \Omega$
$Z_{cs} = 85\ \Omega$	$Z_{cs} = 85\ \Omega$	$Z_{cs} = 85\ \Omega$
$Z_{ca} = Z_{cb} = Z_{cc} = Z_{cd}$	$Z_a = 50\ \Omega$	$Z_c = 30\ \Omega$
$= 70.71\ \Omega$	$\Theta_a = 10°$	$\Theta_c = 5°$
	$Z_b = 72.4\ \Omega$	$Z_d = 71.8\ \Omega$
	$\Theta_b = 68.7°$	$\Theta_d = 75.89°$
	Bandwidth with Insertion Loss Less Than −0.1 dB	
0.47–1.53 GHz	0.48–1.52 GHz	0.48–1.52 GHz
	Bandwidth with Return Loss Less Than −15 dB	
0.46–1.54 GHz	0.43–1.57 GHz	0.44–1.56 GHz

impedance. To reduce the size of the ring filter, CVTs and CCTs are good candidates. Replacing the transmission-line sections by CVTs or CCTs results in the configurations shown in Fig. 11.19, called CVT and CCT ring filters. Design data for three ring filters, CVT and CCT ring filters as well as a conventional ring filter, are given in Table 11.8. Based on the data, the three were designed and simulated, and simulation results are plotted in Fig. 11.20, where Con-ring is the ring filter shown in Fig. 11.17 and CVT-ring and CCT-ring are the CVT and CCT ring filters in Fig. 11.19(a) and (b), respectively. Frequency responses of the insertion losses are plotted in Fig. 11.20(a) and those of the reflection coefficients in Fig. 11.20(b). They are also summarized in Table 11.8. The results show that about the same performances can be obtained from the three ring filters even if the length of each transmission line and the circuit configurations differ from each other. As given in Table 11.8, all three ring filters have more than 100% fractional bandwidths, where insertion losses are below 0.1 dB and reflection coefficients below −15 dB.

Based on the design data, a CCT ring filter were fabricated on a substrate of $\varepsilon_r = 2.88$, $h = 508\ \mu$, $\tan\delta = 0.0064$ at 3 GHz in microstrip technology and its photo is shown in Fig. 11.21. In practical cases, the two short stubs may not be realized equally, which could occur for a sudden phase inversion with a transmission zero. To prevent transmission zero in Fig. 11.21, a circuit for prevention of sudden phase inversion (POSPI) is connected at the two points where two short stubs are connected with the ring. The POSPI can be wire or several ohms to make the voltages at two points the same, and two resistors with several ohms are, for these measurements, connected in series. Simulations were carried out using an EM simulator, Empire, based on the FDTD (finite-difference time-domain) method. The measured and predicted results are plotted in Fig. 11.22, where the insertion losses are compared in Fig. 11.22(a) and the return losses

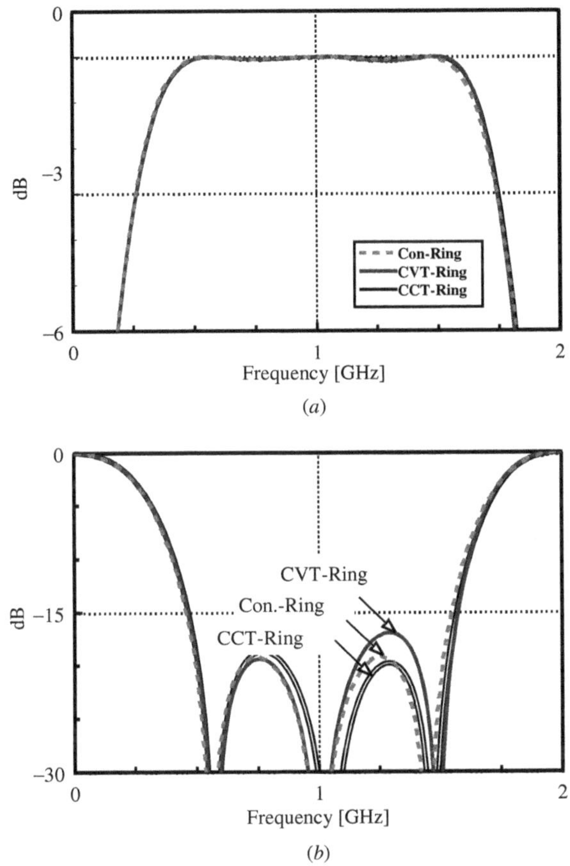

FIGURE 11.20 Simulation results of three ring filters: (*a*) insertion losses; (*b*) return losses.

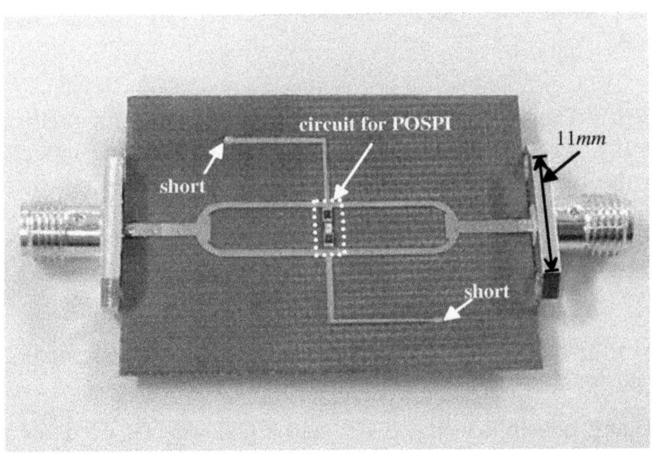

FIGURE 11.21 CCT ring filter.

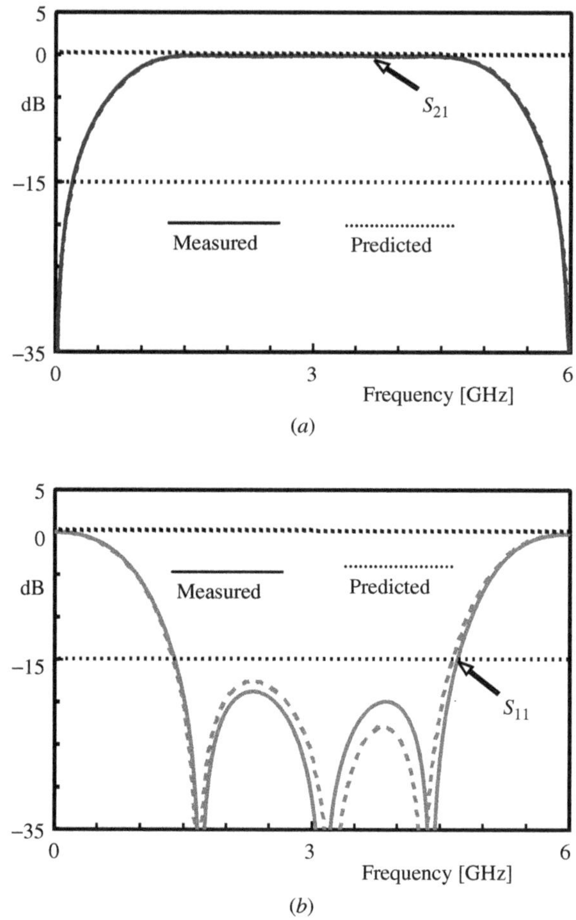

FIGURE 11.22 Measurements and predictions of a CCT ring filter: (*a*) insertion losses; (*b*) return losses.

in Fig. 11.22(*b*). The deviation between measurements and predictions at the band edge in Fig. 11.22(*b*) is due to the stray capacitance and inductance caused by soldering the two resistors on the transmission lines, except that the results measured agree well with the predictions.

EXERCISES

11.1 Set up four tables like Tables 11.2 to 11.4 at intervals of 0.2 when $2 < IR \leq 4$.

11.2 Check any possibility that CVTs and CCTs can be used for a three-port power divider with arbitrary power divisions and arbitrary termination impedances.

11.3 Design the CVT3PD in Fig. 11.10(*a*) when the termination impedances of ports ①, ②, and ③ are 50 Ω, 40 Ω, and 40 Ω, respectively, and $\Theta_b = 20°$.

11.4 Design the CCT3PD in Fig. 11.10(*a*) when the termination impedances of ports ①, ②, and ③ are 50 Ω, 40 Ω, and 40 Ω, respectively, and $\Theta_b = 20°$.

11.5 Design the CCT3PD in Fig. 11.13 so that two output signals at ports ② and ③ are 30° out of phase when the termination impedances of ports ①, ②, and ③ are 50 Ω, 40 Ω, and 40 Ω.

11.6 Design the CVT ring filter in Fig. 11.19(*a*) when it is terminated in 50 Ω and 60 Ω and $\Theta_a = 10°$.

11.7 Design the CCT ring filter in Fig. 11.19(*b*) when it is terminated in 50 Ω and 60 Ω and $\Theta_c = 5°$.

REFERENCES

1. P. H. Smith, *Electronic Applications of the Smith Chart*, McGraw-Hill, New York, 1969, pp. 38–39.
2. G. N. French and E. H. Fooks, The Design of Stepped Transmission Line Transformers, *IEEE Trans. Microwave Theory Tech.*, Vol. 16, October 1968, pp. 885–886.
3. M. A. Hamid and M. M. Yunik, On the Design of Stepped Transmission-Line Transformers, *IEEE Trans. Microwave Theory Tech.*, Vol. 15, September 1967, pp. 528–529.
4. G. D. Vendelin, A. M. Pavio, and U. L. Rohde, *Microwave Circuit Design Using Linear and Nonlinear Techniques*, Wiley, New York, 1990, pp. 163, 172–173.
5. G. N. French and E. H. Fooks, Double Section Matching Transformers, *IEEE Trans. Microwave Theory Tech.*, Vol. 17, September 1969, p. 719.
6. T. Hirota, A. Minakawa, and M. Muraguchi, Reduced-Size Branch-Line and Rat-Race Hybrids for Uniplanar MMIC's, *IEEE Trans. Microwave Theory Tech.*, Vol. 38, March 1990, pp. 270–275.
7. H.-R. Ahn and I. Wolff, Asymmetric Ring Hybrid Phase-Shifters and Attenuators, *IEEE Trans. Microwave Theory Tech.*, Vol. 50, April 2002, pp. 1146–1155.
8. J. Helszajn, *Passive and Active Microwave Circuits*, Wiley, NewYork, 1978, pp. 101–102.
9. E. J. Wilkinson, An *n*-Way Hybrid Power Divider, *IRE Trans. Microwave Theory Tech.*, Vol. 8, January 1960, pp. 116–118.
10. H. Hayashi, H. Okazaki, A. Kanda, T. Hirota, and M. Muraguchi, Millimeter-Wave-Band Amplifier and Mixer MMIC' Using a Broad-Band 45° Power Divider/Combiner, *IEEE Trans. Microwave Theory Tech.,* Vol. 46, June 1998, pp. 811–818.
11. H.-R. Ahn and I. Wolff, Asymmetric Ring Hybrid Phase-Shifters and Attenuators, *IEEE Trans. Microwave Theory Tech.*, Vol. 50, April 2002, pp. 1146–1155.
12. H.-R. Ahn and I. Wolff, Asymmetric Four-Port and Branch-Line Hybrids, *IEEE Trans. Microwave Theory Tech.*, Vol. 48, September 2000, pp. 1585–1588.

13. H.-R. Ahn and I. Wolff, General Design Equations of Three-Port Power Dividers, *IEEE MTT-S Dig.*, Boston, June 2000, pp. 1137–1140.
14. H.-R. Ahn and I. Wolff, Novel Ring Filter as a Wide-Band 180° Transmission-Line, *EUMC Proc.*, Vol. III, October 1999, pp. 95–98.
15. H.-R. Ahn and I. Wolff, Asymmetric Four-Port Hybrids, Asymmetric 3-dB Branch-Line Hybrids, *Asia–Pacific Microwave Conf. Proc.*, Yokohama, Japan, December 1998, pp. 677–680.
16. H.-R. Ahn, I. Wolff, and I.-S. Chang, Arbitrary Termination Impedances, Arbitrary Power Division and Small-Sized Ring Hybrids, *IEEE Trans. Microwave Theory Tech.*, Vol. 44, December 1997, pp. 2241–2247.
17. H.-R. Ahn, I.-S. Chang, and S.-W. Yun, Miniaturized 3-dB Ring Hybrid Terminated by Arbitrary Impedances, *IEEE Trans. Microwave Theory Tech.*, Vol. 42, December 1994, pp. 2216–2221.
18. H.-R. Ahn and I. Wolff, Three-Port 3-dB Power Divider Terminated by Different Impedances and Its Application to MMIC's, *IEEE Trans. Microwave Theory Tech.*, Vol. 47, June 1999, pp. 786–794.
19. H.-R. Ahn and I. Wolff, Asymmetric Three-Port 45° Power Dividers Composed of CVT's and CCT's, *EUMC Dig.*, Vol. 3, September 2001, pp. 165–168.
20. H.-R. Ahn and I. Wolff, General Design Equations of Three-Port Power Dividers, Small-Sized Impedance Transformers, and Their Applications to Small-Sized Three-Port 3-dB Power Divider, *IEEE Trans. Microwave Theory Tech.*, Vol. 49, July 2001, pp. 1277–1288.

APPENDIX A

Symbols and Abbreviations

SYMBOLS

$\boldsymbol{a}$	normalized incident-wave vector
A, B, C, D	elements of an *ABCD* matrix
A_i	vector amplitude of a signal at the i port
A_v	voltage gain
$\boldsymbol{b}$	normalized reflected-wave vector
C	coupling coefficient for coupled lines
d	coupling scaling factor
d_1/d_2	power division ratio for four-port hybrids
D_1, D_2	power division factors
$\boldsymbol{g}$	paraconjugate Hermitian admittance matrix
I_i	incident current
I_r	reflected current
IR_{ij}	impedance transformation ratio between two ports, i and j
k	coupling coefficients for branch-line hybrids
k_e	even-mode impedance scaling factor
k_o	odd-mode impedance scaling factor
$L_1, L_2, L_3, \ldots$	first, second, third, ... loops
NI_b, NI_d	normalized impedance ratios of characteristic impedance of transmission lines to termination impedances R_b and R_d, respectively
NI_{bi}, NI_{di}	normalized impedance ratios with short- and open-circuited loads (i being s or o, symbolizing short or open circuits, respectively)

Asymmetric Passive Components in Microwave Integrated Circuits, by Hee-Ran Ahn

NR_b, NR_d	normalized resistance ratios of resistance to termination impedances R_b and R_d, respectively
$\boldsymbol{P}_{ij}$	partitioned submatrix, i and j being port numbers
$P_1, P_2, P_3, \ldots$	paths for signal flow graph
$\boldsymbol{r}$	paraconjugate Hermitian impedance matrix
R_2, R_3	intermediate resistances of a three-port power divider
R_a, R_b, R_c, R_d	termination real impedances of four- or three-port components
$\boldsymbol{S}$	normalized scattering matrix
$\boldsymbol{S}^I$	current-basis scattering matrix
$\boldsymbol{S}^V$	voltage-basis scattering matrix
$\boldsymbol{T}$	scattering transfer matrix
T_e	complex transmission coefficient of an even-mode network
T_{ij}	complex transmission coefficient from port j to port i
T_o	complex transmission coefficient of an odd-mode network
$\boldsymbol{U}_k$	identity matrix of order k
V_g, V_s, V	sinusoidal voltage sources
V_i	incident voltage
V_r	reflected voltage
$\boldsymbol{Y}$	admittance matrix
Y_{in}	input admittance
Y_L	complex load admittance
$Y_{\text{O/C}}$, $Y_{\text{S/C}}$	even- and odd-mode admittances
Y_{out}	output admittance
$\boldsymbol{Z}$	impedance matrix
Z_a, Z_b, Z_c, Z_d	characteristic impedances of small impedance transformers
Z_{Ad}	arbitrary design Impedance for a three-port power divider
Z_{I1}, Z_{I2}	image impedances at ports ① and ②
Z_{in}	input impedance
Z_L	complex load impedance of small impedance transformer
$Z_{\text{O/C}}$, $Z_{\text{S/C}}$	even- and odd-mode impedances
Z_{out}	output impedance
$z(p)$	reference impedance
Z_r	real reference impedance of small impedance transformer
z_0, Z	characteristic impedance of loads
Z_{0e}	even-mode characteristic impedance of coupled transmission lines
Z_{0i}	characteristic impedances of transmission lines in three-port power dividers (i being from 2 to 5)
Z_{0o}	odd-mode characteristic impedance of coupled transmission lines
Z_0'	characteristic impedance of an impedance transformer of complex impedance into real impedance

Greek Symbols

α	attenuation constant
β	phase velocity
γ	propagation constant per unit length
Γ^I	current basis reflection coefficient
Γ^V	voltage-basis reflection coefficient
Γ_e	complex reflection coefficient of even-mode excitation
Γ_o	complex reflection coefficient of odd-mode excitation
Γ_s	source reflection coefficient
Γ_L	load reflection coefficient
Θ_a^{m1}, Θ_a^{m2}, Θ_c^m	bounded electrical lengths of small impedance transformers
Θ_i	electrical length of transmission lines
ρ^I	current reflection coefficient
ρ^V	voltage reflection coefficient

ABBREVIATIONS

AABH	anti-conventional-direction asymmetric branch-line hybrid
ALERH	asymmetric lumped-element ring hybrid
Al_2O_3	aluminum oxide–Ceramic
CABH	conventional-direction asymmetric branch-line hybrid
CCT	constant-conductance impedance transformer
CCT3PD	CCT 3-dB power divider
CPS	coplanar stripline
CPW	coplanar waveguide
CVT	constant-VSWR impedance transformer
CVT3PD	CVT 3-dB power divider
EMF	electromagnetic force
HIC	hybrid integrated circuit
LET3PD	lumped-element three-port 3-dB power divider
MIC	microwave integrate circuit
MMIC	monolithic microwave integrated circuit
Q-	quality
POSPI	prevention of sudden phase inversion
RF	radio frequency
T3PD	three-port 3-dB power divider
TEM	transverse electric magnetic-
UHF	ultrahigh frequency
VHF	very high frequency
VSWR	voltage standing-wave ratio

APPENDIX B

Conversion Matrices

IMPEDANCE MATRIX → SCATTERING MATRIX

The conversion of an impedance matrix into a scattering matrix is obtained from (2.29*b*), which is given as

$$\boldsymbol{S} = \boldsymbol{h}[\boldsymbol{Z} + z]^{-1}[\boldsymbol{Z} - z_*]\boldsymbol{h}_*^{-1}, \tag{B.1}$$

where

$$\boldsymbol{Z} = \begin{bmatrix} Z_{11} & Z_{12} \\ Z_{21} & Z_{22} \end{bmatrix}, \qquad z = \begin{bmatrix} Z_{01} & 0 \\ 0 & Z_{02} \end{bmatrix},$$

$$\boldsymbol{h} = \begin{bmatrix} \sqrt{Z_{01}} & 0 \\ 0 & \sqrt{Z_{02}} \end{bmatrix}, \qquad \boldsymbol{h}_*^{-1} = \begin{bmatrix} \dfrac{1}{\sqrt{Z_{01}}} & 0 \\ 0 & \dfrac{1}{\sqrt{Z_{02}}} \end{bmatrix}$$

and $z = z_*$ for real Z_{01} and Z_{02}. In more detail it is calculated as

$$S_{11} = \frac{(Z_{22} + Z_{02})(Z_{11} - Z_{01}) - Z_{12}Z_{21}}{\Delta_Z},$$

$$S_{12} = \frac{2\sqrt{Z_{01}}\sqrt{Z_{02}}Z_{21}}{\Delta_Z},$$

$$S_{21} = \frac{2\sqrt{Z_{01}}\sqrt{Z_{02}}Z_{12}}{\Delta_Z},$$

$$S_{22} = \frac{(Z_{11} + Z_{01})(Z_{22} - Z_{02}) - Z_{12}Z_{21}}{\Delta_Z},$$

where $\Delta_Z = (Z_{11} + Z_{01})(Z_{22} - Z_{02}) - Z_{12}Z_{21}$.

SCATTERING MATRIX → IMPEDANCE MATRIX

The conversion of a scattering matrix into an impedance matrix is obtained by rearranging (B.1) and is given as

$$\boldsymbol{Z} = (z\boldsymbol{h}^{-1}\boldsymbol{S}\boldsymbol{h}_* + z_*)(\boldsymbol{U}_n - \boldsymbol{h}^{-1}\boldsymbol{S}\boldsymbol{h}_*)^{-1}, \tag{B.3}$$

where $\boldsymbol{U}$ is an identity matrix of order 2, and

$$\boldsymbol{h}^{-1} = \begin{bmatrix} \dfrac{1}{\sqrt{Z_{01}}} & 0 \\ 0 & \dfrac{1}{\sqrt{Z_{02}}} \end{bmatrix} \quad \text{and} \quad \boldsymbol{h}_* = \begin{bmatrix} \sqrt{Z_{01}} & 0 \\ 0 & \sqrt{Z_{02}} \end{bmatrix}$$

for real Z_{01} and Z_{02}. To reduce calculation errors, $\boldsymbol{h}^{-1}\boldsymbol{S}\boldsymbol{h}_*$ and $(\boldsymbol{U}_n - \boldsymbol{h}^{-1}\boldsymbol{S}\boldsymbol{h}_*)^{-1}$ are first calculated as

$$\boldsymbol{h}^{-1}\boldsymbol{S}\boldsymbol{h}_* = \begin{bmatrix} S_{11} & \dfrac{\sqrt{Z_{02}}}{\sqrt{Z_{01}}}S_{12} \\ \dfrac{\sqrt{Z_{01}}}{\sqrt{Z_{02}}}S_{21} & S_{22} \end{bmatrix}, \tag{B.4}$$

$$(\boldsymbol{U}_n - \boldsymbol{h}^{-1}\boldsymbol{S}\boldsymbol{h}_*)^{-1} = \frac{1}{\Delta_{S_X}} \begin{bmatrix} 1 - S_{22} & \dfrac{\sqrt{Z_{02}}S_{12}}{\sqrt{Z_{01}}} \\ \dfrac{\sqrt{Z_{01}}S_{21}}{\sqrt{Z_{02}}} & 1 - S_{11} \end{bmatrix}, \tag{B.5}$$

where $\Delta_{S_X} = (1 - S_{11})(1 - S_{22}) - S_{12}S_{21}$

The conversion of scattering matrix into the impedance matrix is finally found as

$$\boldsymbol{Z} = (z\boldsymbol{h}^{-1}\boldsymbol{S}\boldsymbol{h}_* + z_*)(\boldsymbol{U}_n - \boldsymbol{h}^{-1}\boldsymbol{S}\boldsymbol{h}_*)^{-1} = \frac{1}{\Delta_{S_X}} \begin{bmatrix} Z_{01}[(1 + S_{11})(1 - S_{22}) + S_{12}S_{21}] & 2\sqrt{Z_{01}}\sqrt{Z_{02}}S_{12} \\ 2\sqrt{Z_{01}}\sqrt{Z_{02}}S_{21} & Z_{02}[(1 + S_{22})(1 - S_{11}) + S_{12}S_{21}] \end{bmatrix}. \tag{B.6}$$

ADMITTANCE MATRIX → SCATTERING MATRIX

The conversion of an admittance matrix into a scattering matrix is obtained from voltage quantities and is given as

$$\boldsymbol{S} = -\boldsymbol{k}[\boldsymbol{Y} + \boldsymbol{y}]^{-1}[\boldsymbol{Y} - \boldsymbol{y}_*]\boldsymbol{k}_*^{-1}, \tag{B.7}$$

where

$$Y = \begin{bmatrix} Y_{11} & Y_{12} \\ Y_{21} & Y_{22} \end{bmatrix}, \qquad y = \begin{bmatrix} Y_{01} & 0 \\ 0 & Y_{02} \end{bmatrix},$$

$$k = \begin{bmatrix} \sqrt{Y_{01}} & 0 \\ 0 & \sqrt{Y_{02}} \end{bmatrix}, \qquad k_*^{-1} = \begin{bmatrix} \dfrac{1}{\sqrt{Y_{01}}} & 0 \\ 0 & \dfrac{1}{\sqrt{Y_{02}}} \end{bmatrix},$$

and $y = y_*$ for real Y_{01} and Y_{02}, $Y_{01} = Z_{01}^{-1}$ and $Y_{02} = Z_{02}^{-1}$. The scattering parameters are again calculated as

$$S = \frac{1}{\Delta_Y} \begin{bmatrix} Y_{01}[(1 - S_{11})(1 + S_{22}) + S_{12}S_{21}] & -2\sqrt{Y_{01}}\sqrt{Y_{02}}S_{12} \\ -2\sqrt{Y_{01}}\sqrt{Y_{02}}S_{21} & Y_{02}[S_{21}S_{12} + (1 - S_{22})(1 + S_{11})] \end{bmatrix} \tag{B.8}$$

where $\Delta_Y = (Y_{11} + Y_{01})(Y_{22} + Y_{02}) - Y_{12}Y_{21}$.

SCATTERING MATRIX → ADMITTANCE MATRIX

The conversion of a scattering matrix into an admittance matrix is obtained by rearranging the matrix in (B.7). To reduce calculation errors, $k^{-1}Sk_*$ is first calculated and then its conversion is computed.

$$k^{-1}Sk_* = \begin{bmatrix} S_{11} & \dfrac{\sqrt{Y_{02}}}{\sqrt{Y_{01}}}S_{12} \\ \dfrac{\sqrt{Y_{01}}}{\sqrt{Y_{02}}}S_{21} & S_{22} \end{bmatrix}, \tag{B.9}$$

$$Y = (y_* - yk^{-1}Sk_*)(k^{-1}Sk_* + U)^{-1} =$$

$$\frac{1}{\Delta_{S_Y}} \begin{bmatrix} Y_{01}[(1 - S_{11})(1 + S_{22}) + S_{12}S_{21}] & -2\sqrt{Y_{01}}\sqrt{Y_{02}}S_{12} \\ -2\sqrt{Y_{01}}\sqrt{Y_{02}}S_{21} & Y_{02}[S_{21}S_{12} + (1 - S_{22})(1 + S_{11})] \end{bmatrix}, \tag{B.10}$$

where $\Delta_{S_Y} = (1 + S_{11})(1 + S_{22}) - S_{21}S_{12}$

ABCD MATRIX → SCATTERING MATRIX

Since an *ABCD* matrix is not directly convertible into a scattering matrix, the conversion of an *ABCD* matrix into an impedance matrix can be used. From Table 2.1,

$$\begin{bmatrix} Z_{11} & Z_{12} \\ Z_{21} & Z_{22} \end{bmatrix} = \frac{1}{C} \begin{bmatrix} A & \Delta \\ 1 & D \end{bmatrix}, \tag{B.11}$$

and its conversion yields

$$\boldsymbol{S} = \boldsymbol{h}[\boldsymbol{A_Z} + \boldsymbol{z}]^{-1}[\boldsymbol{A_Z} - \boldsymbol{z_*}]\boldsymbol{h}_*^{-1}, \tag{B.12}$$

where

$$\boldsymbol{A_Z} = \frac{1}{C}\begin{bmatrix} A & \Delta \\ 1 & D \end{bmatrix}.$$

For the final conversion, $[\boldsymbol{A_Z} + \boldsymbol{z}]^{-1}$ is first calculated and the result is

$$[\boldsymbol{A_Z} + \boldsymbol{z}]^{-1} = \frac{C}{(A + CZ_{01})(D + CZ_{02}) - \Delta}\begin{bmatrix} D + CZ_{02} & -\Delta \\ -1 & A + CZ_{01} \end{bmatrix}, \tag{B.13}$$

where $\Delta = AD - BC$. The final results are given as

$$\boldsymbol{S} = \frac{1}{\Delta_{A_Z}}\begin{bmatrix} AZ_{02} + B - CZ_{01}Z_{02} - DZ_{01} & 2\sqrt{Z_{01}}\sqrt{Z_{02}}(AD - BC) \\ 2\sqrt{Z_{01}}\sqrt{Z_{02}} & -AZ_{02} + B - CZ_{01}Z_{02} + DZ_{01} \end{bmatrix}, \tag{B.14}$$

where $\Delta_{A_Z} = AZ_{02} + B + CZ_{01}Z_{02} + DZ_{01}$.

APPENDIX C

Derivation of the Elements of a Small Asymmetric Ring Hybrid

The design equations of the ring hybrid in Fig. 4.10(b) are

$$Z_1 = \sqrt{\frac{d_1^2 + d_2^2}{d_1^2}}\sqrt{R_a R_b}, \tag{C.1a}$$

$$Z_2 = \sqrt{\frac{d_1^2 + d_2^2}{d_2^2}}\sqrt{R_b R_c}, \tag{C.1b}$$

$$Z_3 = \sqrt{\frac{d_1^2 + d_2^2}{d_1^2}}\sqrt{R_c R_d}, \tag{C.1c}$$

$$Z_4 = \sqrt{\frac{d_1^2 + d_2^2}{d_2^2}}\sqrt{R_d R_a}. \tag{C.1d}$$

The ring hybrid in Fig. 4.10(b) has three transmission-line sections with three-quarter wavelength and one section with a one-quarter wavelength. To replace a three-quarter-wavelength transmission-line section with the one-stage Π-type lumped equivalent circuit in Fig. 3.14(d), any relation of $X_C = Z_i$ and $B_L = 1/Z_i$ can be utilized, i being 1 to 3. Using the relation, the lumped elements in Fig. 4.12 related to the characteristic impedances Z_1, Z_2, and Z_3 in Fig. 4.10 can

Asymmetric Passive Components in Microwave Integrated Circuits, by Hee-Ran Ahn

be derived as

$$\frac{1}{\omega C_a} = Z_1 \rightarrow C_a = \frac{1}{\omega\sqrt{R_a R_b}}\sqrt{\frac{d_1^2}{d_1^2 + d_2^2}}, \tag{C.2a}$$

$$\omega L_a = Z_1 \rightarrow L_a = \frac{\sqrt{R_a R_b}}{\omega}\sqrt{\frac{d_1^2 + d_2^2}{d_1^2}}, \tag{C.2b}$$

$$\frac{1}{\omega C_b} = Z_2 \rightarrow C_b = \frac{1}{\omega\sqrt{R_b R_c}}\sqrt{\frac{d_2^2}{d_1^2 + d_2^2}}, \tag{C.2c}$$

$$\omega L_{bc} = Z_2 \rightarrow L_{bc} = \frac{\sqrt{R_b R_c}}{\omega}\sqrt{\frac{d_1^2 + d_2^2}{d_2^2}}, \tag{C.2d}$$

$$\frac{1}{\omega C_c} = Z_3 \rightarrow C_c = \frac{1}{\omega\sqrt{R_c R_d}}\sqrt{\frac{d_1^2}{d_1^2 + d_2^2}}, \tag{C.2e}$$

$$\omega L_c = Z_3 \rightarrow L_c = \frac{\sqrt{R_c R_d}}{\omega}\sqrt{\frac{d_1^2 + d_2^2}{d_1^2}}, \tag{C.2f}$$

where L_{bc} is an inductance coming from the transmission line connected to ports ② and ③.

To replace the transmission-line section between ports ① and ④ in Fig. 4.10(b) with the one-stage Π-type lumped equivalent circuit in Fig. 3.14(b), a relation of $X_L = Z_4$ and $B_C = 1/Z_4$ is required. So the lumped elements in Fig. 4.12 are derived as

$$\omega L_b = Z_4 \rightarrow L_b = \frac{\sqrt{R_b R_a}}{\omega}\sqrt{\frac{d_1^2 + d_2^2}{d_2^2}}, \tag{C.3a}$$

$$\frac{1}{\omega C_d} = Z_4 \rightarrow C_d = \frac{1}{\omega\sqrt{R_b R_a}}\sqrt{\frac{d_2^2}{d_1^2 + d_2^2}}, \tag{C.3b}$$

$$C_d = C_e. \tag{C.3c}$$

The lumped element L_d in Fig. 4.12 comes from the two transmission lines with Z_1 and Z_2 in Fig. 4.10(b). The inductance L_{bc} is connected in parallel with the inductance L_a, and the result of the parallel connection of two inductances L_{bc} and L_a is L_d. In the same way, the lumped element L_e comes from the two transmissions with Z_2 and Z_3 in Fig. 4.10(b). The inductance L_{bc} is connected in parallel with L_c, and the result of the parallel connection between L_{bc} and L_c

is L_e. Therefore, inductances L_d and L_e in Fig. 4.12 are

$$L_d = L_a \| L_{bc} = \frac{L_a L_{bc}}{L_a + L_{bc}} = \frac{\dfrac{\sqrt{d_1^2 + d_2^2}}{\omega d_1}\sqrt{R_a R_b}\dfrac{\sqrt{d_1^2 + d_2^2}}{\omega d_2}\sqrt{R_c R_b}}{\dfrac{\sqrt{d_1^2 + d_2^2}}{\omega d_1}\sqrt{R_a R_b} + \dfrac{\sqrt{d_1^2 + d_2^2}}{\omega d_2}\sqrt{R_c R_b}}$$

$$= \frac{\dfrac{d_1^2 + d_2^2}{\omega^2} R_b \left(\dfrac{\sqrt{R_a}}{d_1} \dfrac{\sqrt{R_c}}{d_2} \right)}{\dfrac{\sqrt{d_1^2 + d_2^2}}{\omega}\sqrt{R_b}\left(\dfrac{\sqrt{R_a}}{d_1} + \dfrac{\sqrt{R_c}}{d_2} \right)} = \frac{\sqrt{d_1^2 + d_2^2}\sqrt{R_a R_b R_c}}{\omega (d_2\sqrt{R_a} + d_1\sqrt{R_c})}, \quad \text{(C.4}a\text{)}$$

$$L_e = L_{bc} \| L_c = \frac{L_{bc} L_c}{L_{bc} + L_c} = \frac{\dfrac{\sqrt{d_1^2 + d_2^2}}{\omega d_2}\sqrt{R_b R_c}\dfrac{\sqrt{d_1^2 + d_2^2}}{\omega d_1}\sqrt{R_c R_d}}{\dfrac{\sqrt{d_1^2 + d_2^2}}{\omega d_2}\sqrt{R_b R_c} + \dfrac{\sqrt{d_1^2 + d_2^2}}{\omega d_1}\sqrt{R_c R_d}}$$

$$= \frac{\dfrac{d_1^2 + d_2^2}{\omega^2} R_c \left(\dfrac{\sqrt{R_b}}{d_2} \dfrac{\sqrt{R_d}}{d_1} \right)}{\dfrac{\sqrt{d_1^2 + d_2^2}}{\omega}\sqrt{R_c}\left(\dfrac{\sqrt{R_b}}{d_2} + \dfrac{\sqrt{R_d}}{d_1} \right)} = \frac{\sqrt{d_1^2 + d_2^2}\sqrt{R_b R_c R_d}}{\omega (d_1\sqrt{R_b} + d_2\sqrt{R_d})}. \quad \text{(C.4}b\text{)}$$

APPENDIX D

Trigonometric Relations

$$\sin(\alpha \pm \beta) = \sin\alpha\cos\beta \pm \cos\alpha\sin\beta \tag{D.1}$$

$$\cos(\alpha \pm \beta) = \cos\alpha\cos\beta \mp \sin\alpha\sin\beta \tag{D.2}$$

$$\sin\left(\frac{\pi}{2} \pm \alpha\right) = \cos\alpha \tag{D.3}$$

$$\cos\left(\frac{\pi}{2} \pm \alpha\right) = \mp\sin\alpha \tag{D.4}$$

$$\tan\left(\frac{\pi}{2} \pm \alpha\right) = \mp\cot\alpha \tag{D.5}$$

$$\cot\left(\frac{\pi}{2} \pm \alpha\right) = \mp\tan\alpha \tag{D.6}$$

$$\sin\alpha\cos\beta = \tfrac{1}{2}[\sin(\alpha+\beta) + \sin(\alpha-\beta)] \tag{D.7}$$

$$\cos\alpha\sin\beta = \tfrac{1}{2}[\sin(\alpha+\beta) - \sin(\alpha-\beta)] \tag{D.8}$$

$$\cos\alpha\cos\beta = \tfrac{1}{2}[\cos(\alpha+\beta) + \cos(\alpha-\beta)] \tag{D.9}$$

$$\sin\alpha\sin\beta = \tfrac{1}{2}[\cos(\alpha+\beta) - \cos(\alpha-\beta)] \tag{D.10}$$

$$\sin\alpha = 2\sin\frac{\alpha}{2}\cos\frac{\alpha}{2} \tag{D.11}$$

$$\sin 2\alpha = 2\sin\alpha\cos\alpha \tag{D.12}$$

$$\cos\alpha = 2\cos^2\frac{\alpha}{2} - 1 = 1 - 2\sin^2\frac{\alpha}{2} \tag{D.13}$$

$$\cos 2\alpha = 2\cos^2\alpha - 1 = \cos^2\alpha - \sin^2\alpha = 1 - 2\sin^2\alpha \tag{D.14}$$

$$\cos 3\alpha = 4\cos^3\alpha - 3\cos\alpha \tag{D.15}$$

Asymmetric Passive Components in Microwave Integrated Circuits, by Hee-Ran Ahn

$$\cos 4\alpha = 8\cos^4\alpha - 8\cos^2\alpha + 1 \tag{D.16}$$

$$\cos m\alpha = 2^{m-1}\cos^m\alpha - m2^{m-3}\cos^{m-2}\alpha + \frac{m(m-3)}{2!}2^{m-5} \times \cos^{m-4}\alpha + \cdots \tag{D.17}$$

$$1 = \sin^2\alpha + \cos^2\alpha \tag{D.18}$$

$$\sec^2\alpha = \frac{1}{\cos^2\alpha} = 1 + \tan^2\alpha \tag{D.19}$$

$$\csc^2\alpha = \frac{1}{\sin^2\alpha} = 1 + \cot^2\alpha \tag{D.20}$$

$$\tan\alpha = \frac{\sin\alpha}{\cos\alpha} \tag{D.21}$$

$$\sin\alpha = \alpha - \frac{\alpha^3}{3!} + \frac{\alpha^5}{5!} - \frac{\alpha^7}{7!} + \cdots \tag{D.22}$$

$$\cos\alpha = 1 - \frac{\alpha^2}{2!} + \frac{\alpha^4}{4!} - \frac{\alpha^6}{6!} + \cdots \tag{D.23}$$

$$e^{\pm j\alpha} = \cos\alpha \pm j\sin\alpha \tag{D.24}$$

APPENDIX E

Hyperbolic Relations

$$\sinh\alpha = \frac{e^{\alpha} - e^{-\alpha}}{2} = \alpha + \frac{\alpha^3}{3!} + \frac{\alpha^7}{7!} + \cdots \tag{E.1}$$

$$\cosh\alpha = \frac{e^{\alpha} + e^{-\alpha}}{2} = 1 + \frac{\alpha^2}{2!} + \frac{\alpha^4}{4!} + \frac{\alpha^6}{6!} + \cdots \tag{E.2}$$

$$\tanh\alpha = \frac{\sinh\alpha}{\cosh\alpha} = \frac{1}{\coth\alpha} \tag{E.3}$$

$$\sinh(\alpha \pm j\beta) = \sinh\alpha\cos\beta \pm j\cosh\alpha\sin\beta \tag{E.4}$$

$$\cosh(\alpha \pm j\beta) = \cosh\alpha\cos\beta \pm j\sinh\alpha\sin\beta \tag{E.5}$$

$$\sinh(j\alpha) = j\sin\alpha = \frac{e^{j\alpha} - e^{-j\alpha}}{2} \tag{E.6}$$

$$\cosh(j\alpha) = \cos\alpha = \frac{e^{j\alpha} + e^{-j\alpha}}{2} \tag{E.7}$$

Asymmetric Passive Components in Microwave Integrated Circuits, by Hee-Ran Ahn

Index

ABCD matrix, 29–36, 108, 128
  common structures, 32–36
  conversion table with loads, 41
  conversion table without loads, 38
  definition, 29
  input admittance, 31
  input impedance, 31
  output admittance, 31
  output impedance, 31
  relation to admittance matrix, 38
  relation to image parameters, 48–50
  relation to impedance matrix, 37
  relation to scattering matrix, 41
  useful network properties, 30–32
ABCD parameters, 29–36, 64–65, 108, 128, 132–137, 171, 180, 228
  asymmetric ring hybrids, 108
  branch-line hybrids, 128, 132–137
    conventional, 128
    impedance transforming, 132–137
  parallel coupled transmission lines, 64–65
  ring filters, 228
  three-port power dividers, 171, 180
    isolation condition, 17
    for scattering parameters, 180
Actual quantities, 15
Admittance matrix, 36, 108
  common structures, 38
  conversion table with loads, 41
  conversion table without loads, 38
  definition, 36
  input admittance, 31
  input impedance, 31
  output admittance, 31
  output impedance, 31
  relation to ABCD matrix, 38
  relation to image parameters, 48–50
  relation to impedance matrix, 38
  relation to scattering matrix, 41
ALERH, 109–110, 116–119
  asymmetric case, 117–119
  design equation, 110
  symmetric case, 116
Ahn, 1, 6, 97, 106, 155, 206
Arbitrary admittances, 93–94
Arbitrary design impedances, 192–204, 258
  N-way power dividers, 199–202
  three-port power dividers, 192–199
Arbitrary impedances, 106, 109, 258
Arbitrary power divisions, 1–3, 57, 74–75, 93, 101, 106, 109, 122, 139, 156, 170, 192
Arbitrary termination impedances, 1, 6, 75, 93, 122, 192, 193, 258, 265
Asymmetric branch-line hybrids, 139–150
  AABH, 147–151

Asymmetric Passive Components in Microwave Integrated Circuits, by Hee-Ran Ahn

Asymmetric branch-line hybrids (*continued*)
  CABH, 140–146
  relation to four-port hybrids, 139–140
Asymmetric four-port hybrids, 3, 139–147
  branch-line hybrids, 139–147
  ring hybrids, 93–122
Asymmetric ring hybrids, 3, 93–124, 207–209
  arbitrary power divisions, 93, 100, 106, 109, 122
  attenuators, 216–222
  design equations, 96–97
  lumped-element, 106–120
  phase shifters, 206–216
  reflection coefficients, 95
  small, 99–100
  uniplanar, 102–105
  wideband or small, 100–102
Asymmetric ring-hybrid attenuators, 216–222
  asymmetric ring hybrids, 208
  design procedure, 219–220
  magnitude of Γ_{2a}, 217, 219
  magnitude of Γ_{4a}, 217, 219
  microstrip attenuator, 220–222
  simple structures, 218
Asymmetric ring-hybrid phase shifters, 209–216
  asymmetric ring hybrids, 208
  conventional, 211
  design procedure, 215
  phase responses, 211, 215, 216, 221
  scattering matrix (parameters), 207, 208, 209, 210
  simple structures, 213
Asymmetric passive components, 1
Asymmetric three-port power dividers, 5, 155, 160, 170–205
  ABCD parameters, 174, 180
  admittance matrix, 172, 173
  analyses, 173–177
  coplanar T3PD, 196
  design equations, 176, 182, 186–187
  equal power division, 170–180
  equivalent two-port circuits, 180
  even- and odd-modes, 178
  impedance matrix, 178
  LE3PDs, 186–188
  lumped elements, 186
  perfect isolation condition, 171–173
  reflection coefficients, 179, 181, 182, 189
  scattering parameters, 177–184
  transmission line equation, 174

Backward coupler, 3
Baluns, 86–87
  CPW-CPW, 85–86
  CPW-slotline, 85–86
Basis-dependent reflection coefficients, 14
  current-basis, 14
  voltage-basis, 14
Basis-dependent scattering parameters, 14, 16, 17
  current-basis, 14, 16, 17
  voltage-basis, 14, 16, 17
Basis-independent scattering parameters, 17–18
  complex normalized, 18
Branch-line hybrids, 4, 125–151
  asymmetric, 4, 139
  conventional multisection, 127–131
  for impedance transforming, 132–138
    single-section, 135–139
    two-section, 132–135
  multiple, 5, 127–131

CABH, 139–140
CCTs, 2, 6, 237, 240–266
  bounded lengths, 248–251
  impedance transformers, 242–248
    microstrip, 247–248
  mathematical approach, 241–242
  phase angles, 250
  phase shifters, 251–253
  ring filters, 261–266
  three-port power divider, 253–261
    asymmetric 45°, 258–261
    design, 256–258
    isolation circuits, 254–256
CCT3PDs, 2, 253–258
  in-phase circuit, 254
  in-phase impedance or admittance, 254
  isolation circuits, 254–256
  out of phase circuit, 254

out of phase impedance, admittance, 254
responses of isolation circuit, 256
simulated results, 257
CCT-ring filters, 261–266
Characteristic impedances:
asymmetric branch-line hybrid, 139, 140, 142, 145, 147, 150
asymmetric ring hybrid attenuators, 221
asymmetric ring hybrid phase shifters, 207, 209, 211, 213, 216
branch-line hybrids for impedance transforming, 132, 135, 137, 138
coupled transmission lines, 65, 114
N-way arbitrary power dividers, 193, 194, 199, 200, 202
one-port network, 12
ring filters, 225, 227, 230, 235, 237
ring hybrid, 57, 59, 61, 63, 72
small asymmetric ring hybrids, 99
small impedance transformers, 241, 243, 247, 252, 255, 256, 262
three-port power divider, 155, 160, 162, 164, 167, 173, 175, 176, 177, 179, 185, 186
transmission lines, 35, 39, 40, 47
Circuit parameters, 2, 10–55
ABCD, 29–36
admittance, 37–43, 45–47
image, 47–52
impedance, 37–43, 45–47
scattering, 10–29
transfer, 27
Coaxial lines, 3, 56, 61
Combiner, 80, 155
Complex propagation constant, 11, 35, 50, 229, 233, 234
attenuation constant, 11
phase constant, 11
Conjugately matched, 13
Conventional branch-line hybrids, 125–138
ABCD parameters, 129–137
impedance transforming, 132–138
one-stage, 136–138
two-stage, 132–136
five branch lines, 127–129
origin of, 125–127
scattering parameters, 128–130
three-branch lines, 129–130
Conventional circuits, 56–88, 125–139, 154–167
branch-line hybrids, 125–139
impedance transformers, 243
parallel coupled transmission lines, 62–70, 111–115
ring hybrids, 56–88
three-port power dividers, 154–167
Conventional ring hybrids, 3, 4, 56–92
arbitrary power divisions, 56, 74–77
lumped-element, 77–81
mixed small, 80–84
original concept, 57–62
six-fundamental rings, 59
wideband, 71–74
Conventional ring-hybrid phase shifter, 211
Conventional three-port power dividers, 6, 154–167, 196
arbitrary power divisions, 156–160
45° power divider/combiner, 167
introduction, 154
LE3PDs, 186–188
lumped element, 186–188
power divisions, 155, 156, 157
termination impedances, 163–167
complex frequency dependence, 163–167
Conversion matrices, 36–38, 40–43, 272–275
Conversions of parameters, 38, 41, 272–275
Coplanar, 3, 56, 84, 171, 195, 216
CPW feeding lines, 85
ring hybrids, 104, 105
strip, 84, 104
structure, 84
T3PDs, 188–189, 196–197
technology, 96, 102
uniplanar AABH, 148–149
uniplanar CABH, 145–146
uniplanar ring-hybrid phase shifter, 218
Coplanar strip (CPS), 84, 86–87, 103–104
Coupled-line filters, 3, 56, 66–71, 111–121
Coupled slotlines, 84, 85, 86, 88

Coupled transmission lines, 62–71, 111–120
  asymmetrical structures, 111, 112, 114–115
  coupling coefficients, 65, 67, 73, 101, 116
  coupled-line filters, 66–71, 106, 111–112
  equivalent circuits, 66, 111–112
  even-mode impedances, 63, 64, 72, 73
  image impedances, 66, 67
  image propagation constant, 66
  odd-mode impedances, 63, 64
  ring hybrids, 68–70, 72, 111–120
  symmetrical structures, 66–68
Couplers, 3
  backward coupler, 3
  directional couplers, 4, 5
  forward coupler, 3
Coupling coefficients, 65, 67, 73, 101, 116
Current basis scattering matrix, 16, 173
Current quantities, 15–18, 40
Current vectors, 15
CVT3PDs, 2, 253–258
  in-phase circuit, 254
  in-phase impedance or admittance, 254
  isolation circuits, 254–256
  out of phase circuit, 254
  out of phase impedance or admittance, 254
  responses of isolation circuit, 256
  simulated results, 257
CVTs, 2, 6, 237, 240–266
  bounded length, 248–251
  impedance transformer, 242–248
    microstrip, 247–248
  mathematical approach, 241–242
  phase angle, 250
  phase shifters, 251–253
  ring filters, 261–266
  three-port power dividers, 253–261
    asymmetric 45°, 258–261
    design, 256–258
    isolation circuits, 254–256
CVT-ring filters, 261–266

Design equations:
  AABH, 147
  ALERH, 109–118
  asymmetric branch-line hybrids, 142, 147
    AABH, 147
    CABH, 142
  asymmetric power dividers, 176–177
  asymmetric ring hybrids, 96–97, 99–100
    small, 99–100
  CCT3PD and CVT3PD, 253–258
  CCTs and CVTs, 242–247
  coupled-line filters, 66–71, 111–121
  45° power dividers, 259
  mixed small ring hybrids, 80–84
  N-way power dividers, 200
  ring filters, 228
  small ring hybrids, 80–84, 99–100
    asymmetric, 99–100
    conventional, 80–84
  three-port power dividers, 194–198
  three-way power dividers, 202–204
  wideband ring hybrids, 71–72, 100–102
    asymmetric, 100–102
    symmetric, 71–72
Diode, 210
Down-filters, 226, 227–228, 230
Dual networks, 33, 34, 46

Electric wall, 162
Electromagnetic, 2
  boundary, 2
  fields, 11
  force, 24, 86
Equivalent circuits:
  asymmetric ring hybrids, 95, 97, 107
  branch-line hybrids, 128, 129, 133, 136
  coupled-line filters, 66, 111–112
    asymmetric, 112
    symmetric, 66, 111–112
  even- and odd-mode:
    branch-line hybrids, 128, 129, 133, 136
    conventional ring hybrids, 74
    coupled transmission lines, 63–64
    four-port circuits, 44
    L-network, 51
    Π-networks, 45, 78, 80
    ring hybrids, 74
    transmission lines, 45, 82

lumped, 11, 78, 80, 81
asymmetric ring hybrids, 81
T-networks, 78–80
transmission lines, 78, 80
T-networks, 45, 78, 80
Even-mode:
asymmetric three-port power dividers, 172
branch-line hybrids, 128–139
impedance transforming, 132–138
multisection, 128–132
CCT3PDs and CT3PDs, 254–256
conventional ring hybrids, 74–77
conventional three-port power dividers, 160–167
asymmetric, 160–163
complex termination impedance, 163–167
coupled-line filters, 111–121
coupled transmission lines, 64–65
equivalent circuits, 64
4-fort, 44
transmission line sections, 82

Filters, 6, 45, 47, 262
bandpass filters, 225
conventional, 1, 262
coupled-line filters, 3, 56, 66–71, 106, 111–112
ring hybrids, 70
coupled slotline filters, 88, 89
ring filters, 2, 5, 225–238, 261–266
CCT, 2, 237, 261–266
CVT, 2, 237, 261–266
down-filters, 226, 227–228, 230
up-filter, 226, 227–228
tuning filters, 225
wideband filter, 237
Forward bias, 210, 211
Forward coupler, 3

General design equations, 192–204
N-way power dividers, 199–204
three-port power dividers, 192–199
Generalized two-port scattering parameters, 28–29
Group delay, 232, 233

Hermitian matrix, 17, 53

Image impedances, 48, 50–52
common structures, 51
coupled-line filters, 69
definition, 48
symmetric structures, 50–52
Image parameters, 47–52
common structures, 51
relation to *ABCD* matrix, 48, 49
relation to admittance matrix, 50
relation to impedance matrix, 50
Image propagation constants, 47, 49, 68
Impedance matrix, 3, 14, 17, 18, 36
common structures, 38
conversion table with loads, 41
conversion table without loads, 38
coupled transmission lines, 68, 114
definition, 36
input impedance, 39
output impedance, 40
relation to *ABCD* parameters, 37, 38
relation to admittance matrix, 38
relation to image parameters, 50
relation to scattering matrix, 41
Impedance transforming, 1, 2, 6, 145, 148, 158, 160, 165
branch-line hybrids, 132–138
double-section, 132–135
single-section, 135–138
Incident:
current, 12, 13
current vector, 15, 52
quantities, 15
signals, 3
voltage, 12, 13
voltage vector, 15, 52
waves, 1, 11, 12, 14, 20, 24, 26, 29, 30, 69, 94, 181
wave vector, 181
Inherent ring-resonance frequency, 2, 234
In-phase, 43, 75, 88, 199
circuit, 254
mode, 88
power dividers, 199
signal, 43
waves, 75

Input impedances, 36, 39, 48, 74, 95, 157, 158, 176, 177, 193, 262
Isolation circuits, 256

LET3PDs, 186–187
Linear passive networks, 10
L-network, 51, 52
Loss, 106, 110, 234
Lossless, 5, 12, 53, 85, 194
  ABCD parameters, 30, 128
  branch-line hybrids, 130, 132, 133, 134
  ring filters, 226, 230
  T-junctions, 85, 194
  transmission lines, 12, 160, 177
  symmetric networks, 43
  unitary property, 95
Lumped-element, 2, 62, 78, 84, 106
  power divider, 164, 166, 170, 186–188
  ring hybrids, 4, 122
    asymmetric, 106–107, 109–110, 113–114, 117–118
    symmetric, 81
  transmission lines, 11, 78–80

Magnetic plane, 58
Magnetic wall, 160
Magnitude of reflection loads for attenuators, 210, 216, 219
Mason's rule, 26–27
Matching networks, 1, 4, 10, 39, 47, 93, 170
Mathematical approach, 241
Matrices, 10–54
  ABCD, 29–36
  admittance, 36–40
  conversion tables, 38, 40–43
  impedance, 36–40
  image, 47–52
  scattering, 11–29
  scattering transfer, 27–28
Microstrip, 66, 84, 106, 116, 156, 192, 199, 225
  asymmetric ring hybrid, 100–102
  attenuator, 220–222
  CCT-ring filter, 265
  CCTs and CVTs, 247–248
  inherent ring resonance frequency, 230–236
  ring filters, 230–231
  three-way power dividers, 202–203
Mixed small ring hybrids, 80–84
MMICs, 2, 3, 4, 56, 83, 84, 85, 93, 106, 167, 258
Modified ring hybrids, 106, 107
Multiport networks, 19, 21, 22, 210

Normalized, 74
  conductance, 242
  current, 243
  frequency, 234
  impedances, 29, 207, 241, 242
  incident waves, 18, 23, 24, 26, 69
  reflected waves, 18, 23, 24, 26, 69
  scattering parameters (matrix) 17–18
  susceptance, 249
  voltage, 243
  wave vector, 18, 181
N-port network, 11, 14, 17–19
  incident current vector, 15, 16
  incident voltage vector, 15
  scattering matrix, 14, 18
    current basis, 16
    normalized, 17, 18
    voltage basis, 16
N-way power dividers, 192–204
  arbitrary design impedance, 192, 194, 197, 198, 202
  general design equations, 200
  isolation circuit, 194, 200
  low characteristic impedance, 203
  three-port, 192–199
    coplanar, 196–197
    general design equations, 194
  three-way power dividers, 202

Odd mode:
  asymmetric three-port power dividers, 172
  branch-line hybrids, 128–139
    impedance transforming, 132–138
    multisection, 128–132
  CCT3PDs and CT3PDs, 254–256
  conventional ring hybrids, 74–77
  conventional three-port power dividers, 160–167

asymmetric, 160–163
complex termination impedance, 163–167
coupled-line filters, 111–121
coupled transmission lines, 64–65
equivalent circuits, 64
4-fort, 44
transmission line sections, 82
Off-state, 210
One-port network, 12, 13, 14
On-state, 210
Open-circuit, 10, 32, 36, 66, 211, 213, 243
impedance parameters (matrix), 14, 36
Out of phase, 94, 96, 99, 130, 139,142, 154, 182, 219, 226, 230

Phase angle circles, 250
Phase shifters, 1, 2, 4, 6, 56, 206–216
asymmetric phase shifters, 206–216
asymmetric ring hybrids, 207–210
conventional, 211
definition, 210
design method, 215
phase responses, 214
scattering parameters, 208, 209, 210, 212
simple structures, 213
Para-Hermitian part, 17, 53
admittance, 17
impedance, 17, 53
Perfect isolation, 1, 65, 154, 155, 159
Perfect isolation condition:
three-port power divider, 171–173
Π-networks, 34, 38, 45, 51, 78, 80
Power dividers:
branch-line hybrids, 125–151
asymmetric, 139–147
AABH, 147–150
CABH, 140–145
conventional, 125–132
directional couplers, 4, 64
N-way, 192–204
ring hybrids, 56–122
asymmetric, 93–122
conventional, 56–91
three-ports, 154–204
asymmetric, 170–190
conventional, 154–168

Q-factor, 250
Quadrature, 154

Reciprocal, 1, 30, 32
network, 17, 36, 53
property, 43, 49, 77, 108
Reduced multiports, 18, 19, 21
Reference impedances, 12, 13, 24
Reflected, 26, 125, 181, 226, 234
amplitudes, 128, 129
coefficients, 132
components, 24
currents, 11, 12, 13
current vector, 16, 17
loads, 20, 21
powers, 125, 235
power waves, 181
scattering parameters, 64, 132, 143
signals, 3
symmetry, 170
variables, 24
voltages, 12, 13
voltage vectors, 15, 17
waves, 3, 11, 14, 20, 24–26, 29, 30, 52, 53, 69, 125, 134
wave vector, 18, 94, 181
Reflection coefficients, 14
current-basis, 12, 14
normalized, 17–18
voltage-basis, 12, 14
Relation, 2, 12, 14, 18, 27, 29, 33, 36, 37, 39, 110, 117
between circuit parameters, 2, 38, 41, 43, 46, 50, 108, 134, 171, 173
between coupling and image impedance, 69, 72
between ring and branch-line hybrids, 139–140
between voltages and currents, 12, 13, 34, 39, 75
of $NI_{do} \cdot NI_{bs}$, 216
Ring filters, 225–237, 261–266
ABCD parameters, 228
analyses, 226
CCT and CVT, 261–266
characteristic impedances, 228
down-filter, 226–227
introduction, 225
up-filter, 226–227

Ring hybrids, 56–122
  asymmetric, 93–122
  conventional, 56–88
  coupled-line filters, 62–70, 111–120
  lumped-element, 77–80
    asymmetric, 106–116
    conventional, 77–80
  mixed small, 80–83
  original concept, 57–62
  symmetric, 74–78
  wideband, 71–73
  uniplanar, 84–90
Ring resonance frequency, 2, 230, 233–237
  group delay, 232
  with loss, 234
  lossless, 232

Scattering matrix (parameters), 10–29
  basis-dependent, 16, 18
  basis-independent, 17
  complex normalized, 18
  conversions, 38, 41
  current-basis, 14, 16
  definition, 14, 16, 17, 18
  generalized, 28
  input admittance, 31
  input impedance, 31
  n-port network, 14
  one-port network, 12
  output admittance, 31
  output impedance, 31
  reduced multiport, 18–21
  two-port networks, 22, 28
  voltage gain, 25
  voltage-basis, 14, 16
Scattering transfer parameters, 27
Series inductance per unit length, 11
Series resistance per unit length, 11
Signal flow graph, 25, 26
  branches, 26
  first-order loop, 26
  loop, 27
  nodes, 26
  path, 26
  second-order loop, 26
Shunt conductance per unit length, 11
Shunt capacitance per unit length, 11
Strip line, 3, 56
Symmetric networks, 43–47, 50–52
  ABCD matrix, 48
  admittance matrix, 46
  even and odd modes, 44–47
  image impedances, 48, 50
  impedance matrix, 46
  Π-networks, 45, 51
  T-networks, 45, 51
  transmission lines, 35, 38, 45, 47, 51, 78, 82

T-junctions, 3, 5, 56, 57, 85, 86
  CPW, 87
  CPW-slotline, 87
T-networks, 32, 34, 39, 45, 46, 52, 114
Three-port networks, 22, 85
Three-port power dividers, 2, 5, 6, 81, 154–205
  asymmetric, 170–190
  conventional, 154–168
Transitions, 86
  CPW-to-coupled slot lines, 86
  CPW-to-CPS, 86
  CPW-to-slotline, 86
Transfer parameters, 27
Transmission line theory, 10, 11–12
  current-basis reflection coefficient, 12
  series inductance per unit length, 11
  series resistance per unit length, 11
  shunt capacitance per unit length, 11
  shunt conductance per unit length, 11
  propagation constant, 11
    attenuation constant, 11
    phase constant, 11
  voltage and current amplitudes, 11
  voltage-basis reflection coefficient, 12
Two-port networks, 22, 23–43
  ABCD parameters, 29–35
  analyses, 23–25
  generalized scattering parameters, 28
  input impedances, 31, 39
  output impedances, 31, 39

scattering transfer parameters, 27
signal flow graph, 25–27
voltage gain, 25

Uniplanar baluns, 86
wideband uniplanar, 86
Uniplanar ring hybrids, 84–90
Uniplanar structures, 84
baluns, 86
CPW-CPW, 85–86
CPW-slotline, 85–86
coplanar strip lines, 84, 85
coplanar wave guide(CPW), 84
coupled slotlines, 84
ring hybrids, 88–89
slotlines, 84
T-junctions, 85
transitions, 86
Uniplanar technique, 3, 57, 62
Unitary property, 108, 132, 134, 137, 209

Voltage-basis quantities, 17
Voltage-basis scattering matrix, 12, 16, 17, 171, 173
Voltage gain, 25
Voltage quantities, 18, 42
Voltage maximum, 43, 57–59, 128
Voltage minimum, 43, 57–59, 128

Wavefront, 62
Wave-guide T-junctions, 56

WILEY SERIES IN MICROWAVE AND OPTICAL ENGINEERING

KAI CHANG, Editor
Texas A&M University

FIBER-OPTIC COMMUNICATION SYSTEMS, Third Edition • *Govind P. Agrawal*

COHERENT OPTICAL COMMUNICATIONS SYSTEMS • *Silvello Betti, Giancarlo De Marchis, and Eugenio Iannone*

PHASED ARRAY ANTENNAS: FLOQUET ANALYSIS, SYNTHESIS, BFNs, AND ACTIVE ARRAY SYSTEMS • *Arun K. Bhattacharyya*

HIGH-FREQUENCY ELECTROMAGNETIC TECHNIQUES: RECENT ADVANCES AND APPLICATIONS • *Asoke K. Bhattacharyya*

COMPUTATIONAL METHODS FOR ELECTROMAGNETICS AND MICROWAVES • *Richard C. Booton, Jr.*

MICROWAVE RING CIRCUITS AND ANTENNAS • *Kai Chang*

MICROWAVE SOLID-STATE CIRCUITS AND APPLICATIONS • *Kai Chang*

RF AND MICROWAVE WIRELESS SYSTEMS • *Kai Chang*

RF AND MICROWAVE CIRCUIT AND COMPONENT DESIGN FOR WIRELESS SYSTEMS • *Kai Chang, Inder Bahl, and Vijay Nair*

MICROWAVE RING CIRCUITS AND RELATED STRUCTURES, Second Edition • *Kai Chang and Lung-Hwa Hsieh*

MULTIRESOLUTION TIME DOMAIN SCHEME FOR ELECTROMAGNETIC ENGINEERING • *Yinchao Chen, Qunsheng Cao, and Raj Mittra*

DIODE LASERS AND PHOTONIC INTEGRATED CIRCUITS • *Larry Coldren and Scott Corzine*

RADIO FREQUENCY CIRCUIT DESIGN • *W. Alan Davis and Krishna Agarwal*

MULTICONDUCTOR TRANSMISSION-LINE STRUCTURES: MODAL ANALYSIS TECHNIQUES • *J. A. Brandão Faria*

PHASED ARRAY-BASED SYSTEMS AND APPLICATIONS • *Nick Fourikis*

FUNDAMENTALS OF MICROWAVE TRANSMISSION LINES • *Jon C. Freeman*

OPTICAL SEMICONDUCTOR DEVICES • *Mitsuo Fukuda*

MICROSTRIP CIRCUITS • *Fred Gardiol*

HIGH-SPEED VLSI INTERCONNECTIONS: MODELING, ANALYSIS, AND SIMULATION • *A. K. Goel*

FUNDAMENTALS OF WAVELETS: THEORY, ALGORITHMS, AND APPLICATIONS • *Jaideva C. Goswami and Andrew K. Chan*

HIGH-FREQUENCY ANALOG INTEGRATED CIRCUIT DESIGN • *Ravender Goyal (ed.)*

ANALYSIS AND DESIGN OF INTEGRATED CIRCUIT ANTENNA MODULES • *K. C. Gupta and Peter S. Hall*

PHASED ARRAY ANTENNAS • *R. C. Hansen*

MICROSTRIP FILTERS FOR RF/MICROWAVE APPLICATIONS • *Jia-Sheng Hong and M. J. Lancaster*

MICROWAVE APPROACH TO HIGHLY IRREGULAR FIBER OPTICS • *Huang Hung-Chia*

NONLINEAR OPTICAL COMMUNICATION NETWORKS • *Eugenio Iannone, Francesco Matera, Antonio Mecozzi, and Marina Settembre*

FINITE ELEMENT SOFTWARE FOR MICROWAVE ENGINEERING • *Tatsuo Itoh, Giuseppe Pelosi, and Peter P. Silvester (eds.)*

INFRARED TECHNOLOGY: APPLICATIONS TO ELECTROOPTICS, PHOTONIC DEVICES, AND SENSORS • *A. R. Jha*

SUPERCONDUCTOR TECHNOLOGY: APPLICATIONS TO MICROWAVE, ELECTRO-OPTICS, ELECTRICAL MACHINES, AND PROPULSION SYSTEMS • *A. R. Jha*

OPTICAL COMPUTING: AN INTRODUCTION • *M. A. Karim and A. S. S. Awwal*

INTRODUCTION TO ELECTROMAGNETIC AND MICROWAVE ENGINEERING • *Paul R. Karmel, Gabriel D. Colef, and Raymond L. Camisa*

MILLIMETER WAVE OPTICAL DIELECTRIC INTEGRATED GUIDES AND CIRCUITS • *Shiban K. Koul*

MICROWAVE DEVICES, CIRCUITS AND THEIR INTERACTION • *Charles A. Lee and G. Conrad Dalman*

ADVANCES IN MICROSTRIP AND PRINTED ANTENNAS • *Kai-Fong Lee and Wei Chen (eds.)*

SPHEROIDAL WAVE FUNCTIONS IN ELECTROMAGNETIC THEORY • *Le-Wei Li, Xiao-Kang Kang, and Mook-Seng Leong*

ARITHMETIC AND LOGIC IN COMPUTER SYSTEMS • *Mi Lu*

OPTICAL FILTER DESIGN AND ANALYSIS: A SIGNAL PROCESSING APPROACH • *Christi K. Madsen and Jian H. Zhao*

THEORY AND PRACTICE OF INFRARED TECHNOLOGY FOR NONDESTRUCTIVE TESTING • *Xavier P. V. Maldague*

OPTOELECTRONIC PACKAGING • *A. R. Mickelson, N. R. Basavanhally, and Y. C. Lee (eds.)*

OPTICAL CHARACTER RECOGNITION • *Shunji Mori, Hirobumi Nishida, and Hiromitsu Yamada*

ANTENNAS FOR RADAR AND COMMUNICATIONS: A POLARIMETRIC APPROACH • *Harold Mott*

INTEGRATED ACTIVE ANTENNAS AND SPATIAL POWER COMBINING • *Julio A. Navarro and Kai Chang*

ANALYSIS METHODS FOR RF, MICROWAVE, AND MILLIMETER-WAVE PLANAR TRANSMISSION LINE STRUCTURES • *Cam Nguyen*

FREQUENCY CONTROL OF SEMICONDUCTOR LASERS • *Motoichi Ohtsu (ed.)*

WAVELETS IN ELECTROMAGNETICS AND DEVICE MODELING • *George W. Pan*

SOLAR CELLS AND THEIR APPLICATIONS • *Larry D. Partain (ed.)*

ANALYSIS OF MULTICONDUCTOR TRANSMISSION LINES • *Clayton R. Paul*

INTRODUCTION TO ELECTROMAGNETIC COMPATIBILITY, Second Edition • *Clayton R. Paul*

ADAPTIVE OPTICS FOR VISION SCIENCE: PRINCIPLES, PRACTICES, DESIGN AND APPLICATIONS • *Jason Porter, Hope Queener, Julianna Lin, Karen Thorn, and Abdul Awwal (eds.)*

ELECTROMAGNETIC OPTIMIZATION BY GENETIC ALGORITHMS • *Yahya Rahmat-Samii and Eric Michielssen (eds.)*

INTRODUCTION TO HIGH-SPEED ELECTRONICS AND OPTOELECTRONICS • *Leonard M. Riaziat*

NEW FRONTIERS IN MEDICAL DEVICE TECHNOLOGY • *Arye Rosen and Harel Rosen (eds.)*

ELECTROMAGNETIC PROPAGATION IN MULTI-MODE RANDOM MEDIA • *Harrison E. Rowe*

ELECTROMAGNETIC PROPAGATION IN ONE-DIMENSIONAL RANDOM MEDIA • *Harrison E. Rowe*

HISTORY OF WIRELESS • *Tapan K. Sarkar, Robert J. Mailloux, Arthur A. Oliner, Magdalena Salazar-Palma, and Dipak L. Sengupta*

SMART ANTENNAS • *Tapan K. Sarkar, Michael C. Wicks, Magdalena Salazar-Palma, and Robert J. Bonneau*

(List continues on the inside back cover)

NONLINEAR OPTICS • *E. G. Sauter*

APPLIED ELECTROMAGNETICS AND ELECTROMAGNETIC COMPATIBILITY • *Dipak L. Sengupta and Valdis V. Liepa*

COPLANAR WAVEGUIDE CIRCUITS, COMPONENTS, AND SYSTEMS • *Rainee N. Simons*

ELECTROMAGNETIC FIELDS IN UNCONVENTIONAL MATERIALS AND STRUCTURES • *Onkar N. Singh and Akhlesh Lakhtakia (eds.)*

FUNDAMENTALS OF GLOBAL POSITIONING SYSTEM RECEIVERS: A SOFTWARE APPROACH, Second Edition • *James Bao-yen Tsui*

RF/MICROWAVE INTERACTION WITH BIOLOGICAL TISSUES • *André Vander Vorst, Arye Rosen, and Youji Kotsuka*

InP-BASED MATERIALS AND DEVICES: PHYSICS AND TECHNOLOGY • *Osamu Wada and Hideki Hasegawa (eds.)*

COMPACT AND BROADBAND MICROSTRIP ANTENNAS • *Kin-Lu Wong*

DESIGN OF NONPLANAR MICROSTRIP ANTENNAS AND TRANSMISSION LINES • *Kin-Lu Wong*

PLANAR ANTENNAS FOR WIRELESS COMMUNICATIONS • *Kin-Lu Wong*

FREQUENCY SELECTIVE SURFACE AND GRID ARRAY • *T. K. Wu (ed.)*

ACTIVE AND QUASI-OPTICAL ARRAYS FOR SOLID-STATE POWER COMBINING • *Robert A. York and Zoya B. Popović (eds.)*

OPTICAL SIGNAL PROCESSING, COMPUTING AND NEURAL NETWORKS • *Francis T. S. Yu and Suganda Jutamulia*

SiGe, GaAs, AND InP HETEROJUNCTION BIPOLAR TRANSISTORS • *Jiann Yuan*

ELECTRODYNAMICS OF SOLIDS AND MICROWAVE SUPERCONDUCTIVITY • *Shu-Ang Zhou*

ASYMMETRIC PASSIVE COMPONENTS IN MICROWAVE INTEGRATED CIRCUITS • *Hee-Ran Ahn*

Lightning Source UK Ltd.
Milton Keynes UK
UKHW030600130619
344257UK00003B/97/P

9 780471 737483